MODERN ASPECTS OF ELECTROCHEMISTRY

No. 45

Modern Aspects of Electrochemistry

Topics in Number 44 include:

- The basic mathematical models which arise in electrochemical processes regarding the galvanic corrosion phenomena
- Near field transducers and their applications to nanochemistry and electrochemistry
- The role of symmetry considerations in modern scientific computation
- The methods used in the analysis of electrochemical systems, including computer simulations and quantum and statistical mechanics
- The techniques used to investigate ionic and solvent transfer and transport in electroactive materials
- Monte Carlo simulations of the underpotential deposition of metal layers on metallic substrates
- The models applied to localized corrosion processes and a review of recent work on the modeling of pitting corrosion
- Descriptions of the conceptual structure of Density-Functional Theory and its practical applications
- Scanning acoustic microscopy's applications to nanoscaled electrochemically deposited thin film systems
- The processes and equations underlying the modeling of the current distribution in

Topics in Number 43 include:

- Mathematical modeling in electrochemistry using finite element and finite difference methods
- Modeling atomic systems of more than two bodies
- Modeling impedance of porous electrodes, in order to permit optimal utilization of the active electrode material
- Multi-scale mass transport in porous silicon gas
- Physical theory, molecular simulation, and computational electrochemistry for PEM fuel cells, with emphasis on fundamental understanding, diagnostics, and design
- Modeling of catalyst structure degradation in PEM fuel cells
- Modeling water management in PEM fuel cells
- Modeling electrochemical storage devices for automotive applications

For other volumes in this series, go to
www.springer.com/series/6251

MODERN ASPECTS OF ELECTROCHEMISTRY

No. 45

Series Edited by

RALPH E. WHITE
Series Editor
University of South Carolina
Columbia, SC, USA

CONSTANTINOS G. VAYENAS
Series Editor
University of Patras
Patras,Greece

and

MARIA E. GAMBOA-ALDECO
Managing Editor
Superior, CO, USA

Springer

Editor
Ralph E. White
Department of Chemical Engineering
University of South Carolina
Columbia, SC 29208
USA

Series Editors
Ralph E. White
Department of Chemical Engineering
University of South Carolina
Columbia, SC 29208
USA

Constantinos G. Vayenas
Department of Chemical Engineering
University of Patras
Patras 265 00
Greece

Managing Editor
Maria E. Gamboa-Aldeco
Superior, CO 80027
USA

ISSN 0076-9924
ISBN 978-1-4899-8312-1 ISBN 978-1-4419-0655-7 (eBook)
DOI 10.1007/978-1-4419-0655-7
Springer Dordrecht Heidelberg London New York

Printed on acid-free paper

Springer is part of Springer Science+Business Media (www.springer.com)

Preface

This volume contains five chapters. The topics covered are the cathodic reduction of nitrate, for which the key issues are product selectivity and current efficiency; a discussion of ionic liquids known as molten or fused salts in key industrial processes; the properties of nanowires made of metals and semiconductors by means of electrodeposition into porous templates and template synthesis; and an overview of a range of alloys used in electroplating, and trends in ammonium electrolysis. Also, included is a review of the applications of synchroton x-ray scattering to the electrochemical interphase.

Chapter 1 by Celia Milhano and Derek Pletcher reviews the cathodic reduction of nitrate as a "truly fascinating reaction" since processes for nitrate removal from drinking water and effluents and reducing nitrates in nuclear waste is an essential step in nuclear waste disposal. The authors also review electrochemical membrane technology not involving the cathodic reduction of nitrate.

In Chapter 2, the authors discuss substances called ionic liquids, and the fact that this is really a new more modern term for molten or fused salts. However, the label of ionic liquids is usually applied to the liquid state of salts at or near room temperature, which is a way to differentiate low-melting salts from higher melting cousins. These molten or fused salts form the basis of several key industrial processes, and this chapter is primarily devoted to the non-haloaluminate ionic liquids as reviewed by Tetsuya Tsuda and Charles L. Hussey.

Reginald M. Penner reviews and discusses in Chapter 3 the properties of nanowires composed of metals and semiconductors. He focuses on the step edge decoration method which is horizontal instead of vertical.

Gerardine S. Botte and Madhivanan Muthuvel review some of the trends in ammonium electrolysis in Chapter 4 as a renewable source of fuel.

In Chapter 5, Zoltan Nagy reviews the usefulness of synchrotron x-ray scattering to a wide range of electrode phenomena by reviewing the physics of x-ray scattering and describing applications of the technique in response to a variety of electrochemical phenomena.

R.E. White
University of South Carolina
Columbia, South Carolina

C.G. Vayenas
University of Patras
Patras, Greece

Contents

Chapter 1

THE ELECTROCHEMISTRY AND ELECTROCHEMICAL TECHNOLOGY OF NITRATE

Clelia Milhano and Derek Pletcher

I. Introduction 1
II. Thermodynamics of Nitrate Reduction 2
III. Mechanisms for Nitrate Reduction 5
IV. The Influence of Electrolysis Conditions 13
1. Acidic Media 13
2. Neutral Media 26
3. Alkaline Media 29
V. Applications of Electrochemical Nitrate Reduction 33
1. Analysis 33
2. Electrosynthesis 36
3. Treatment of Nuclear Waste 38
4. Removal of Nitrate from Natural Waters and Effluents 44
VI. Conclusion 55
References 55

Chapter 2

ELECTROCHEMISTRY OF ROOM-TEMPERATURE IONIC LIQUIDS AND MELTS

Tetsuya Tsuda and Charles L. Hussey

I. Introduction 63
II. Synthesis and Purification of Room-Temperature Ionic Liquids 75
1. Dialkylimidazolium Chlorides 75

2. Dialkylimidazolium Salts with Fluorohydrogenate Anions 76
3. Dialkylimidazolium Salts with Fluorocomplex Anions 77
4. Tetraalkylammonium Salts with Bis[(trifluoromethyl)-sulfonylimide Anions 78
III. Fundamental Properties of Room-Temperature Ionic Liquids 79
1. Thermal Stability 79
2. Water Contamination 82
3. Chloride Ion Contamination 84
4. Oxygen Contamination 85
5. Conductance Measurements at High Frequencies 88
6. Other Considerations 89
IV. General Electrochemical Techniques 89
1. Electrodes and Experimental Considerations 90
2. An Overview of the Techniques Used for Electrochemical Analysis in Ionic Liquids 99
(*i*) Cyclic Voltammetry 99
(*ii*) Hydrodynamic Voltammetry 103
(*iii*) Chronoamperometry 107
(*iv*) Sampled Current Voltammetry 112
V. Electrochemical Applications 113
1. Surface Finishing 121
(*i*) Metal Halide–Organic Halide Salt Binary Systems 132
(*ii*) RTILs with Fluoroanions 133
(*iii*) Room-Temperature Melts (RTMs) 134
2. Fuel Cells 134
(*i*) Hydrogen Electrode Reaction 136
(*ii*) Oxygen Electrode Reaction 137
(*iii*) Fuel Cell Systems Based on RTILs 137
3. Lithium and Lithium-Ion Batteries 142
4. Nuclear Waste Treatment 148
(*i*) Removal of $^{137}Cs^{+}$ and $^{90}Sr^{2+}$ from Spent Reactor Fuel 149
(*ii*) Recovery of Actinides with RTILs 153
Symbols (Usual Units) 153
References 154

Chapter 3

ELECTROCHEMICAL STEP EDGE DECORATION (ESED): A VERSATILE TOOL FOR THE NANOFABRICATION OF WIRES

Reginald M. Penner

I. Introduction 175
II. Metals 178
  1. Canonical ESED of Metal Nanowires 179
  2. Coinage and Noble Metals 182
  3. Base Metals 182
  4. Nanowire Nucleation with Seeds Prepared Using Physical Vapor Deposition (PVD) 185
III. Compound Semiconductors 189
  1. Stoichiometric Electrodeposition of CdSe and Bi_2Te_3 Nanowires using Cyclic Electrodeposition-Stripping in Concert with ESED 190
  2. Electrochemical/Chemical Synthesis of CdS and MoS_2 Nanowires 195
  3. Oxides 197
IV. Reducing the Diameter of ESED Nanowires by Kinetically-Controlled Nanowire Electrooxidation 200
V. Summary 202
Acknowledgements 204
References 204

Chapter 4

TRENDS IN AMMONIA ELECTROLYSIS

Madhivanan Muthuvel and Gerardine G. Botte

I. Introduction 207
II. Hydrogen Production Methods 211
  1. Steam Reforming 211
  2. Partial Oxidation Reforming 212
  3. Coal Gasification 212

4. Water Electrolysis 213
5. Efficiency for Hydrogen Production Methods 215
III. Ammonia 217
1. Sources 217
2. Ammonia Cracking 219
3. Ammonia Electrolysis 219
4. Comparison of Ammonia Electrolysis and Ammonia Cracking 221
IV. Development of Components 223
1. Anode Electrode..223
2. Cathode Electrode 234
3. Electrolyte 235
4. Separator Membrane 238
V. Ammonia Electrolyzer: A Prototype 239
VI. Summary 242
References 243

Chapter 5

APPLICATIONS OF SYNCHROTRON X-RAY SCATTERING FOR THE INVESTIGATION OF THE ELECTROCHEMICAL INTERPHASE

Zoltán Nagy and Hoydoo You

I. Introduction 247
II. Theory and Practice of X-Ray Scattering........................250
1. General Description of Surface-X-Ray Scattering (SXS)..250
2. X-Ray Reflectivity and Crystal Truncation Rods 253
3. Resonance Anomalous Surface X-Ray Scattering 259
III. Metal Surface Preparation, Restructuring, and Roughening 259
1. Relaxation/Reconstruction of Metal Surfaces 260
(*i*) Relaxation of Platinum Surfaces 260
(*ii*) Reconstruction of Platinum Surfaces 262
(*iii*) Reconstruction of Gold Surfaces......................263
2. Surface Roughness Measurements 264
(*i*) Platinum Oxidation/Reduction 264

IV. Double Layer Structure Studies 270
1. Water at the Silver Surface.......................................272
2. Water at the Ruthenium Dioxide Surface..................273
3. Ionic Distribution in the Double Layer 276
V. Adsorption/Absorption at Electrode Surfaces 277
1. Adsorption of Bromide on Gold Surfaces.................277
2. Adsorption of Carbon Monoxide on Platinum Surfaces...280
3. Absorption of Oxygen below Platinum Surfaces 282
VI. Kinetics/Mechanism and Electrocatalysis of Charge Transfer Reactions..283
1. Oxygen Reduction Reaction on Gold Surfaces.........283
2. Oxygen Reduction Reaction on Platinum Surfaces...286
VII. Metal Deposition at the Submonolayer, Monoloyer, and Multilayer Level..288
1. Deposition of Lead on Silver Surfaces......................288
2. Deposition of Palladium on Platinum Surfaces.........290
3. Deposition of Thallium on Silver Surfaces 293
4. Deposition of Copper on Gold Surfaces....................298
VIII. Oxide Film Formation on Metals and Passivation 299
1. Oxidation of Ruthenium Surfaces.............................299
2. Oxidation of Platinum Surfaces 302
IX. Corrosion and Metal Dissolution......................................308
1. Pitting Corrosion of Copper Surfaces 308
X. Porous Silicon Formation...308
XI. Investigation of Interphases other than Metal/Electrolyte..313
1. Oxidation of Ruthenium Dioxide Surfaces 313
XII. Concluding Remarks...315
Appendix..315
1. X-Ray-Electrochemical Cell Designs 316
(*i*) Reflection-Geometry Cell Design...................317
(*ii*) Transmission-Geometry Cell Design 318
(*iii*) Hanging-Drop Transmission-Geometry Cell Design...319
(*iv*) Comparison of Cell Designs...........................321
2. Problems with Solution Impurities at Beamlines......322
Acknowledgement..326
References..326

List of Contributors, MAE 45

Gerardine G. Botte
Department of Chemical and Biomolecular Engineering
Ohio University
Athens, OH 45701
Ph: 740-593-9670
Fax: 740-593-0873
botte@ohio.edu

Charles L. Hussey
Departments of Chemistry and Biochemistry
The University of Mississippi
University, Mississippi 38677-1848
chclh@chem1.olemiss.edu

Celia Milhano
School of Chemistry, The University
Southampton S019 1BJ, England

Madhivanan Muthuvel
Department of Chemical and Biomolecular Engineering
Ohio University
Athens, OH 45701

Zoltan Nagy
Materials Science Division
Argonne National Laboratory
Argonne, Illinois 60439
Present address:
Department of Chemistry
Campus Box 3290
The University of North Carolina at Chapel Hill
Chapel Hill, North Carolina 27599-3290)
nagyz@email.unc.edu

Reginald M. Penner
Institute for Surface and Interface Science
Department of Chemistry
University of California
Irvine, California 29697-2025
rmpenner@uci.edu

Derek Pletcher
School of Chemistry, The University
Southampton S019 1BJ, England
dp1@soton.ac.uk

Tetsuya Tsuda
Departments of Chemistry and Biochemistry
The University of Mississippi
University, Mississippi 38677-1848
ttsuda@olemiss.edu

Hoydoo You
Materials Science Division
Argonne National Laboratory
Argonne, Illinois 60439

1

The Electrochemistry and Electrochemical Technology of Nitrate

Clelia Milhano and Derek Pletcher

School of Chemistry, The University, Southampton SO19 1BJ, England

I. INTRODUCTION

The cathodic reduction of nitrate is a truly fascinating reaction. It can lead to at least eight different products in reactions involving the transfer of one to eight electrons per nitrate ion, see Fig. 1. A mixture of products is usual and it is never obvious whether products are being formed in a sequence of electron transfer steps or by parallel reduction mechanisms. Two consequences follow. Firstly, simply the determination of the average number of electrons involved in the reduction is a very poor indicator of the likely products. Secondly, it is apparent that the product spectrum is likely to depend on a number of experimental parameters including the concentration of nitrate and pH as well as cathode material, electrode potential (or current density) and charge passed. All the reduction reactions of nitrate involve protonation and hence the local pH at the cathode surface can be quite different from the bulk value and it can vary during an experiment, e.g. during a potential sweep. Particularly with high nitrate concentrations, (and/or high current density), the reaction layer will only be buffered in

R.E. White (ed.), *Modern Aspects of Electrochemistry, No. 45*, Modern Aspects of Electrochemistry 45, DOI 10.1007/978-1-4419-0655-7_1,

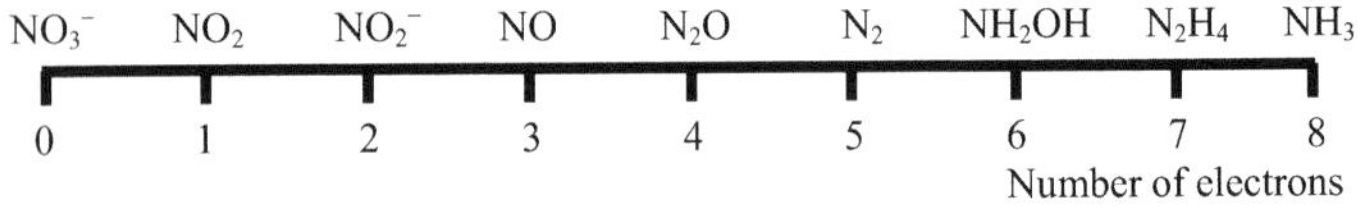

Figure 1. Possible products from the reduction of nitrate ion.

solutions at the extremes of pH; buffers added to the solution of nitrate have a strong influence on the reduction of nitrate, even killing the response entirely. Moreover, experience shows that the presence of both cations and anions (including nitrite) at a trace or impurity level can lead to changes in the product spectrum; hence, the purity of chemicals and water used in the preparation of solutions is an issue in all experimental studies. Certainly, product selectivity is a concern in all applications involving the reduction of nitrate ion. Also, hydrogen evolution is usually a competing reaction producing multiple challenges for electrocatalysts—enhancing the rate and selectivity of nitrate reduction while minimising hydrogen evolution. In addition, perhaps not surprisingly, the mechanisms for nitrate reduction remain very poorly understood; most perplexing, the mechanisms are least understood when the voltammetry is simplest!

Despite these difficulties, there are several reasons why nitrate reduction is of interest to electrochemical technology. Processes have been developed for the removal of nitrate from both, drinking water and effluents, for the reduction of nitrate as an essential step in nuclear waste disposal and for the manufacture of hydroxylamine while both electroanalytical procedures and sensors for nitrate have been described. Some of these objectives have also been approached by electrochemical membrane technology using approaches not involving the cathodic reduction of nitrate and the final section of this chapter will also review such methods.

We are not aware of any extensive reviews of the electrochemistry of nitrate since the chapter by Plieth published in 1978.[1]

II. THERMODYNAMICS OF NITRATE REDUCTION

As noted above, a number of products (namely nitrite ion, oxides of nitrogen, nitrogen, hydroxylamine, hydrazine and ammonia)

Table 1
Formal Potentials vs. the Saturated Calomel Electrode for the Reduction of Nitrate to Various Products in Acid and Alkaline Solutions.[1]

Reaction	E_e^0 vs. SCE/V		Eq.
	pH 0	pH 14	
$NO_3^- + 2H^+ + e^- \rightarrow NO_2 + H_2O$	+0.53	−1.14	(1)
$NO_3^- + 2H^+ + 2e^- \rightarrow NO_2^- + H_2O$	+0.70	−0.13	(2)
$NO_3^- + 4H^+ + 3e^- \rightarrow NO + 2H_2O$	+0.72	−0.39	(3)
$2NO_3^- + 10H^+ + 8e^- \rightarrow N_2O + 5H_2O$	+0.88	−0.16	(4)
$2NO_3^- + 12H^+ + 10e^- \rightarrow N_2 + 6H_2O$	+1.01	+0.01	(5)
$NO_3^- + 7H^+ + 6e^- \rightarrow NH_2OH + 2H_2O$	+0.49	−0.48	(6)
$2NO_3^- + 16H^+ + 14e^- \rightarrow N_2H_4 + 6H_2O$	+0.58	−0.37	(7)
$NO_3^- + 9H^+ + 8e^- \rightarrow NH_3 + 3H_2O$	+0.64	−0.30	(8)

have been observed during the cathodic reduction of nitrate ion. The overall reactions involved in their formation are listed in Table 1 together with their formal potentials in both acid and alkaline aqueous solution,[2,3] for easy comparison and practical use, these formal potentials are quoted versus a saturated calomel electrode.

In aqueous acid solution, nitrate ion is a strong oxidising agent. Even the oxidation of water by nitrate ion to give oxygen and nitrogen is thermodynamically favourable in very concentrated acid although it has only a small negative free energy of reaction. Certainly, the reaction between many metals used as cathodes (e.g., Cu, Sn, Zn) and nitrate ion in acidic solution should be spontaneous; corrosion of the metals should occur and any of the products listed in the table could be formed. In practice, the kinetics of nitrate reduction is universally poor and nitrate reduction is seldom observed. Hence, nitrate never reacts with water and spontaneous reaction with metals is generally only observed in very acid media; many metals are stable to oxidation by nitrate at room temperature at pH > 0 provided no impurities are present. The literature is, however, complicated by strong evidence that several species including nitrite, oxides of nitrogen, chloride and ferric ion, even when present in low concentration, can catalyse the oxidation of metals by nitrate.[4] The importance of such effects is, of course, magnified by the fact that several possible reduction products from nitrate are included in the list of potential catalysts. Taking the example of nitrite ion, the chemical reaction

$$NO_3^- + NO_2^- + 2\,H^+ \rightarrow 2\,NO_2 + H_2O \quad (9)$$

is thermodynamically favourable under appropriate conditions and, in consequence, the presence of nitrite ion can introduce the possibility of entirely new reaction pathways. Indeed, since the reduction

$$2\,NO_2 + M \rightarrow 2\,NO_2^- + M^{2+} \quad (10)$$

can also be thermodynamically favourable and, indeed, kinetically facile, the chemical (and electrochemical) reductions of nitrate can be autocatalytic since reaction (9) followed by (10) leads to a net increase in the concentration of nitrite ion. Hence, the purity of chemicals and water in laboratory studies and the exact composition of feed solutions to commercial processes is an important factor. It should also be noted that the formal potentials for most of the reactions, (1)-(8) are quite similar and hence it is not surprising that mixtures of products are common and, small variations in reaction or electrolysis conditions can lead to different products.

In alkaline solutions, nitrate is thermodynamically more stable. Reaction with water is always unfavourable. From a thermodynamic viewpoint, the likelihood of reactions with metals remains although the driving force for reaction depends on the strength of complexation of the resulting metal ions with hydroxide. Again, in practice, kinetic factors in the reduction of nitrate and passivating oxide layers on the surface of the metals generally prevent spontaneous reduction of nitrate by the most metals. At pH 14, the formal potentials are more spread out than in acid media but again there is little difference in the thermodynamics of several of the reactions.

Table 1 only considers the overall reduction of nitrate to the eight different products. Clearly, some of the products could be discrete intermediates in the conversion to more reduced products. The thermodynamics of all the possible transformations is set out in the book edited by Bard, Parsons and Jordan.[2] It should also be emphasised that

(a) the reduction can occur in a sequence of discrete electron transfer steps involving relatively stable intermediate products, and

(b) a number of *reproportionation reactions* could occur as coupled reactions in homogeneous solution.

For example, the reaction

$$3\,H^+ + 3\,NO_3^- + 5\,NH_3 \rightarrow 4\,N_2 + 9\,H_2O \qquad (11)$$

has a negative free energy in both acid and alkaline solution; at pH 0, $\Delta G = -114$ kJ for reaction (11) as written.

The hydrogen evolution reaction has a formal potential of –0.24 V vs. SCE at pH 0 and a formal potential of –1.07 V vs. SCE at pH 14. Hence, from a thermodynamic viewpoint, the cathodic reduction of nitrate would always be favoured compared to hydrogen evolution. Experimentally, it has always been found that the reduction of nitrate is a very irreversible reaction and usually it is observed to occur at potentials where hydrogen evolution can occur. The extent of hydrogen evolution is determined by the kinetics of the hydrogen evolution reaction on the cathode surface being employed. The current efficiency for nitrate reduction will depend on the thermodynamics and kinetics of both nitrate reduction and hydrogen evolution.

Hence, throughout this review of cathodic nitrate reduction, key issues will be the product selectivity and the current efficiency for the reactions under study.

III. MECHANISMS FOR NITRATE REDUCTION

At this early stage in this review, it needs to be recognised that in most situations there is little definitive evidence to support detailed mechanisms for the cathodic reduction of nitrate. Many papers have sought to define *reaction pathways*, i.e., to identify intermediate oxidation states on the route to the final product, but even here it is not always clear whether compounds are formed in series or parallel reactions and it is important that conclusions take into account the way in which the product spectrum changes with charge consumed. A few cathode materials support the reduction of nitrate to nitrite with little further reduction. It is, however, more commonly found that the kinetics of nitrite reduction (as well as the reduction of some oxides of nitrogen) are substantially faster than the kinetics of nitrate reduction and, hence, once nitrate reduction is initiated, multi-electron transfer can take place. It is therefore not surprising that the products from nitrate reduction depend on the concentration of nitrate, pH, other species in solution, current density, mass transport conditions, conversion, etc.

It is, however, the initial steps in the nitrate reduction reaction that remain a mystery. Voltammograms for nitrate reduction can have a very simple form. For example, at a copper disc in acidic media, well-formed sigmoidal waves are recorded and the limiting currents are proportional to the square root of the rotation rate of the disc confirming mass transfer control in the plateau region, see Fig. 2. Even so, an electron transfer as the first step seems unlikely. Few people are willing to write the first step as

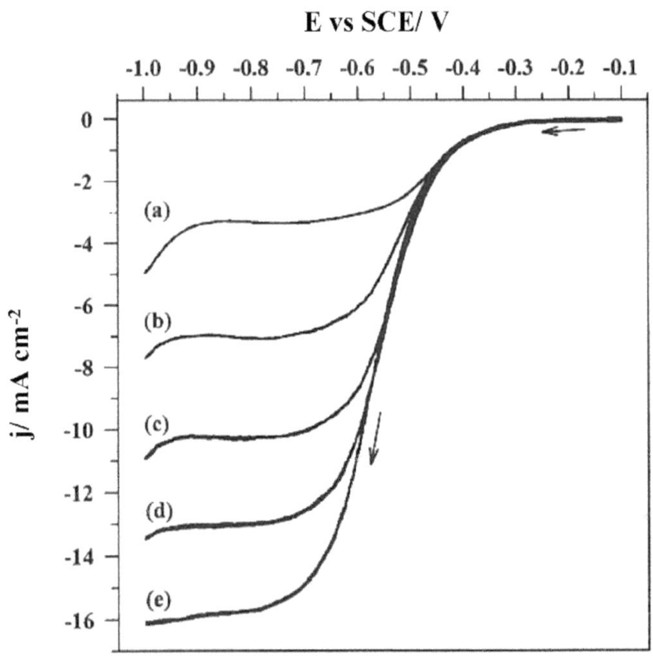

Figure 2. Voltammograms for 5 mM $NaNO_3$ in 0.5 M Na_2SO_4 pH 2.7 at a polished Cu rotating disc electrode (a) 100 (b) 400 (c) 900 (d) 1600 (e) 2500 rpm. Potential scan rate 10 mV s^{-1}.

$$NO_3^- + e^- \rightarrow NO_3^{2-} \qquad (12)$$

or

$$HNO_3 + e^- \rightarrow HNO_3^- \qquad (13)$$

Hence, how does the cleavage of the N-O bond occur? There is a general belief, in some situation supported by good evidence, that the adsorption of nitrate is a critical first step but the next is never clearly defined. Moreover, it should be emphasised that the same uncertainties can be found in the literature on the chemical reduction of nitrate and, indeed, a recent review[5] of the chemical reduction of nitrate makes little mention of the mechanism for the initial stages of nitrate reduction. The chemistry of the nitrate ion, however, confirms that it is generally a poor ligand and the kinetics of its reduction is universally poor.

Perhaps the clearest evidence for a mechanism for a nitrate reduction reaction comes from investigations of naturally occurring enzymes. Such enzymes have a metal centre, usually molybdenum, capable of forming a bond to an oxygen atom in a nitrate ion. Several papers[6–8] conclude that the nitrate reductase enzymes containing a Mo centre convert nitrate into nitrite by an oxygen transfer mechanism. The key steps are formation of an enzyme-substrate complex

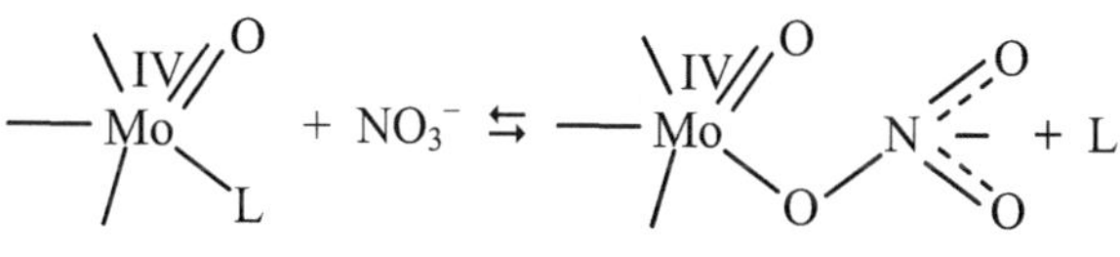

(14)

followed by a rate determining step involving oxygen transfer and conversion of a Mo(IV) centre to a Mo(VI) centre,

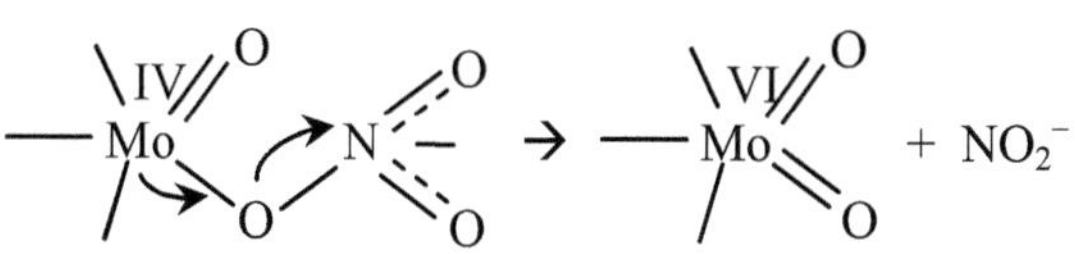

(15)

Mechanisms involving such oxygen transfer reactions are likely to apply to some indirect cathodic reductions with mediators such as molybdate[9-11] and the uranyl cation.[12] On the other hand, such mechanisms have seldom been discussed for the direct cathodic reduction of nitrate. In principle, oxygen transfer to the cathode surface followed by reduction of a metal-oxygen bond is a possible mechanism but there seems little correlation between the strength of the M-O bond and the effectiveness of the material as a nitrate reduction cathode.

Nadjo and coworkers[13-16] have sought to design heteropolyanions as homogeneous catalysts for nitrate reduction in acidic media. Significant catalytic activity was seen only with heteropolyanions substituted by Cu(II) and Ni(II) with Fe(III) substituted anions showing a small activity. It was found that the catalytic activity increased with the number of Cu(II) or Ni(II) centres in the catalyst and this led[16] to the development of the heteropolyanion, $[Cu_{20}Cl(OH)_{24}(H_2O)_{12}(P_8W_{48}O_{184})]^{25-}$; studies at pH 0 and 5 were reported. This heteropolyanion shows peaks for the Cu(II)/Cu(I) and Cu(I)/Cu couples between 0 and –0.3 V vs. SCE and for tungsten couples at more negative potentials. The reduction of nitrate is only catalysed negative to –0.8 V where the tungsten centres are electroactive but currents of 0.25 mA cm^{-2} were achieved at pH 5 with 2.4 mM nitrate and 0.04 mM catalyst. Nitrite ion is reduced at the much more positive potential where the Cu(II)/Cu(I) couple is active.

The cathodic reduction of nitrate at precious metal cathodes (bulk polycrystalline metals,[17–20] single crystals,[21-24] metal blacks[25–27] and dispersed on carbon supports[28–31]) has been investigated in both acid and alkaline solution. In these systems, the potential for nitrate reduction generally corresponds to that where hydrogen adsorption takes place. Moreover, the rate of reduction reflects qualitatively the real surface area of the precious metal exposed to the solution. Hence, it appears that the mechanism is analogous to that for the heterogeneous reduction with hydrogen gas[32-35] with adsorbed hydrogen atoms as key intermediates, i.e.,

$$H^+ + M + e^- \rightarrow M\text{–}H \qquad (16)$$

or

$$H_2O + M + e^- \rightarrow M\text{–}H + OH^- \quad (17)$$

followed by a reaction such as

$$2\,M\text{–}H + NO_3^- \rightarrow NO_2^- + H_2O + 2\,M \quad (18)$$

It is tempting to believe that adsorption of the nitrate ion is essential to such mechanisms. Indeed, it has been reported that at precious metal surfaces, the reduction of nitrate becomes inhibited by a high coverage of adsorbed hydrogen[32] reducing the number of sites available for the weaker adsorber, the nitrate anion. It is also found that the rate of reduction and even the products of reduction depend on the anion of the supporting electrolyte and this results from competitive adsorption of the nitrate and electrolyte anion onto the surface. Koper and coworkers[19,36] reported that there was a higher activity for nitrate reduction at Pt in perchloric acid compared to sulphuric acid especially at lower nitrate concentrations and explained this observation by adsorption of sulfate competing with nitrate for surface sites. At very high nitrate concentration (> 1 M) this difference in rate was much diminished as nitrate became more competitive with sulfate for surface sites. The addition of chloride ion, even in low concentrations, to either perchloric or sulphuric acids was also found to inhibit the reduction of nitrate at both platinised[37] and rhodised[38] electrode surfaces; chloride ion is, of course, recognised as a strongly adsorbing anion. The sensitivity of nitrate reduction to submonolayer coverages of other species also implies that adsorption of nitrate is critical to its rate of reduction. For example, the underpotential deposition of germanium onto Pt, Pd and Pt/Pd alloys leads to an increase in the rate of nitrate reduction.[39,40] The rate increases with germanium coverage and it was proposed that nitrate absorbed on germanium sites and could then react with hydrogen adsorbed on precious metal sites.

Parallel evidence is available to suggest that nitrate adsorption also is a step in its reduction mechanism at metal electrodes where hydrogen adsorption is weak (e.g., Cu and Au) and mechanisms involving absorbed hydrogen seem inappropriate. Nitrate reduction at copper in acid is inhibited by halide ions. Radovici et al.[41] showed that all halides can inhibit nitrate reduction at copper in sulphuric acid solution; even μM concentrations of iodide and bromide diminished the peak current on cyclic voltammograms for solutions containing mM nitrate, while the peak almost disappeared with 10 μM I^- or 25 μM Br^-. In a more limited study, Pletcher and Poorabedi[42] found that chloride ion added to perchloric acid led to a shift in the peak for nitrate reduction to more negative potential and some diminution in peak current density. Again, in contrast, underpotential deposition of metals enhances nitrate reduction; this has been demonstrated for cadmium on both gold and silver.[43-45]

The voltammetric response for nitrate reduction has been found to be sensitive to the state of the copper surface and this also implies a key role for surface chemistry in the reduction. The reduction of nitrate at polished bulk copper in acid solution gives a well formed wave, see Fig. 2. But at a freshly deposited Cu surface, produced by the presence of trace Cu(II) in the electrolyte so that in situ deposition of copper occurs continuously, the reduction wave is steeper and shifted to more positive potentials.[42,46–50] More recently, it has been demonstrated that similar voltammetry can be achieved by the in situ deposition of copper onto a diamond surface.[51,52] Similarly, it is possible to activate a copper surface for nitrate reduction by cycling the potential of a copper electrode so that it first corrodes and then copper is replated; such effects have been demonstrated, for example, for hydroxide[53] and sulfate[54] solutions. Burke et al.[55] have more clearly applied the concepts of special sites on the copper surface to the mechanism of nitrate reduction. Applying their general theory[56,57] of the importance of *metastable metal surface states* to the reaction, they propose the existence of a low coverage of copper sites (e.g., adatoms, step sites or other special sites) where the copper atoms are less stabilised than in the bulk lattice and hence the potential of the Cu/Cu(I) couple is shifted to more negative potentials, i.e., the copper atoms at such sites are stronger reducing agents than bulk copper. On copper subjected to abrasion with fine emery paper, these authors have identified a low current density voltammetric peak whose potential coincides with that for nitrate reduction and suggest the following mechanism for nitrate reduction

$$Cu^* + NO_3^- \rightarrow Cu(I)_{surface} + \text{intermediate} \qquad (19)$$

$$Cu(I)_{surface} + e^- \rightarrow Cu^* \qquad (20)$$

$$\text{intermediate} \xrightarrow{\text{further reduction}} \text{product} \tag{21}$$

involving a catalytic cycle with the active copper centre, Cu^*, as the key reducing species. Roue et al.[58] have investigated the influence of ball milling copper powder on the activity for nitrate reduction in basic media. Ball milling for 60 minutes leads to a large increase in activity and, more interestingly, ball milling in air produces a much larger effect than ball milling in argon. The role of the ball milling is explained in terms of an increase in the number of active surface defects.

Understanding the mechanism of the electrochemical reduction of nitrate is further complicated by the number of possible homogeneous chemical reactions involving nitrate ion. The nitronium ion, NO_2^+, is well known as the intermediate in the nitration of organic molecules[59] in strongly acidic media, e.g. mixtures of concentrated nitric acid and sulphuric acid. Moreover, it has been identified spectroscopically and some salts have been isolated. Hence, in acidic electrolytes, a mechanism involving the pre-equilibrium

$$HNO_3 + H^+ \leftrightarrows NO_2^+ + H_2O \tag{22}$$

followed by electron transfer

$$NO_2^+ + e^- \rightarrow NO_2 \tag{23}$$

should always be considered even when the equilibrium concentration of the nitronium ion may be low.

In addition as noted above, the thermodynamic data in Table 1 would indicate that equilibria such as

$$2\,HNO_3 \leftrightarrows 2\,NO_2 + H_2O + \tfrac{1}{2}\,O_2 \tag{24}$$

are possible. Again, even if the equilibrium lies to the left, they could be significant routes to nitrate reduction[3,4,36] if the following electron transfer step is facile. The likelihood of the involvement of homogeneous chemical reactions is further enhanced by the fact that a large number of reproportionation reactions are thermodynamically possible (see Table 1), including for example

$$HNO_3 + HNO_2 \leftrightarrows 2\,NO_2 + H_2O \qquad (25)$$

Such reactions also introduce the possibility that reduction reactions of nitrate are autocatalytic and maybe the products change with time—i.e., the formation of particular products could initiate new reaction pathways. Certainly, the mechanism and kinetics of nitrate reduction could be strongly influenced by trace quantities of nitrite ion. This was demonstrated more than 50 years ago by Vetter[60] and a recent paper has shown that for a platinised titanium electrode, the addition of some nitrite ion leads to a very substantial increase in electroactivity for the reduction of nitrate to nitrite[61]. This could be one reason why the reduction chemistry of nitrate is influenced by its concentration in solution; impurity levels in the nitrate used for preparation of solutions will increase proportionately with the concentration of nitrate. Moreover, since equilibria such as (22) and (24) may be influenced by both heat and light, the precise experimental conditions may be important. Certainly, from a practical point of view, the purity of chemicals and of the water used in studies of the mechanism of nitrate reduction is an important issue – nitrite and chloride ion, common contaminants, both have a strong influence on the reaction. It is not surprising that the literature is not entirely consistent!

All the reduction reactions of nitrate and, indeed, the reductions of almost all stable intermediates (NO_2, NO_2^-, NO, etc.) in the reduction sequences involve net protonation. This may occur via protonation on the nitrate ion or the intermediates prior to electron transfer or protonation of intermediates following electron transfer. The likelihood and rate of all such reactions will be a function of pH. In addition, in solutions that are unbuffered (or insufficiently buffered), particularly neutral and weakly acidic solutions, the electrochemistry will lead to upwards swing in pH in the reaction layer close to the electrode surface. The extent of the pH shift will depend on the bulk nitrate concentration, current density and the mass transport regime as well as the initial pH and buffer regime. For example, in acidic solutions, the products and rate of reactions can depend strongly on the ratio of proton and nitrate fluxes to the cathode surface; products are likely to be different in conditions where

(a) there is sufficient proton present that the reduction of nitrate does not lead to a significant pH swing at the surface, or
(b) the reduction of nitrate causes depletion of the proton close to the surface and the pH increases.

The conclusions to this section are clear. Firstly, the mechanism for the early steps in nitrate reduction remains a matter for speculation. Secondly, the mechanism and products will depend on the exact conditions of the experiment (or process operation) and quite subtle changes to conditions or impurity levels could lead to significant changes in reaction kinetics, the reaction pathway and final products.

IV. THE INFLUENCE OF ELECTROLYSIS CONDITIONS

In discussing the influence of electrolysis conditions on the cathodic reduction of nitrate, two important factors must be recognised. Firstly, it must be stressed that the products, rate of reduction and mechanism can all depend on a large number of experimental parameters including the electrode material, the concentration of nitrate, pH, other deliberately added ions/molecules or impurities, electrode potential (or current density), mass transfer regime, charge passed and temperature. Secondly, the experiments carried out and reported in the literature have been strongly influenced by the target application, e.g. the range of nitrate concentration when the objective is an analytical procedure or a water treatment is quite different from that for a synthesis or met in nuclear waste treatment. A consequence is a lack of uniformity in the experiments reported making comparisons hazardous. In this review, we have selected pH and electrode material as the principle parameters in this discussion.

1. Acidic Media

At a mercury cathode, the direct reduction of nitrate only occurs in very concentrated acidic solutions. These conditions are, however, important because they have been used to develop a process for the electrosynthesis of hydroxylamine. An early paper[62] describes the reduction of nitric acid in 50% sulphuric acid at a mercury surface and a current efficiency of 60-70% was achieved. This was followed up by workers at the Olin Corporation[63-66] who by using concentrated nitric acid as the feed and a high current density of 350 mA cm^{-2} were able to produce hydroxylammonium nitrate with a current efficiency of 85%. This chemistry was scaled up to a small commercial process as will be described later. Mercury appears to be the only practical cathode material for the manufacture of hydroxylamine although Bard et al.[67] have described a modification of a vitreous carbon surface that leads to good yields of hydroxylamine. The procedure involves in situ formation of a polymer on the surface by including 50 mM p-phenylenediamine in the nitric acid catholyte.

The reduction of nitrate at cadmium in acid solutions has also been reported.[68,69] The voltammograms are affected by the anodic dissolution of the metal but it is possible to define a potential range where the reduction is mass transfer controlled. In this potential region and with an acid concentration above 0.1 M, the major product is ammonia.[70] Nitrate also gives similar voltammograms at tin and zinc cathodes.[70] All three electrodes are likely to corrode rapidly in preparative conditions employing high concentration of nitrate.

Pletcher and Poorabedi[42] report a rather complete study of nitrate reduction at a copper rotating disc electrode using low concentrations of nitrate (1–5 mM) in perchlorate and sulfate solutions pH 1–3. The voltammograms are straightforward and well formed, sigmoidal waves are recorded (at pH 1 in perchlorate medium, $E_{1/2}$ = –430 mV vs. SCE). For pH $\leq$ 3, the limiting currents are proportional to the square root of the rotation rate of the disc electrode confirming that in the plateau region, the current is mass transport controlled, see Fig. 2; the currents are consistent with an $8e^-$ reduction. Moreover, coulometry confirms an $8e^-$ reduction and the yield of ammonia (as determined by a spectroscopic method) at a potential in the limiting current region is consistent with ammonia as the only product and there is no evidence for intermediate products, see Fig. 3. Hence, in these conditions the reduction of nitrate is

$$NO_3^- + 9\,H^+ + 8\,e^- \rightarrow NH_3 + 3\,H_2O \qquad (8)$$

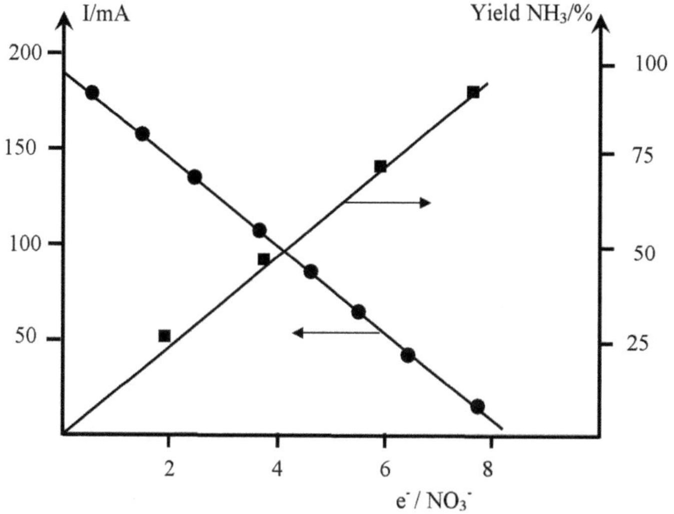

Figure 3. Electrolysis of 5 mM nitrate in 0.1 M $HClO_4$ + 0.9 M $NaClO_4$ at a Cu gauze electrode (9 cm^2) using a potential of - 0.60 V vs SCE.

When the pH was increased to 3, the limiting currents were diminished indicating that the number of electrons involved in the reaction had decreased. This demonstrates the importance of the ratio of nitrate and proton concentrations to the electrochemistry. An excess of protons at the electrode surface is essential to obtain complete reduction as the nitrate reduction itself consumes proton leading to depletion at the electrode surface. The conclusions from this paper were later used to develop analytical methods for the determination of nitrate.[49,50] Carpenter et al.[49] described a procedure for the determination of nitrate in drinking water that also uses the observation that the sensitivity may be enhanced by the inclusion of a low concentration of copper (II) in the analyte so that a fresh, high surface area copper surface was continuously deposited. Davis et al.[50] also used electrodeposited, high surface area copper surfaces to give a lower detection limit suitable for the determination of nitrate in lettuces. Dima et al.[19] have compared the activity of several cathode materials for nitrate reduction using 0.1 M nitrate in 0.5 M perchloric acid or sulphuric acid and conclude that copper is the most active. Using differential electrochemical mass spectroscopy, they identified some nitric oxide product in these conditions. However, the amount of nitric oxide is not quantified and a low percentage of this partially reduced product could arise because of the small excess of protons in these solutions. Kaczur[71] used a three dimensional copper felt cathode for the efficient removal of nitrate ion (typically 1.3 mM) at pH 2.26 and 3.05. He detected no ammonia and considered nitrogen to be the major product (some nitrite was formed at intermediate stages in the depletion). The partially reduced product may be explained by the very low current density used (the cell current was 1 A with a cathode total surface area of 5460 cm^2), well below the mass transport limited value, cf. Poorabedi and Pletcher[42] who obtained ammonia in mass transport controlled conditions. Kaczur[71] also reports that the copper cathode gave a much higher rate of nitrate reduction than either graphite or stainless steel in these conditions.

Although in 1.0 M perchloric acid, nitrate ions are not reduced before hydrogen evolution on nickel, alloying of copper with nickel has been shown to lead to a substantial positive shift in the reduction wave for nitrate at the NiCu alloy compared to pure copper.[72,73] It is suggested that the catalysis results from a mechanism where nitrate adsorbs on copper and hydrogen atoms on the nickel. Practically, alloying with nickel decreases the contribution from background currents resulting from hydrogen evolution in the plateau region for nitrate reduction, thereby lowering the detection limit in analytical procedures. With the alloy $Cu_{25}Ni_{75}$, the product is reported to be ammonia but the limiting currents are lower with $Cu_{50}Ni_{50}$ alloy indicating that only partially reduced products are formed.

In strongly acidic solutions, nitrate reduction does not occur at gold cathodes; only hydrogen evolution is observed.[19,74] With solutions with pH > 1.6 and a high nitrate concentration (0.5 M), however, at least at low current densities compared to mass transport control, the reduction of nitrate becomes the predominant cathode reaction and a mixture of nitrite and ammonia is found.[74] As expected from the concept of proton deficiency, the ratio of nitrite/ammonia increases with increasing pH. Interestingly, changing the cation of the electrolyte from Na^+ to Cs^+ influences the experimental observations. The authors explain this in terms of *underpotential deposition of these group I metals*. This seems unlikely in an acidic aqueous solution and a double layer effect seems more probable. Silver is also a poor cathode material in strong acid although some reduction is observed at potentials just positive to hydrogen evolution.[19]

Many studies employ precious metal cathodes and low current density reactions are generally observed. The reports of the reduction of nitrate at platinum are many. There is general agreement that the reduction of nitrate can occur via a *direct* mechanism[17-19,37,75–78] and an *indirect* mechanism[59, 79–82] and present knowledge of these reactions is discussed thoroughly in a recent paper by de Groot and Koper.[36] The *direct* mechanism can occur at all nitrate and acid concentrations while the *indirect* mechanism is only observed in concentrated nitric acid in the presence of nitrite, either

(i) added deliberately,
(ii) present as an impurity in the chemicals,
(iii) formed by the chemical decomposition of the nitric acid, or
(iv) formed by cathodic reduction.

Figure 4 shows a typical cyclic voltammogram at Pt, in fact for a solution containing 0.1 M $NaNO_3$ + 0.5 M $HClO_4$. On the forward scan to more negative potentials, the *direct* reduction is seen as a large peak at ~ +0.1 vs. RHE. A substantial cathodic peak is also seen on the potential sweep towards more positive potentials over the same range of potentials as the forward scan. In general, the *direct* reduction always occurs at potentials negative to +400 mV vs. SHE, i.e., in the potential range where hydrogen adsorption occurs. The rate of the *direct* reduction depends on the coverages of the surface by both adsorbed hydrogen and adsorbed nitrate and hence is influenced by potential, concentration of nitrate and the presence of other species, e.g., sulfate, capable of adsorption. This explains the sharp reduction peak in Fig. 4, as negative to the peak the coverage by hydrogen approaches one and nitrate is not able to adsorb and continue the reduction. It also explains the cathodic peak on the reverse scan; as the coverage by adsorbed hydrogen decreases, the nitrate can again adsorb and nitrate reduction recommences. It must be strongly emphasised that the rate of reaction is also always slow, orders of magnitude below the mass transport controlled value. Current densities are < 1 mA cm^{-2} even for > 1 M nitrate ion. The major product of reduction over the range +400 mV to 0 mV is always ammonia but differential elec-

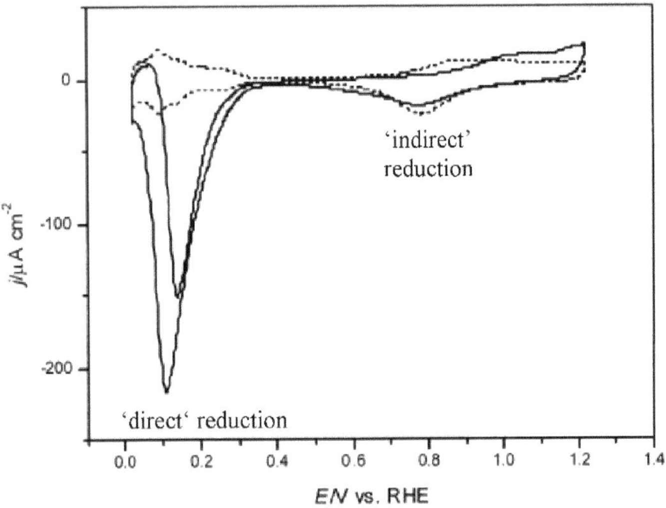

Figure 4. Cyclic voltammogram of polycrystalline Pt in 0.5 M $HClO_4$ with (solid line) and without (dashed line) 0.1 M $NaNO_3$. Potential 20 mV s^1. Reprinted from Ref. 19, Copyright (2003) with permission from Elsevier.

trochemical mass spectroscopy detects small amounts of NO, N_2O and N_2. It can also be shown by applying a potential of ~ 350 mV vs. SHE (in the foot of the reduction peak) to the Pt electrode in the solution containing nitrate, removing and washing the electrode with water and then recording a voltammogram in nitrate free acid that NO adsorbs quite strongly on the Pt surface. Coverages by NO estimated from the voltammograms can reach 0.4. The NO_{ads} leads to an increase in cathodic charge in the hydrogen adsorption region, showing that at more negative potentials, the adsorbed NO reduces further to ammonia. Hence, the mechanism of the *direct* reduction could be written

$$NO_3^- \rightarrow (NO_3^-)_{ads} \tag{26}$$

$$H^+ + e^- \rightarrow H_{ads} \tag{27}$$

$$(NO_3^-)_{ads} + 3\,H_{ads} + H^+ \rightarrow NO_{ads} + 2\,H_2O \tag{28}$$

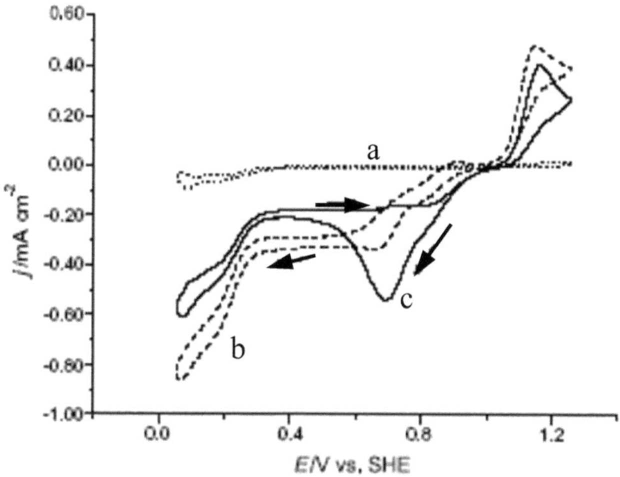

Figure 5. Cyclic voltammograms for a polished Pt electrode in (a) 1 M $NaNO_3$ (b) 25 mM $NaNO_2$ (c) 1 M $NaNO_3$ + 25 mM $NaNO_2$. in 2 M $HClO_4$ + 0.5 M H_2SO_4. Potential scan rate 10 mV s^{-1}. Reprinted from Ref. 36, Copyright (2004) with permission from Elsevier.

$$NO_{ads} + 5\,H_{ads} + H^+ \rightarrow NH_4^+ + H_2O \qquad (29)$$

although the later reduction steps could be electron transfers rather than reaction with adsorbed hydrogen atoms. The *indirect* reduction gives rise to the smaller symmetrical peak at ~ 700 mV vs. RHE on cyclic voltammograms, the current dropping down towards zero before the *direct* reduction commences at more negative potentials. Figure 5 shows cyclic voltammograms that illustrate responses obtained for the *indirect* reduction. By comparing the voltammogram (c) with (a) and (b), it can be seen that the presence of nitrite is leading to a reduction current for nitrate and a cathodic peak at ~ 700 mV. The voltammograms are also showing waves for both nitrate and nitrite at less positive potentials for the *direct* reduction. The peak for the *indirect* reduction is, however, much more prominent in a more acidic solution containing a

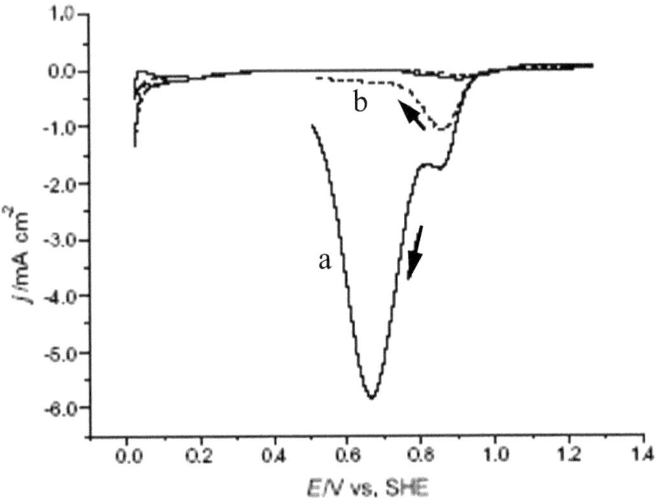

Figure 6. Voltammograms for 5 M HNO_3 + 0.5 M H_2SO_4 at a polished Pt disc electrode (a) stationary and (b) rotating (49 rps) electrode. Potential scan rate 15 mV s^{-1}. Reprinted from Ref. 36, Copyright (2004) with permission from Elsevier.

higher nitrate concentration as is illustrated with the voltammograms in Fig. 6. While the current densities are still low compared to the mass transport limited value for the nitrate in solution, it can be seen that in appropriate conditions the peak current density can be significantly bigger than that observed for the *direct* reduction. Moreover, the peak current density is also enhanced by increasing the nitrite concentration or the temperature. The *indirect* reduction peak is only observed when both nitrate and nitrite (as an impurity or a deliberate addition) are present in the electrolyte and the peak current density can be a function of time since nitrite is not stable in strong acid media (products such as NO and NO_2 are formed). Also,

(a) The reaction is autocatalytic. For example, in cyclic voltammetry, the second cycle can show a peak substantially bigger than on the first cycle and the peak current also increases as the potential scan rate is decreased,

(b) When explored with a rotating Pt disc electrode, the currents are found to decrease with increasing rotation rates. In Fig. 6 an intermediate key to the *indirect* reduction mechanism is being swept away from the electrode surface by the convection resulting from rotation.

Cyclic voltammetry also indicates that NO is not adsorbed at the potentials of the *indirect* reduction peak. Differential electrochemical mass spectroscopy identifies NO as the major gaseous product together with traces of NO_2 in the potential region of the *indirect* peak. At slightly more negative potentials, N_2O becomes a major product together with some N_2. These experimental results lead to a possible mechanism. The *indirect* reduction is considered to occur by the sequence

$$HNO_2 + H^+ \leftrightarrows NO^+ + H_2O \tag{30}$$

$$NO^+ + e^- \rightarrow NO \tag{31}$$

$$2\,NO + HNO_3 + H_2O \rightarrow 3\,HNO_2 \tag{32}$$

with the reduction of NO^+ as the electron transfer reaction. The reaction is autocatalytic because the consumption of two molecules leads to the formation of three molecules of nitrous acid. Also removal of the nitrous acid from the electrode surface by convection leads to a decrease in current density. The sharp, symmetrical reduction peak then arises because at slightly more negative potentials, the NO is removed from the catalytic cycle by reduction to N_2O in a reaction such as

$$2\,NO + 2\,H^+ + 2\,e^- \rightarrow N_2O + H_2O \tag{33}$$

Nitrogen is then thought to result from the further reduction of N_2O. Balbaud et al.[83] have extended the study to 373 K and both the *direct* and *indirect* mechanisms are still observed. While different chemical reactions are proposed, the general concepts are the same. Of course, much higher current densities are obtained. Higher current densities are also obtained by using platinised platinum surfaces[17,25,27,37,84] although they remain very low for practical application. Studies of the *direct* reduction confirm the role of adsorbed hydrogen and the need for the adsorption of nitrate.

Finally, the reduction of nitrate has been investigated at single crystal Pt surfaces.[21-24,85] The cyclic voltammetric response is sensitive to the structure of the platinum surface but this structure sensitivity is considered largely to arise from sensitivity of the adsorption of hydrogen (critical to the *direct* reduction) and other anions such as sulfate (competing for sites with nitrate) rather than a direct influence on the behaviour of nitrate. At the Pt(111) surface[23] in 0.1 M $HClO_4$ solution containing nitrate the first cycle of a voltammogram shows two reduction peaks, a sharper peak at +0.80 V that overlaps the peak for OH desorption and a broader peak at +0.33 V vs. RHE associated with a mechanism involving adsorbed hydrogen. On later cycles the more positive peak is absent. Experiments at [n(111) x (111)] stepped Pt surfaces[23] showed that the rate of nitrate reduction in the hydrogen adsorption region increases markedly with the step density and it was concluded that nitrate was able to adsorb at (111) monoatomic steps but not on (111) terrace sites. At the Pt(110) surface[24] in the same solution, the response is dominated by a symmetrical peak at ~ +0.2 V vs. RHE in the hydrogen adsorption region. Taguchi and Feliu[23,24] conclude that the data at single crystal Pt electrodes is compatible with a Langmuir-Hinshelwood mechanism involving adsorbed hydrogen and adsorbed nitrate.

Dima et al.[19] have reported voltammetry for nitrate in acid media using several precious metal cathodes and conclude from comparison of the current densities at polished electrodes that the activity decays in the order Rh > Ru > Ir > Pt > Pd. The differences are substantial; at Rh, steady state reduction currents are clearly seen positive to hydrogen evolution while at Pd the activity is negligible. At all precious metal cathodes, however, the current densities for the reduction of 0.1 M nitrate are low and any practical application would require the use of a high area dispersed metal centres on a substrate such as carbon. Indeed, a recent paper[31] employing highly dispersed palladium, mean particles size 10.5 nm, on high surface area carbons still showed only very low current densities for the reduction of nitrate (0.5 M) in 1.0 M $HClO_4$ although the activity did increase with temperature. It was also possible to confirm reduction of nitrate by differential electrochemical mass spectroscopy and both N_2O and NO were detected as products. At rhodised Rh, the current/potential characteristic again shows a relatively high rate of reduction and it is sigmoidal in shape so that there is no evidence for blocking of the surface by adsorbed hydrogen atoms.[38] On the other hand, nitrate ion reduction is inhibited by the adsorption of chloride ion[38] and other chemisorbed species[86] and there is both differential electrochemical mass spectroscopy and FTIR evidence for the involvement of NO as an intermediate.[18,19] Hence, it may be assumed that the mechanism for nitrate reduction on rhodium is similar to the *direct* reduction on platinum. There is also some evidence that the major product on rhodium is ammonia. There are also preliminary studies of platinum/rhodium[18] and platinum/iridium[87] alloys and evidence is presented that the Pt/Ir alloy is more active than either of the single metals.

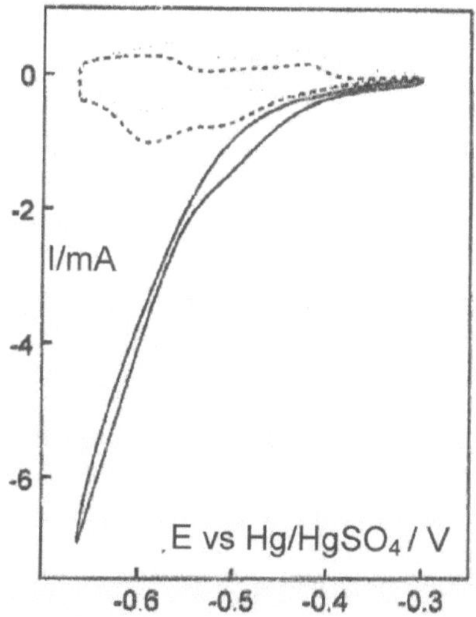

Figure 7. Cyclic voltammogram for a clean platinised Pt (dashed curve) and platinised Pt covered with submonolayer Ge (coverage ~ 0.15). Solution 0.5 M H_2SO_4 + 0.1 M KNO_3. Scan rate 20 mV s^{-1}. Reprinted from Ref. 39, Copyright (1997) with permission from Elsevier.

Sub-monolayer deposition of many metals increases the response of platinum and platinised platinum for nitrate reduction[39,40,88–94] with germanium[39,40] and tin[91–94] showing the largest effect. Figure 7 shows the voltammetry for 0.1 M NO_3^- + 0.5 M H_2SO_4 at a platinised platinum electrode, with and without a sub-monolayer coverage by germanium.[39] In the presence of the germanium, the current increases continuously as the potential is scanned more negative and on reversing the potential scan there is little hysteresis. The response for 10 mM nitrate is increased by a factor of 70 by a low germanium coverage although the current density remains much lower than that expected for mass transport control. The presence of sub-monolayer germanium also prevents inhibition by adsorbed hydrogen and leads to a change of product; hydroxylamine is formed alongside ammonia and oxides of nitrogen; the selectivity to hydroxylamine increases to about 60% as the germanium coverage is increased to 0.3. Later studies[91] have concluded that the germanium on the surface is present as a hydride species and its role is to prevent the formation/adsorption of inhibiting species rather than positively promoting the reduction on nitrate. Very recently, FTIR spectroscopy has confirmed[40] that, at least in the potential range for nitrate reduction, the germanium is present on the surface as a hydride species. It is suggested that the change of major product from ammonia to hydroxylamine is a *third body* effect. The germanium hydride on the surface limits the number of available neighbouring Pt sites required for the cleavage of a N-O bond. Probably, the NO intermediate must lie flat on the surface in order for the final N-O bond to break and ammonia to result and this required two neighbouring sites. Shimazu et al.[91] first reported the strong positive influence of tin adatoms on the Pt surface on the reduction of nitrate. Using a Sn coverage of 0.34 and slow scan voltammetry, they were able to obtain current densities in excess of 5 mA cm^{-2} for a 10 mM nitrate solution in 0.1 M $HClO_4$. This rapid catalysis was confirmed by others[92-94] but Kim et al.[94] found that loss of activity was a problem. They suggested two mechanisms; the first involves $Sn(OH)_2$ formation during the fabrication of the Sn-modified surface and its later dissolution while the second results from diffusion of the Sn atoms into the Pt lattice. Nitrogen is the major product formed at Sn-modified platinum.

Underpotential layers of transition metals on the surface of other precious metals can likewise have a dramatic effect on the activity for nitrate reduction. Such enhancements have been reported for Cd on gold and silver,[17,35–37] Cu on palladium,[95,96] Ge on palladium,[39,97] Cd on gold,[43] Pb on gold,[98] and Ag and Pb on Au(111).[99] Sn on palladium[92,100] is a particularly catalytic surface and current densities > 10 mA cm^{-2} could be achieved with a tin coverage of 0.65 for 10 mM nitrate solution in 0.1 M $HClO_4$. The major product is nitrous oxide in contrast to nitrogen at Sn/Pt. In general, it appears that these increases in rate result from the avoidance of poisoning species on the surface rather than a direct catalysis of nitrate reduction. Long term stability of such surface would also be a general concern in many applications.

The activity of the precious metals for the cathodic reduction of nitrate in acid media can also be enhanced by alloying. Precious metal alloys studied include Pt/Rh[18] and Pt/Ir.[87] More interesting alloys are those of Pd-Sn. Casella and Contursi[101] electrodeposited a series of Pd-Sn alloys and found that, although neither palladium or tin were active for nitrate reduction in 50 mM H_2SO_4, some of the alloys gave well-formed, mass transport controlled peaks for the reduction of nitrate. The most active alloy was $Pd_{33}Sn_{67}$.

Two papers consider the influence of deposition of palladium onto copper[102] and of copper onto palladium[95] and claim enhancement in the activity for nitrate reduction in acid solution compared to the individual elements. Both also stress the inhibition of the catalysis by sulphate and chloride ion and conclude that this provides evidence for the adsorption of nitrate as a critical step in its reduction. The paper by de Vooys et al.[95] considers in detail the influence of copper coverage on the reaction. It is shown that the current density at fixed potential increases linearly with coverage and leads to a change in product selectivity. Nitrogen is the major product at low Cu coverage while nitrous oxide dominates as a monolayer of copper is approached. The behaviour of a monolayer of copper is, however, not the same as the bulk metal. It is concluded that copper activates the first electron transfer while the palladium directs the selectivity towards nitrogen. The trend is also towards nitrous oxide with increasing nitrate concentration.

In all its forms (graphite, vitreous carbon, diamond), carbon has been studied as a cathode material for nitrate reduction at higher pH. In acidic solutions, the overpotential for nitrate reduction appears to be very high and the application of carbon is limited to that of substrate for other materials. The active materials include copper[49-51] and metal complexes.[103]

2. Neutral Media

While the electrochemical reduction of nitrate in neutral media is of interest for the development of analytical procedures and sensors as well as in water and effluent treatment (where the preferred product is nitrogen), reliable, fundamental studies of mechanism and kinetics are seldom achievable. All reductions of nitrate, e.g., reactions (1)–(8) consume protons and in unbuffered, neutral solutions this will lead to a rapid rise in pH close to the electrode surface and the extent of this rise will depend on the nitrate concentration and current density as well as the mass transport regime. So for example, the local pH at the cathode surface will change during a potential sweep through the peak for nitrate reduction. Indeed, the fact that nitrate reduction in neutral solutions leads to the formation of hydroxide ion at the cathode surface has long been recognised and applied as a procedure for the controlled formation of metal hydroxide/oxide layers on a substrate surface. The best known example is the deposition of a nickel hydroxide layer by reduction of an aqueous nickel nitrate solution.[104,105] In general, the use of buffers has not been followed up because it is to be expected that they would change markedly the mechanism and kinetics of the cathode reaction, for example by adsorption of the buffer ions on the electrode surface. To some extent, the impact of the pH increase is reduced in many technological situations where the concentration of nitrate is low and/or the use of a flowing solution sweeps the hydroxide away from the cathode. Even so, the conclusions from mechanistic and kinetic studies in *neutral media* really apply to an ill-defined slightly alkaline pH.

The most active metal cathodes for nitrate reduction, as in mildly acidic solutions, are freshly plated copper[51,52] and the precious metals, particularly iridium[28,29,106] and rhodium[30,107,108] that are reported to give higher rates of nitrate reduction than platinum or palladium. The activity of rhodium centres deposited onto pyrolytic graphite depends on the deposition conditions (and correlates with centre size) but the product is always a mixture of nitrite and ammonia.[107,108] Peel et al.[30] investigated the application of a commercial Rh on carbon cloth electrode for the removal of nitrate from groundwater. They confirmed the activity of Rh as a catalyst and showed that the electrode material was capable of decreasing nitrate levels to < 50 mg dm^{-3}, as required in the production of potable water but a mixture of products were formed.

Politades and Kyriacou[109] compared the products from the reduction of 0.1 M nitrate in neutral sulfate at Cu, Zn, Al, Pb, C felt and two alloys ($Cu_{60}Zn_{40}$, $Sn_{85}Cu_{15}$); nitrate could be reduced at all these materials but the highest selectivity for nitrogen was found at aluminium and the tin/copper alloy.

A preliminary note[110] has reported a high reduction activity for a copper/palladium surface. Copper/platinum surfaces have been investigated by Vila et al.[111] This group modified a carbon surface by cathodic reduction of a diazonium salt to produce a carbon surface covered with 4-sulfophenyl groups. These were then complexed with Cu(II) and the Cu(II) reduced cathodically or by reaction with sodium borohydride. Finally, these surfaces were dipped into a K_2PtCl_6 solution for various periods. The resulting Cu/Pt surfaces gave good voltammetric peaks in 2 M sodium nitrate but the peak height decreased as the Cu atoms were replaced by platinum.

Dash and Chaudhari[112] investigated graphite, iron, aluminium and titanium as cathode materials at pH 7 and 9. With the metal cathodes, they report effective removal of nitrate at a level of 100 mg dm^{-3} using a rather high current density of 0.14 A cm^{-2}. At iron and aluminium, ammonia was formed in almost quantitative yields but at titanium, nitrogen was said to be the major product as other products could not be identified.

A Greek group[113,114] have used tin cathodes at very negative potentials for the reduction of nitrate in potassium sulphate and demonstrate an interesting result. Using a potential of –2.9 V vs. Ag/AgCl, they found the major product to be nitrogen with a yield of 92% with some ammonia (~ 8%) and a trace of nitrite. At –2.4 V, the yields of nitrogen and ammonia were almost equal (35-40%). At both potentials, the rate of conversion was high and the current efficiencies were reasonable, initially 60% at –2.9 V. The base electrolyte also influences the rate of reduction. At –1.8 V, the rate of nitrate reduction was highest in the presence of multi-charged cations and followed the sequence, $La^{3+} > Ca^{2+} > NH_4^+ > Cs^+ > K^+ > Na^+ > Li^+$ and also, for halides, followed the order, $F^- >$

Cl^- > Br^- > I^-. The latter was thought to be a double layer effect while the influence of the anions is similar to that observed for, for example, copper in acid solution.[42]

Diamond cathodes have been investigated.[115,116] The latter paper[116] compares three different preparations of diamond and finds that

(i) the surface structure of the diamonds as observed by electron microscopy differ markedly,
(ii) the voltammograms also show differences but all give substantial current densities for nitrate reduction negative to –1.50 V vs. SCE and these can reach > 50 mA cm^{-2} with 1 M KNO_3 as the medium, and
(iii) the ratio of the products is independent of the diamond preparation; the major products are always nitrite and nitrogen containing gases in almost equal amounts.

Taniguchi et al.[117] have investigated the reduction of 0.1 M nitrate at a series of cathodes (Hg, Pb, Cu and Ag) in the presence and absence of CoL^{3+} and $Ni(II)L^{2+}$ where L is the macrocyclic ligand 1,4,8,11-tetra-azacyclotetradecane. The metal complexes are effective electrocatalysts leading to substantial reductions in overpotential for nitrate reduction. Moreover, the complexes appear to be stable and give high turnover numbers and selective products. Interestingly, the presence of the complexes leads to a change in product from ammonia to hydroxylamine. Moreover, the concentration of catalyst required is very low (e.g., 20 μM) and it is thought that the active catalyst species are adsorbed on the cathode surfaces. Modified electrodes based on polymerised nickel and copper tetraamino-phenylporphyrin entities[98] have also been shown to be active for nitrate reduction in neutral solution when the major product is nitrite.

A very surprising result was reported by Zhang et al.[118] They report that, in terms of the potential, polypyrrole was the most active electrocatalyst for nitrate reduction yet reported; in 0.1 M nitrate, pH 7, they found that reduction commenced at +600 mV vs. SCE. It would be interesting to see this system investigated further.

3. Alkaline Media

In alkaline solutions, a major driving force for investigations has been the wish to remove nitrate from various waste streams in the nuclear industry and a large number of cathode materials have been investigated in appropriate conditions for this application. In addition, academic papers have sought to probe fundamental aspects of electrocatalysis but such work is more limited than that in acid solution. There is little overlap between these two strands and it is therefore difficult to integrate the conclusions. In general, in alkaline media, the potential for nitrate reduction is shifted to substantially more negative potentials and the electroactivity is less in alkaline than in acidic aqueous solutions. One reason is the greater chemical stability of nitrate itself (as well as its reduction products) in alkaline solutions and the intervention of homogeneous chemical reactions is much less likely.

Copper remains a popular choice of cathode material[58,74,119-122] for the reduction of nitrate at alkaline pH. There is general agreement that the activity of copper for nitrate reduction depends on the history of the surface and its pre-treatment, supporting the view that special sites on the surface are important to nitrate reduction on copper. Cattarin[119] reports variation of the cyclic voltammetric response with the number of cycles, the potential limits and the direction of scan. A recent paper[58] has investigated in more detail the changes to the copper surface produced by potential cycling (either stepping or scannining) in 1 M NaOH and has shown that high surface area and highly structured surfaces can be produced. Figure 8 shows SEM images of copper surfaces produced by anodisation and subsequent reduction under controlled conditions and it can be seen that the copper layer is made up of nanowires. The same paper goes on to show that there is a good correlation between the structure of the Cu layer and the potential for nitrate reduction in 1 M NaOH; with the high surface area structure, there is a positive shift of ~ 100 mV compared to smooth copper. Paidur, et. al.[120] report that a low concentration of Cu(II) in the electrolyte prevents deactivation of the surface and this probably results from the continuous deposition of a high surface area form of copper.

The most dramatic effects resulted from ball milling.[58] Ball milling in air and argon led to quite different forms of copper powder. While both increase the activity for nitrate reduction, ball

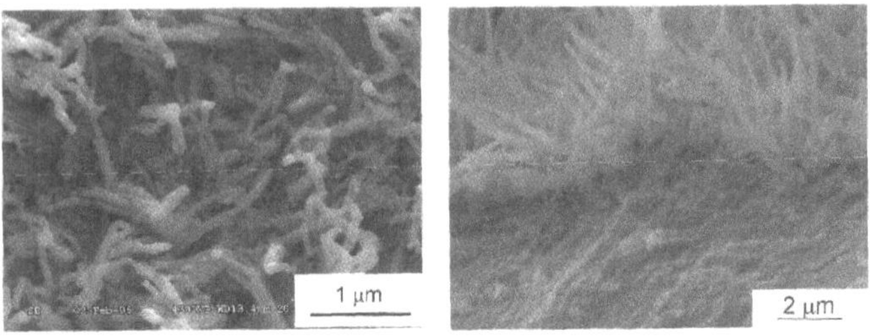

Figure 8. SEM images of copper deposits produced by anodising the copper at –100 mV vs Hg/HgO for 900 s followed by reduction with 20 cycles between –450 mV and –1650 mV, both in 1 M NaOH. (a) top view (b) cross section. Reproduced from Ref. 197, Copyright (2007) with permission of the ECS—The Electrochemical Society.

milling in air produces the more active surface for nitrate reduction in base. During cyclic voltammetry, ball milling in air leads to large increases in current density at all potentials negative to –0.90 V vs. Hg/HgO (more than a factor of ten) and this is shown not to be a surface area effect. It is suggested that the increased activity results from the creation of *active sites* (grain boundaries and/or surface defects) during the ball milling. In prolonged electrolyses, ball milling increased the coulombic efficiency for nitrate reduction to ammonia.

In general, cyclic voltammetry with a copper electrode in 1 M NaOH, nitrate gives a well formed reduction peak at ~ –1.30 V vs. SCE. The current efficiency for nitrate removal at copper is high and the major products are nitrite and ammonia. The literature gives the impression that reduction occurs in two stages since the selectivity to ammonia increases with charge passed and as the potential is made more negative (current density is increased). Homeric, et. al.[74] report a surprising variation in the products with the cation of the electrolyte at pH in the range 10-13. With Na^+, the current efficiency for ammonia increases with pH from 55% to 75% while the nitrite remains constant at ~ 27%, while with Cs^+, nitrite is the major product with the ratio of NH_3/NO_2^- in the range 0.36-0.48. Reduction of nitrate also occurs at silver in alkaline solutions;[119] a reduction peak is seen on voltammograms at ~ –1.2 V vs. SCE but at this metal, reduction appears to stop at the stage of nitrite ion.

Casella and Gatta[123] report that the addition of thallium to copper increases the electrocatalytic activity for nitrate reduction with $Cu_{45}Tl_{55}$ giving the highest response. They noted that the copper/thallium alloys formed by electrodeposition had a higher roughness than copper and also suggested that thallium may promote the adsorption of nitrate. Milhano and Pletcher[124] have compared the catalysis of nitrate reduction in alkaline solution at Pd-Cu alloys prepared by electrodeposition. Palladium itself is a very poor catalyst but with increasing palladium content of the Pd-Cu alloy, the reduction wave shifts positive by > 300 mV.

The Canadian Group has described the preparation of both Cu-Ni and Cu-Pd alloys by ball milling. Cu-Ni alloys[125] with bulk nickel contents of 20, 40 and 80% were prepared from the elements. When examined as catalysts for nitrate reduction in 1 M NaOH, the changes in performance were small; there was a small decrease (20%) in the rate of nitrate destruction and a small increase in the selectivity for the formation of ammonia (92 to 98%) as the nickel content was increased from 20 to 80%. The Cu-Pd catalysts[126] were prepared in a two stage process. Copper was initially ball milled for 6 hours before the addition of 1 at % palladium followed by further ball milling for 5-30 minutes. The objective was to minimise the palladium usage by limiting the alloy formation to the surface of the Cu particles. Palladium surface concentrations over the range 3-62% were produced. In line with the studies described above, the Pd-Cu were very active catalysts for nitrate reduction in 1 M NaOH. Again compared to copper, a surface alloy with ~ 70% Pd showed a positive shift of ~ 100 mV for nitrate reduction and also increased the selectivity to ammonia. It is suggested that the Cu atoms in the surface are responsible for the initial steps in nitrate reduction while the palladium participates in the later steps leading to ammonia. These recent papers are consistent with the earlier literature describing the strong catalysis by Pd/Cu surfaces produced by underpotential deposition of copper onto palladium.[95] Pd-Cu is indeed a most promising catalyst for nitrate reduction in alkaline media. Moreover, we understand that it will be the subject of several further papers from the Canadian group in the coming year.[127]

The electrochemical reduction of nitrate in alkaline solution has also been studied at hydrogen storage alloys,[128] $Mm(NiAlMnCo)_5$ where Mm is a mixture of La, Ce and Pr. It was shown that the hydrogen loaded alloys react chemically with nitrate and the alloys also act as cathodes, the major product being ammonia. Ball milling has also proved to be a useful way to fabricate alloys for formulating into cathodes.

Hobbs and coworkers[129-131] were interested in removing nitrate from nuclear waste streams and have investigated several cathode metals using solutions typically containing 3 M NaOH and 0.25 M Na_2CO_3. Early experiments[129,130] were conducted at controlled potential and low nitrate concentrations. Zinc, lead, iron, nickel and platinum cathodes were all found to give reasonable current efficiencies for nitrate removal with the efficiency decreasing along the series Zn > Pb > Fe > Ni. At zinc and iron, the selectivity to ammonia was high while at nickel and platinum, nitrogen was the major product. At lead, the ratio of ammonia/nitrogen was dependent on both, potential (or current density) and initial nitrate concentration. Later[131] much higher nitrate concentrations were employed and electrolyses were carried out at constant current. Nine cathode materials (Pb, Cd, Pb alloy, 316 stainless steel, Ni, Cu, porous Ni, graphite, Medicate ES6) were examined at room temperature for the reduction of a mixture of 1.95 M nitrate and 0.6 M nitrite. When the current density was 0.14 A cm^{-2}, all these cathodes led to significant removal of both nitrate and nitrite with nitrogen as a major product; lead and cadmium performed best. The study was continued in a flow cell using an even higher current density (0.5 A cm^{-2}) and lead and nickel were selected as the cathode materials. With both materials, a large amount of hydrogen evolution was obtained and early in the electrolyses, nitrogen and nitrous oxide were the main gaseous products from the nitrate and nitrite, with ammonia dominating when higher charges have been passed. The current efficiency was higher at lead when significant conversion of nitrate to nitrite occurred in the early stages of the electrolysis. Although these studies led to the conclusion that the conditions employed were effective for decreasing the nitrate and nitrite content of the nuclear waste, the current densities are very high compared to those observed during voltammetry and presumably correspond to rather negative potentials. Thus, for example a voltammogram at Ni shows a low current density, symmetrical peak at ~ –1.3 V vs. SCE and the response implies inhibition of the surface for nitrate reduction by a high coverage of adsorbed hydrogen atoms (as in acidic solutions).[121]

The precious metals have been less studied in alkaline solutions and the impression is that they are rather poor electrocatalysts in alkaline media. Hirani and Rizmayer[25] investigated platinised platinum and report a low current density peak on voltammograms in the potential range where hydrogen adsorption occurs. Tucker et al.[20] report similar behaviour for high area rhodium surfaces and also describe an activation of the rhodium surface for nitrate reduction by cycling the potential in the alkaline nitrate solution. Palladium shows a very low activity for nitrate reduction[95,132] but it is increased by underpotential deposition of copper.[95]

Several papers have addressed[128,133-135] the catalysis of cathodic nitrate reduction by macrocyclic complexes in alkaline conditions. None have found application but the most promising was $Co(1,4,8,11\text{-tetraazacyclotetradecane})^{3+}$ where significant nitrate reduction peaks were observed on voltammograms with low concentrations of nitrate.[128,134]

V. APPLICATIONS OF ELECTROCHEMICAL NITRATE REDUCTION

1. Analysis

The reliable and rapid analysis of nitrate is important in many fields. For health reasons, the intake of nitrate by humans must be controlled. Hence, throughout the world, the maximum level of nitrate in drinking water is fixed by legislation (in the UK the maximum permitted level is 50 mg/litre, i.e., ~ 0.8 mM). Similarly, the level of nitrate in food needs to be monitored while the concentration of nitrate in physiological systems is of interest to the health profession. The use of nitrates as fertilisers in agriculture, as a component in household detergents and in many industrial processes also makes the concentration of nitrate in natural waters and effluents a concern. The mapping of nitrate concentrations (within the range 0-40 μM) in the oceans is a routine procedure for defining ocean circulation and this is important to the understanding of the physics and chemistry of the oceans, weather forecasting, etc. Nitrate and nitrite are often found together and, indeed, nitrite is the more toxic. Hence, analytical procedures that distinguish nitrate and nitrite are particularly attractive although in many practical situations, the nitrite level is much lower than that of nitrate and nitrite is therefore not necessarily a serious interference to nitrate determination. Moorcroft et al.[136] have reviewed the analytical procedures, including amperometric and voltammetric methods, for nitrate and nitrite analysis.

Since the direct reduction of nitrate is generally not observed on mercury, polarographers needed to find indirect approaches in order to establish methods for nitrate analysis. Two ways forward were defined. The first used a soluble mediator capable of reducing nitrate and hence leading to a catalytic cycle where the limiting current plateau for the mediator was enhanced by the presence of nitrate. Suitable ions include molybdate[9-11] and the uranyl cation;[12] all give well formed polarographic waves well positive to the negative potential limit with significant current enhancements dependent on the concentration of nitrate in solution. The second[137] employs a highly charged cation such as La^{3+} or Ce^{3+}. The presence in the electrolyte of such ions leads to a polarographic wave when nitrate is present in solution. It is thought that such highly charged cations ion-pair with the nitrate and the positively charged ion pair is able to enter the double layer at the mercury drop surface at negative potentials. With both approaches, however, the relationship between limiting current and nitrate concentration is non-linear and this makes the methods unattractive.

Solid electrodes give more promising results partly because of the greater sensitivity achievable at Cu and Cd when the reduction wave results from an $8e^-$ reduction, see reaction (8). Early experiments used cadmium[68] and copper[42] rotating disc electrodes but a major advance occurred when it was found that the quality of the response was much enhanced by the addition of Cu^{2+} and/or Cd^{2+} to the analyte so that a fresh metal deposit was formed continuously during the analytical determination of nitrate. This approach was introduced by Bodini and Sawyer[69] who deposited in situ both copper and cadmium onto a pyrolytic graphite cathode. By varying the concentrations of Cu^{2+} and Cd^{2+} in solution, they were able to obtain linear calibration plots between reduction peak current on cyclic voltammograms and nitrate concentration over ranges between 0-10 μM and 0-10 mM. They also applied their procedure to the determination of nitrate in irrigation waters and airborne particulate samples. Johnson and Sherwood[138,139] used a variation of this procedure where they in situ plated copper onto a cadmium

RDE and adapted the method to design a detector for nitrate for use in high performance liquid chromatography.

Later papers recognised the importance of a fresh surface of metal on the detection electrode but consider copper alone to be adequate. Albery and coworkers[46] described a packed bed wall jet dual electrode where the Cu^{2+} could be generated from a bed of copper placed before the jet of a normal wall jet electrode; this creates a highly reproducible fresh copper on the wall disc electrode. Fogg and coworkers[48] designed and developed a capillary fill sensor fabricated using screen printing but used in situ copper plating onto the screen printed carbon working electrode to obtain a reproducible response. Davis et al.[50] reported that freshly plated electrodes could give separate peaks for nitrate and nitrite and analysed for nitrate in lettuces and sewage outfall waters. The group in Oxford[51] also report that boron doped diamond is a good substrate for copper deposition in these analytical procedures. These procedures are all based on potential sweep methods where the precise measurement of peak height can be a problem. Carpenter and Pletcher[49] therefore devised a potential step sequence whereby fresh copper was plated in situ onto a copper disc during a first pulse and the nitrate reduction current was measured following a step to a more negative potential. The analysis required only the measurement of a single current and the potential sequence was optimised to give the maximum response. They applied the method to the determination of nitrate in local utility water. A Turkish group[140] further improved the lowest detection limit by using square wave voltammetry at a glassy carbon disc on the analyte following a small addition of Cu^{2+}. Nitrate is seen as a well formed and easily measured peak negative to that for copper plating and a lowest detection limit of ~ 1 μM is claimed. Bertotti et al.[141] demonstrated that the Cu(II) required to plate an active surface could be formed in situ in the analyte. Prior to each differential pulse voltammogram used for an analysis, a polished and acid washed copper electrode was polarised at +0.50 V for 10 s (to drive copper dissolution) and then –0.25 V for 15 s (to redeposit a fresh copper surface). This pre-treatment led to a dramatic improvement in the peak size and shape for low concentrations of nitrate. Moreover, the peak height for nitrate reduction is proportional to nitrate concentration over the range 0.1-2.5 mM. The procedure was illustrated by the analysis of a mineral water (nitrate determined as 14.7 ppm) and a sausage meat (nitrate determined as 26.4 ppm).

Another way to improve the sensitivity is to use a microelectrode array when a sigmoidal current/voltage response is obtained and Ward-Jones et al.[52] have used in situ Cu plating onto a boron doped diamond microelectrode array to determine nitrate. They also claim a lowest detection limit of ~ 1 μM and report excellent analyses of natural waters containing 35 μM, 194 μM and 532 μM nitrate. Finally, Davis et al.[142] investigate the influence of ultrasound on these in situ copper deposition procedures for the determination of nitrate.

2. Electrosynthesis

In principle, the cathodic reduction of nitrate could be a route to a number of compounds, see Table 1. Indeed, some would be technologically feasible, for example, the reduction at copper in acid solution could be developed into a selective route to ammonia. The only process that is economically attractive, however, is the reduction to hydroxylamine nitrate

$$2\,NO_3^- + 8\,H^+ + 6\,e^- \rightarrow NH_3OHNO_3 + 2\,H_2O \qquad (34)$$

Hydroxylamine nitrate is a component of weapons and rocket propellants but it is also used in automobile air bag inflators and as a stripper in the manufacture of electronic components. A process for its manufacture was developed by the Olin Company in the USA.[63-66] No alternative to a mercury cathode was found and hence a cell was developed, see Fig. 9. It had horizontal electrodes with a 6 m^2 mercury cathode separated from the platinum clad niobium anode by a Nafion membrane. The catholyte feed was 5 M nitric acid and the anode reaction was oxygen evolution, also with a concentrated nitric acid electrolyte. The electrolysis was carried out at ~ 0.3 A cm^{-2} and operated with a current efficiency of 85%. The cell is capable of manufacturing 70 tons/year of 13 M hydroxylamine nitrate. The product stream from the cell contains unreacted nitric acid and this is neutralised with hydroxylamine. For this reason part of the product stream is converted to free hydroxylamine using an ion exchange column and cycled back

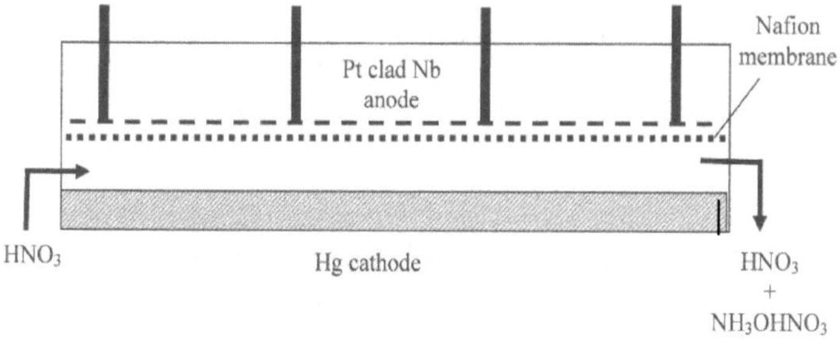

Figure 9. Schematic of the Olin cell for the manufacture of hydroxylamine nitrate.

into the anolyte exit stream. The stability of the product in all streams is greatly decreased by the presence of trace metal ions in solution and hence priority was given to pure feedstocks and corrosion resistant materials for the cell and auxiliary equipment.

Rutten et al.[61] have described a specific situation where the cathodic reduction of nitric acid to nitrous acid could be of interest. Manganese dioxide (γ-MnO_2) can be electrodeposited by the anodic oxidation of an acidic manganese nitrate solution. In a divided cell, the cathode reaction could be used to prepare nitrous acid for further extraction of manganese from its crude ore. Using a platinised titanium cathode and an initial catholyte feed containing 1 M HNO_3 + 1.2 M NO_3^- + 0.01 M HNO_2 at 353 K and a current density of 0.15 A cm^{-2}, the initial rise in nitrous acid concentration was rapid. Unfortunately, the concentration later reached a constant value (~ 0.06 M) because the nitrous acid was not completely stable.

A much more ambitious synthesis is the cathodic conversion of nitrate and carbon dioxide to urea as the first step in the production of organic molecules from two simple inorganic starting materials. Shibata et al.[143] have demonstrated this reaction using gaseous carbon dioxide fed to a gas diffusion electrode in contact with a solution containing 0.2 M $KHCO_3$ + 20 mM KNO_3 at 298 K. They investigated a number of metals as the cathode catalyst in the gas diffusion electrode and concluded that metals that supported the reduction of nitrate to ammonia and also carbon dioxide to carbon monoxide were the best catalysts for synthesising urea; indeed a good linear correlation was noted. The highest current efficiency to urea was ~ 35% using a potential of –1.75 V vs. SCE and zinc as catalyst. This is a promising result and it would be interesting to investigate the reaction further, particularly the influence of the electrolyte composition.

The cathodic reduction of nitrate also has a role in the electrodeposition of oxide and hydroxide layers on conducting substrates. All the reduction reactions of nitrate in unbuffered media lead to an increase in the pH at the cathode surface and it has been shown that the local pH can increase from 7 to > 12, allowing the deposition of for example, zinc oxide.[144] The local pH and hence deposition rate will depend on a number of parameters including the nitrate concentration, the current density, bulk pH and temperature and the properties of the deposits can be further controlled by potential modulation.[145] This controlled increase in pH at the surface can be used to deposit layers with good adhesion and defined properties. Initially, the electrolysis of an aqueous nickel nitrate solution was used for the deposition of nickel hydroxide onto nickel or steel cathodes during the manufacture of nickel/cadmium batteries.[104,105] More recently, the procedure has been modified and used for the deposition of nickel hydroxide onto Pt and Au microelectrode arrays in the fabrication of pH sensitive microelectrochemical sensors.[146] In addition the concept has been used for the electrodeposition of other oxide/hydroxide layers often onto indium-tin oxide coated glass as components of optical or light sensitive devices. These include titanium dioxide,[147] zinc oxide,[144,145,148,149] cuprous oxide,[150] $BaTiO_3$[151] and mixed Cu/Zn oxide.[152] The procedure has also been extended to nitrate reduction in a non-aqueous solvent, isopropanol in order to deposit a precursor to the high temperature superconductor, $YBa_2Cu_3O_{7-x}$; the electrodeposition largely produces a mixture of hydroxides that, on heating in oxygen, are converted to the superconductor.[153]

3. Treatment of Nuclear Waste

Hobbs and collaborators[131,154-158] have carried out extensive studies of the electrochemical treatment of low-level nuclear waste. This is a solution that remains after the removal of almost all the radioactive elements and it is a complex brew typically containing 1.95 M $NaNO_3$, 0.6 M $NaNO_2$, 1.33 M NaOH as well as lower concentrations of several other sodium salts (aluminate, sulfate, carbonate, chloride, fluoride, chromate, phosphate, silicate and tetraphenylborate) and traces of radionuclides and metal ions such as ruthenium and mercury. The objectives of the treatment would be

(a) to reduce the level of hazardous materials including nitrate, nitrite and the radionuclides,
(b) to reduce the corrosiveness of the waste – nitrate is an aggressive ion leading to stress corrosion cracking, pitting and crevice corrosion in the carbon steels employed in the nuclear industry, and
(c) to minimise the volume of the waste requiring disposal.

Recycle of chemicals, particularly the sodium hydroxide, within the waste would be a considerable benefit.

The cathodic reduction of the nitrate and nitrite to gaseous products is an obvious way to lower the nitrogen content of the waste. Following earlier experiments in glass cells,[130,131,133,134] lead and nickel were selected for extended testing and small pilot plant systems were constructed around two flow cells,[134] a FM01 LC (ICI, electrode areas 64 cm^2) and a MP cell (Electrocell, electrode areas 100 cm^2). The cells had a platinum anode and were operated both with and without a Nafion 417 membrane separator. A number of 50 hour electrolyses were carried with a simulant mix at 343-353 K at a current density of 0.5 A cm^{-2} using a charge sufficient to remove 67% of all the nitrate and nitrite. The membrane improved performance (avoiding oxidation of nitrite to nitrate at the anode) and lead was superior to nickel as the cathode since it gave less hydrogen as a co-product; moreover, contrary to expectations, the lead also appeared to be stable in these electrolysis conditions. At both cathodes, nitrogen, nitrous oxide and ammonia were the major gaseous products and the ammonia built up gradually until it predominated at the end of the electrolyses. At lead, nitrite initially builds up before declining but no such increase is seen at nickel. At lead the overall destruction efficiency was ~ 75%. Also the technicium and ruthenium contents in the waste are decreased as TcO_2 and Ru metal were deposited onto the cathode. To further demonstrate the concept, two 1000 hour tests were completed using 70 dm^3 batches of simulant mix in a divided cell with a lead cathode and a Pt coated anode at 343 K. The electrolyses led to > 99% removal of the nitrate and nitrite with a destruction efficiency that dropped from 70% to 55% as the concentra-

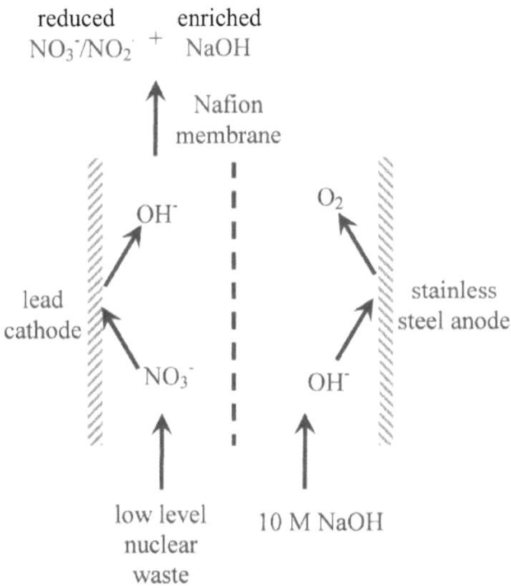

Figure 10. Preferred cell configuration for the removal of nitrate and nitrite from low level nuclear waste.

tions of nitrate/nitrite dropped. Some anode corrosion was noted during the electrolyses (using a sulphuric acid anolyte). As a result a further 100 hour test was carried out with a 10 M NaOH anolyte and a stainless steel anode and this change seemed to avoid anode corrosion without affecting the cathode performance. In this configuration, the sodium hydroxide concentration in the catholyte increased and it was possible to recover solid NaOH by crystallisation from the treated waste. The preferred cell configuration is sketched in Fig. 10. It should be noted that all these electrolyses were carried out where the current densities are very much higher than in all academic studies of electrode materials (ca. 500 mA cm^{-2} vs. < 1 mA cm^{-2}); the efficiency is surprising and different mechanisms for nitrate reduction must be expected. These flow cell electrolyses were also successfully modelled to allow prediction of the process performance with change in operating conditions.[155] In another study, the oxygen evolving anode chemistry was replaced by hydrogen consuming gas diffusion electrode[156] and this change was not found to change the ability of the cathode to remove nitrate and nitrite. The cell was operated with both acidic and alkaline anolytes in membrane divided cells and the emphasis was on devising conditions where the sodium hydroxide concentration in the waste (catholyte) built up and could be recovered. Acceptable performance was obtained with current densities in the range 0.15-0.25 A cm^{-2} and the cell voltage (hence, the cell energy consumption) was significantly reduced by using the hydrogen anode. No long term electrolyses to test the stability of the gas diffusion electrodes were, however, reported. Bockris and Kim[159,160] report the voltammetry of the species in the simulated nuclear waste using a nickel cathode and then show that the nitrate/nitrite, mercury, chromate and ruthenium can all be removed in cells with packed bed cathodes. The influence of the bed metal (Ni, Pb or Fe) and particle size were defined.

Hobbs[154] described a different approach leading to the recovery of pure sodium hydroxide and pure nitric acid solutions suitable for recycle back into the treatment sequence for the nuclear waste. The three compartment flow cell is shown schematically in Fig. 11. The low level nuclear waste is fed to the central compart-

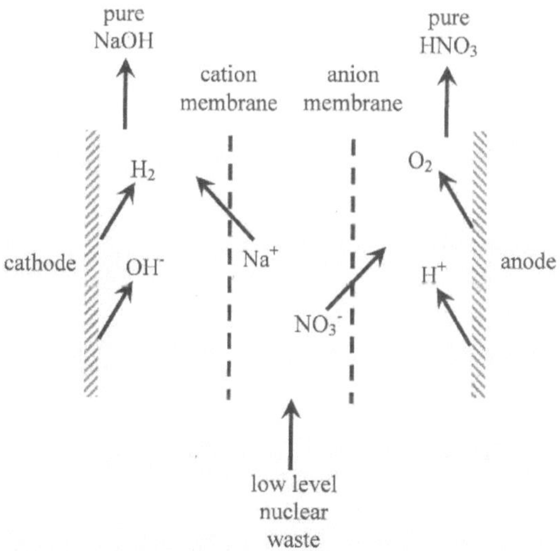

Figure 11. Schematic of the three compartment cell for recovery of both sodium hydroxide and nitric acid from the low level nuclear waste.

ment and water electrolysis is carried out at the two electrodes; use of an anion permeable membrane and a cation permeable membrane allows the specific transport of nitrate and sodium ions from the central compartment to the anolyte and catholyte respectively. A number of anion and cation membranes were tested. Anion membranes generally show poor stability to strong alkali and nitrite but the combination of Neosepta AM1 anion permeable membrane and a Nafion cation membrane gave satisfactory results. It was possible to remove almost all the nitrate and nitrite from the waste stream and a considerable reduction in waste volume was also achieved since water accompanies the ions through the membranes. Theoretically, a 75% volume reduction is possible. Although this approach seems attractive, it was not followed up, probably because of insufficient stability of the anion membrane for long term operation and the high cost of three compartment electrolysis cells.

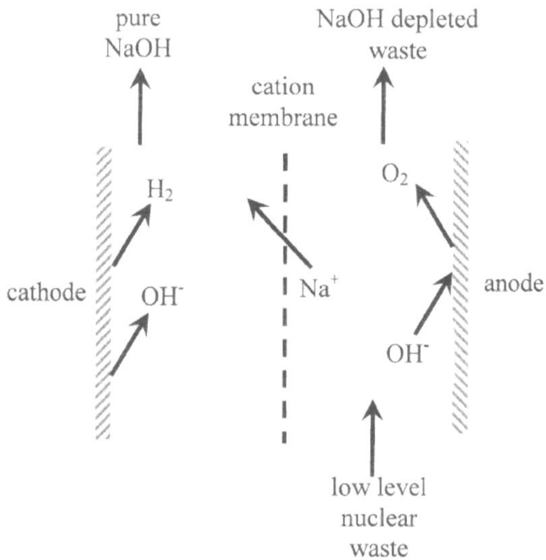

Figure 12. Schematic of the two compartment cell for recovery of high quality sodium hydroxide from the low level nuclear waste.

More recently, Hobbs[157] has described an electrolytic process that recovers sodium hydroxide from the low level nuclear waste but without removing the nitrate and nitrite. A schematic of the cell is shown in Fig. 12. The low level waste is fed to the anode of a cell carrying out water electrolysis. Hydroxide is converted to oxygen at the anode but is reformed at the cathode leading to an increase in concentration of a pure NaOH stream. This is a simpler cell than that described above and avoids the use of the anion permeable membrane. Hobbs describes a number of electrolyses in a MP cell with a nickel cathode, either nickel or platinised titanium anode and either a Nafion polymer membrane or a NASD ceramic membrane. Satisfactory performance was obtained with all the components tested although the ceramic membrane had to be operated at a lower current density. The current efficiency for hydroxide formation at the cathode was always high and it was possible to prepare 14% NaOH. The process also leads to a reduction in the volume of the waste as water accompanies the sodium ions through the membrane. With a Nafion membrane, but not the ceramic membrane, there was some contamination by radioactive caesium present in the waste. With a Pt/Ti anode, nitrite in the waste was oxidised to nitrate and with both Ni and Pt/Ti anodes, it was necessary to limit the removal of hydroxide, since below ~ 0.4 M the aluminium in the waste precipitated as a solid. In longer term testing, the nickel anode was found to corrode. But a FM01 cell with a nickel cathode, Pt/Ti anode and a Nafion 350 membrane was successfully operated for 1079 hours[158] although it was necessary to carry out several maintenance activities during this time. The test was carried out with a current density of 0.4 A cm^{-2} at a temperature of 313 K. The cell consistently produced 13.8% NaOH and there was no sign of electrode corrosion. The membrane did eventually show evidence for precipitation within its structure and this was thought to be aluminium and silicon species but it was concluded that with appropriate management, the membrane life would be > 5000 hours in this process environment.

Hirose and coworkers, working in the laboratories of Hitachi in Japan, have developed procedures for the removal of ammonium nitrate from the effluent of a process sequence to recover and recycle uranium in a plant for manufacture of LWR fuel.[161,162] The effluent contains 1.3 M ammonium nitrate and two approaches were considered. The first involved cathodic reduction of nitrate to nitrite in a divided cell with a lead cathode following adjustment of the effluent pH to 9 with ammonia. The product ammonium nitrite was then thermally decomposed at 343 K,

$$NH_4NO_2 \rightarrow N_2 + 2\,H_2O \quad (35)$$

Essentially, complete reduction of nitrate was possible and the current efficiency for nitrate reduction approached 100%. The current efficiency for nitrite formation was > 70% so that the process looked attractive especially since no by-products were formed. Unfortunately, the lead cathode deactivated with cell operation and attempts to reactivate led to lead corrosion. The second approach used a three compartment cell similar to that in Fig. 11 to split the ammonium nitrate into nitric acid and ammonia and it was possible to generate 6 M HNO_3 and 8 M NH_4OH using a current density of 0.3 A cm^{-2}. At the time of the presentation it was intended to scale up the system to 4500 moles/day. A four compartment cell process to concentrate the 6 M HNO_3 to 10.6 M HNO_3 was also described but this is likely to be uneconomic in practice.

Another opportunity in the nuclear industry is in *acid killing*. In the treatment of the aqueous phase resulting from the solvent extraction of Fast Breeder Reactor fuel reprocessing, it is necessary to reduce the nitric acid concentration in a waste stream from > 4 M to almost neutral.[163] Mallika, et. al.[164] have looked at precious metal anodes in an electrolytic process for this application. They investigated electroplated platinised platinum, both with and without diffusion annealing and a mixed metal oxide coating on titanium. Their preferred anode coating was electroplated 3 μm Pt and this appeared stable when operated at 80 mA cm^{-2} for 110 hours.

4. Removal of Nitrate from Natural Waters and Effluents

Throughout the world, pollution of ground water, rivers and lakes by nitrate is a problem. The nitrate largely arises from present agricultural techniques, particularly the intensive use of fertilisers, and is a risk to human health. Nitrate in drinking water has been identified as a cause of methemoglobenemia or blue baby syndrome and is also implicated in some forms of cancer, diabetes and birth defects. As a result, legislation generally limits the nitrate level in water and maximum concentrations in the range 10-50 ppm are enforced in most countries. Nitrate is also found in industrial effluents, sometimes at a relatively high level, and it then makes sense to remove it before it enters the natural water system. Consequently, technology to remove nitrate from waters is now essential. Successful approaches include biological denitrification, reverse osmosis, ion exchange, electrodialysis, catalytic hydrogenation and electrolytic methods[165,166] although it should be noted that reverse osmosis, ion exchange and electrodialysis are only concentration technologies. Either the technology must produce sufficiently concentrated nitrate solutions for them to be recycled or the technology is producing a secondary effluent that must be subjected to further treatment before disposal. This review will only consider methods for nitrate removal based on electrochemistry but it must always be remembered that, in practice, they are in competition with the non-electrochemical technologies.

The nitrate levels of interest in water treatment are generally low and certainly the target will be to remove the nitrate to $<< 1$ mM. Also, the volumes to be treated are usually large. Hence, the technology will be built either around a three dimensional electrode or the treatment process will have two stages, a preconcentration step, usually ion exchange, coupled to electrolytic destruction of nitrate from the concentrate. Paidur et al.[120] investigated the influence of cell design on the removal of nitrate from the regenerant solution for a strongly basic ion exchange resin, i.e., 1 M sodium bicarbonate containing low levels of chloride and sulfate as well as 1 g dm^{-3} nitrate. They selected copper as the cathode material and compared cells with a plate cathode, a plate with a fluidised bed of inert glass beads in the interelectrode gap, a packed bed of copper particles and a vertically moving copper particle bed. As expected, nitrate removal occurred in all the cells and ammonia was the major product with some nitrite early in the electrolyses. Moreover, in all cells the continuous creation of a freshly deposited copper surface, e.g., by addition of Cu(II) to the solution, increased the rate of removal. Even in the simple plate cell, reduction in the nitrate level from 1000 ppm to < 20 ppm was possible at low current densities (2 mA cm^{-2}) although the current efficiency fell to < 20% towards the end of the electrolysis. Introducing the bed of inert particles improved the mass transport and a higher rate of removal was possible; the current efficiency for 90%

nitrate removal was ~ 30% even with a current density of 32 mA cm^{-2}. While the rate of nitrate removal and the current efficiency was highest with the moving bed of copper particles, the performance of the three dimensional copper electrodes was not as good as predicted by theory and the authors believe that this is the result of insufficient control of the potential and current distributions through the electrodes. Kaczur[167] describes the use of a cell with a cathode (30 cm x 7.6 cm x 0.3 cm) of compressed copper fibres (60 μm x 75 μm, specific surface area 74 cm^2 cm^{-3}) for the treatment of both 85 mg dm^{-3} and higher (22-123 g dm^{-3}) nitrate concentrations. A very low current density was employed (the cell current was only 1.01 A) and the solutions used were acidified to pH 2.5 and 1.5 respectively. With the dilute solution up to 90% removal of the nitrate was achieved in a single pass through the cell at low flow rates and the current efficiency was ~ 15%. With increasing flow rate, the conversion decreased but the current efficiency improved. Little nitrite was formed and it is suggested that the major product is nitrogen. This is contrary to most of the literature on copper cathodes since ammonium ion is expected but nitrogen formation could result from proton depletion at the cathode surface. With the more concentrated solutions, the electrolyte was recycled through the cell many times and a copper felt cathode was used. The current efficiency for nitrate removal was very good but the products were not determined.

A number of papers have sought to employ metal catalysts dispersed over three dimensional carbon substrates. The metal catalysts include iridium,[28,29,103] rhodium,[20,30,168] Sn[113,114] and Pd/Sn[169] alloy and all appear effective in removing nitrate from neutral solutions. The academic literature, however, does not feature attempts to place such cathodes in engineered cells. On the other hand, Upscale Water Technologies Inc has marketed the N-Forcer™ for maintaining the nitrate level in closed loop fish tank systems at an acceptable level.[168] The key component is a cathode formed from rhodium catalysed carbon cloth. The three-dimensional cathode cell had a cathode surface area/catholyte compartment volume ratio of 7900 cm^2 cm^{-3}. Hence, a large fraction of nitrate is removed in a single pass. Systems were developed for both salt and fresh water and for 200 and 400 gallon tanks. Much larger systems able to handle up to 100,000 gallons/day were also being developed. These might have up to 160 cells in parallel and were designed to remove 80% nitrate in a single pass with an inlet nitrate concentration of 200 ppm.

Ionex Ltd has developed an integrated ion exchange column/electrochemical cell system for the removal of nitrate ion from household water.[20, 170-174] The overall objective is to convert the nitrate in the water to nitrogen gas with the minimum addition of chemicals or water or formation of waste product. The feed water typically has a nitrate level in the range 50-100 ppm and this is removed on ion exchange resin. The ion exchange columns are regenerated with a 2 M KCl solution with a residual level of nitrate and it leaves the columns containing ~ 20 g dm^{-3} nitrate. This is the feed to an electrolysis cell designed for simplicity and low cost. It is an undivided tank cell with a volume of ~ 400 dm^3 and fitted with 100 titanium bipolar electrodes. The cathode surfaces are electrodeposited rhodium that has been activated by potential cycling in the cell feed.[20] This cycling procedure leads to a restructuring of the rhodium to give a surface increased in real area to ~ 230 times the planar area along with morphological modifications; this activation procedure has to be repeated periodically during the cell operation. The anode surfaces are standard DSA RuO_2 coatings and in the cell operating conditions the anode reaction is oxygen evolution with the protons formed largely maintaining a pH balance in the cell. The cell operates at a current density of 100 mA cm^{-2} (based on superficial geometric area) and the feed is recycled through the cell to achieve a 90% reduction of the nitrate concentration. Nitrogen, together with hydroxide ion, is the only product formed at the cathode. After treatment in the electrolytic cell, the pH of the solution is readjusted to neutral with hydrochloric acid and it is then ready for recycle to the ion exchange columns. The technology is now licensed to BOC Ltd and is presently being tested within two UK water utilities, each providing 3000-4000 m^3/day of water to customers. To date the technology is performing well.

This technology has replaced an earlier concept[173] where the ion exchange columns were replaced by a three compartment cell with two anion permeable membranes and the centre compartment was filled with ion exchange beads. The water to be treated passed through the central compartment and the anion exchange resin acted as an instantaneous *trap* for the nitrate in the feed water. Electromigration induced by the cell current was used to drive the nitrate ion out from this central compartment into the anode compartment before it was transferred to a destruction cell. Clearly, this is a more complex arrangement and, in any case, the available anion membranes were not completely stable. Van Velzen and Langenkamp[175] described another approach to the removal of nitrate from chloride based ion exchange column regenerant. They used a commercial DEM cell (dished electrode membrane cell) with parallel plate electrodes. It had a Hastelloy C steel cathode, a DSA anode and a Nafion cation permeable membrane. This cathode material led to substantial amounts of ammonia as a product but this was removed by circulating the catholyte exit stream through the anode compartment where hypochlorous acid was formed, allowing the reaction

$$3\ HClO + 2\ NH_3 \rightarrow N_2 + 3\ H_2O + 3\ HCl \qquad (36)$$

In fact, several papers have suggested combining the generation of ammonia at a cathode with destruction of the ammonia by hypochlorous acid formed at an anode in a chloride electrolyte. Hiro, et. al.[176,177] used a Cu/Zn cathode and a precious metal anode in an undivided cell and treated solutions containing nitrate, nitrite and ammonium ion and were able to achieve 90% conversion of the nitrate to nitrogen gas. An Italian group[178] used a membrane cell with a Pd/Co oxide coated titanium cathode but found that the ammonia formed was transported through the membrane to be destroyed at Pt/Ir on Ti anode.

Cheng and coworkers[179] have described the application of a solid polymer electrolyte cell to the removal of nitrate in a bicarbonate based medium. The cell was based on a membrane electrode assembly (MEA) fabricated by hot pressing a Pd-Rh coated Ti cathode mesh and Pt/Ti anode mesh onto a Nafion 117 membrane. The cell was operated with a current density between 1 and 20 mA cm^{-2} giving a steady state cell voltage of 1.6 V to 7.4 V at a temperature of 333 K. It was possible to achieve almost complete removal of nitrate ion although the current efficiency dropped substantially as the nitrate level dropped. Nitrogen was the major product at lower current densities. The same cell has also been used[180] for the simultaneous removal of nitrate and ammonia at the cathode and anode respectively. Again a bicarbonate based medium was employed and it was possible to drop the levels of both nitrate and ammonia to a low level with nitrogen as the main product. The current efficiency for ammonia removal was, however, poor.

Dziewinski[181,182] has described a totally different approach to the removal of nitrates that is suited to both low level nitrate treatment of water and high level nitrate in industrial effluent. The electrolytic step is the reaction

$$Zn^{2+} + 2e^{-} \rightarrow Zn \tag{37}$$

carried out in a rotating cathode in conditions where the zinc is formed as a very fine and active powder. The zinc powder is then transferred to a chemical reactor where it is reacted with the nitrate containing solution after slight acidification to pH 2-4 and addition of sulfamic acid. The following reactions are rapid and selective,

$$Zn + 2\,H^{+} + NO_3^{-} \rightarrow Zn^{2+} + NO_2^{-} + H_2O \tag{38}$$

$$NO_2^{-} + H_2NSO_3H \rightarrow N_2 + SO_4^{2-} + H^{+} + H_2O \tag{39}$$

thus, ensuring that the nitrate is converted only to nitrogen gas. The exit stream from this reactor then passes through the electrolysis cell for the removal of the zinc cation from the water and the recycle of the zinc powder. The process may be illustrated by treatments of water containing ~ 220 mg dm^{-3} nitrate; the nitrate was decreased to < 1 mg dm^{-3} and the treated water contained only ~ 5 mg dm^{-3} ammonium ion and a trace of sulfamic acid. Clearly this approach consumes a chemical, sulfamic acid, and requires careful control. It is, however, rapid and capable of handling nitrate concentrations over the range < 50 mg dm^{-3} to > 1 M and sulfamic acid is a cheap chemical. A recent paper[183] has reported the influence of the reaction conditions on the rate and efficiency of nitrate removal.

Electrodialysis[184] has proved to be a very successful technology for the removal of nitrates from drinking water. A schematic of the cell used for the removal of nitrate by electrodialysis is shown in Fig. 13. For clarity, the cell is shown with only three pairs of anion/cation permeable membranes; a commercial cell will have 100-1000 membrane pairs between the two electrodes. In electrodialysis, the ionic separation occurs at the membranes and the role of the electrodes (usually carrying out water electrolysis) is only to create the electric field to drive the ionic migration through the membranes, anions towards the anode and cations towards the cathode. It can be seen in Fig. 13 that the cell has two streams fed to alternate intermembrane gaps, the water to be treated and a product stream, and the sodium nitrate passes out of the water stream into a product stream that becomes a concentrated sodium nitrate solution (for recycle or a small volume requiring further treatment).

In Europe, the development of the nitrate removal from drinking water technology has been led by Eurodia, a company in France.[185,186] Their technology is based on:

(a) The availability of an anion permeable membrane (Tokuyama Coup, Neosepta ACS) that shows considerable selectivity for the transport of nitrate versus other ions found in natural waters, particularly bicarbonate, chloride and sulfate. The exact selectivity depends on the ratio of the anions in the solution feed, current density, etc. but the ACS membrane performs significantly better than other commercial membranes; for example, for a 20% total dissolved salt reduction from a sample of water containing equal concentrations of the anions, the anions removed could be 50% nitrates, 40% chlorides, 10% bicarbonates, and 0% sulfates.

(b) An improved design of the spacer in each intermembrane gap (not shown in the Fig. 13). The main role of the spacer is to act as a turbulence promoter and hence to increase the flux of nitrate ions to the membrane surface. This increases the rate and selectivity of nitrate removal without degrading the other performance characteristics (e.g., power consumption).

(c) the development of a compact stack with 700-membrane pairs, each membrane with an active area of 0.4 m^2. The stack is capable of handling 45 m^3/hour, giving 90% removal of nitrate and 93-98% recovery of water. Typically, the membrane current density is ~ 2 mA cm^{-2}. The stack is also programmed for current and hydraulics reversal to avoid blocking of the membranes.

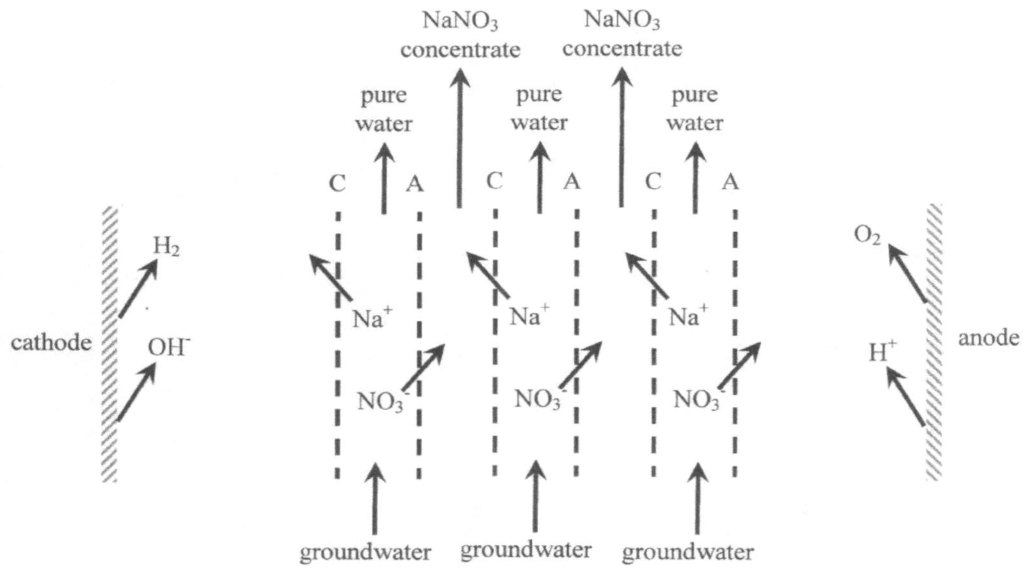

Figure 13. Schematic of an electrodialysis stack for the removal of $NaNO_3$ from groundwater. A = anion permeable membrane, C = cation permeable membrane.

A number of commercial plants have been commissioned and are operating around Europe to provide potable water meeting legal requirements. Table 2 lists these plants with scales of operation and some performance data. The plant in Austria has been described in some detail[187] and there is also further information on the web.[188]

GE Ionics also offer technology for nitrate removal from groundwater.[189-190] In the USA, the nitrate problem is slightly different; high nitrate is usually accompanied by a high total dissolved salt level and the GE Ionics technology has usually sought to reduce the level of other salts as well as nitrate. The membranes, the cell stack technology, and the current reversal approach has all been developed within the company and few details have been published. Details of some of the plants installed are given in Table 3 where water recovery is generally > 90%. Most of these plants have been operational for more than ten years, some for twenty years.

Table 2
Eurodia Electrodialysis Plants for the Removal of Nitrate Ion for Potable Water

Site	Volume treated (m^3 day^{-1})	Membrane area (m^2)	Feed Water: Total salt (ppm)	Feed Water: Nitrate, (ppm)	Product Water: Total salt (ppm)	Product Water: Nitrate (ppm)
Montfano, Italy	1000	660		92		34
Haraucourt, France	250	340				
Bern, Switzerland	1200	780				
Amsterdam, Holland	120	130				
Imola, Italy	600	440		52		9
Kleylehof, Austria	3500	1344	704	120	550	41
Jaunay-Clan, France	3900	1680	715	25	517	7.5

Three compartment, two membrane cells (similar to that sketched in Fig. 11) have been used for splitting ammonium nitrate into reusable nitric acid and ammonia[191,192] and such a cell is well suited to coupling to an electrodialysis unit to produce zero effluent technology.[193] The cell had a stainless steel cathode and a DSA anode, an AW anion permeable membrane (Solvay) and a Nafion cation permeable membrane. The ammonia formed at the cathode was stripped continuously while it was possible to form up to 8 M HNO_3 at the anode with a current efficiency of ~ 55%. The cell voltage was ~ 8 V with a current density of 0.2 A cm^{-2}. The diluted ammonium nitrate solution from the central compartment was then concentrated by electrodialysis and the concentrated stream returned to the electrolysis cell. Typically the Eurodia EU-2P-10 electrodialysis unit with CMV/AMV membranes increased the ammonium nitrate concentration from ~ 0.4 M to 1.1 M with the diluate stream having only trace amounts. The system was run for 1000 hours in 24-hour cycles.

A bipolar membrane electrodialysis stack[184] is similar to an electrodialysis stack except three membranes—an anion permeable membrane, a cation permeable membrane and a bipolar membrane—are alternated through the stack of 100-1000 membranes, see Fig. 14. There are now three streams, the sodium nitrate feed and the nitric acid and sodium hydroxide product streams. The bipolar membrane[184] has two layers composed of an anionic and a cationic polymer respectively. Under the influence of a potential field, water is split into proton and hydroxide and these migrate in

Table 3
Ionics Electrodialysis Plants for the Removal of Nitrate Ion for Potable Water

Site	Volume treated (m^3 day^{-1})	Feed Water		Product Water	
		Total salt (ppm)	Nitrate, (ppm)	Total salt (ppm)	Nitrate (ppm)
Delaware, USA	2300	114	61	11	5
Bermuda	1600	1614	66	278	8
Milan, Italy	11700	1012	120	474	37
Donnington, UK	600				
Kazusa, Japan	150		80		27
Arizona, USA	62		100		
Safaria, Israel	2250		100		45

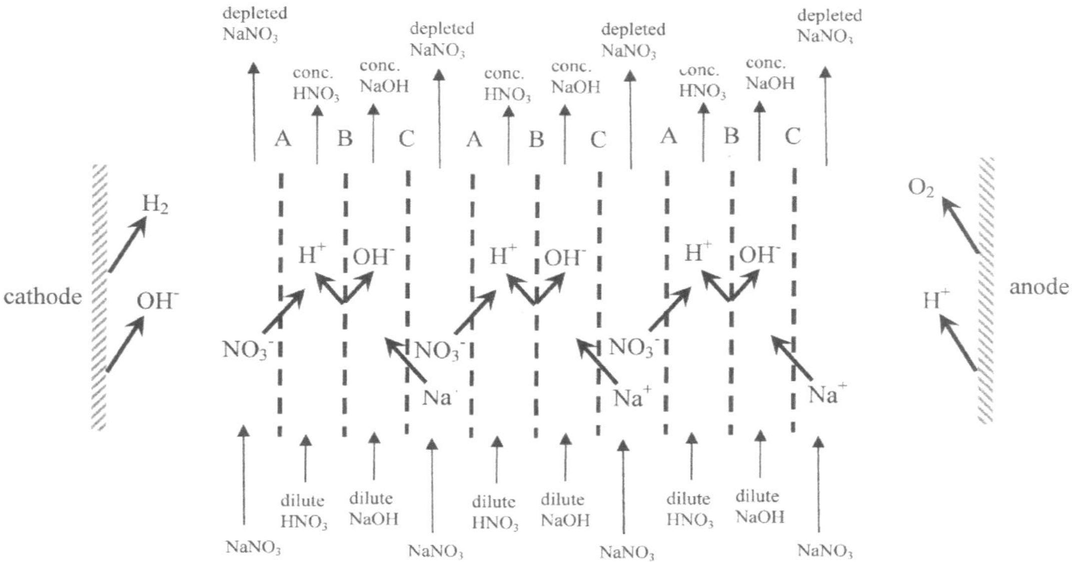

Figure 14. Schematic of a bipolar membrane electrodialysis stack cell for the splitting of sodium nitrate to nitric acid and sodium hydroxide. A = anion permeable membrane, B = bipolar membrane, C = cation permeable membrane.

opposite directions through the cation permeable and anion permeable layers of the bipolar membrane. The nitric acid is formed by combination with a nitrate ion that has migrated through an anion permeable membrane while the sodium hydroxide results from sodium ions migrating through a cation permeable membrane into the stream receiving protons. Bar and Lutin[194] have described a trial to investigate the possibility of converting an industrial effluent stream containing 200 g dm^{-3} KNO_3 into potassium hydroxide and nitric acid for reuse internally within the facility. The pilot plant used a small stack with five sets of membranes each with an area of 0.2 m^2 and it was operated continuously for 1085 hours with a current density of 0.1 A cm^{-2}. It proved possible to generate 2 to 3 M NaOH and 2 M nitric acid suitable for the applications envisaged and the current efficiency for the acid was 60-70% (losses are due to proton transport back through the anion permeable membrane). The economics of a plant to handle 3200 tons of sodium nitrate/year within the waste stream were considered promising. Similar bipolar membrane processes[195,196] have been described for the conversion of ammonium nitrate to nitric acid and ammonia although the properties of the bipolar membranes limit the concentration of nitric acid produced (~ 2 M).

VI. CONCLUSION

The cathodic reduction of nitrate is a reaction finding diverse applications. Despite an extensive literature, however, there are many fundamental questions still to be answered and improving the product selectivity remains a challenge to all who wish to apply the reduction of nitrate.

REFERENCES

[1]W.J. Plieth in *Encyclopedia of Electrochemistry of the Elements, Volume VIII,* ed. A.J. Bard, Marcel Dekker, 1978.
[2]J.T. Maloy in *Standard Potentials in Aqueous Solution,* eds. A.J. Bard, R. Parsons and J. Jordan, Marcel Dekker, 1985.
[3]K. Jones in *Comprehensive Inorganic Chemistry, Volume 2,* eds. J.C. Bailor, H.E. Emeleus, R. Nyholm and A.F. Trotman-Dickenson, Pergamon, 1973.

[4]J.W. Mellor, *Comprehensive Treatise on Inorganic and Theoretical Chemistry, Part VIII,* Longmans Green, 1928.
[5]J. C. Fanning, *Coord.Chem. Rev.*, **199** (2000) 159.
[6]K. Wieghardt, M. Woeste, P.S. Roy, and P. Chaudhuri, *J. Am. Chem. Soc.*, **107** (1985) 8276.
[7]J.A. Craig, and R.H. Holm, *J. Am. Chem. Soc.*, **111** (1989) 2111.
[8]R. Hille, *Chem. Revs.*, **96** (1996) 2757.
[9]M.G. Johnson and R.J. Robinson, *Anal. Chem.,* **24** (1952) 366.
[10]I.M. Kolthoff and I. Hodura, *J. Electroanal. Chem.,* **5** (1963) 2.
[11]T.E. Edmonds, *Anal. Chim. Acta,* **116** (1980) 323.
[12]L.M. Kolthoff, W.E. Harris, and G. Matsuyama, *J. Am. Chem. Soc.*, **66** (1944) 1782.
[13]B. Keita, E. Abdeljalil, L. Nadjo, R. Contant and R. Belgiche, *Electrochem. Commun.*, **3** (2001) 56.
[14]B. Keita, I-M. Mbomekalle, L. Nadjo and R. Contant, *Electrochem. Commun.*, **3** (2001) 267.
[15]B. Keita, I-M. Mbomekalle and L. Nadjo, *Electrochem. Commun.*, **5** (2003) 830.
[16]D. Jabbour, B. Keita, L. Nadjo, U. Kortz and S. S. Mal, *Electrochem. Commun.,* **7** (2005) 841.
[17]O.A. Petrii and T.Y. Safonova, *J. Electroanal. Chem.,* **331** (1992) 897.
[18]M. Da Cunha, J.P.I. De Sousa and F.C. Nart, *Langmuir,* **16** (2000) 771.
[19]G.E. Dima, A.C.A. de Vooys, and M.T.M. Koper, *J. Electroanal. Chem,* **554/5** (2003) 15.
[20]P.M. Tucker, M.J. Waite, and B.E. Hayden, *J. Applied Electrochem.,* **34** (2004) 781.
[21]X.K. Xing, D.A. Scherson, and C. Mak, *J. Electrochem. Soc.,* **137** (1990) 2166.
[22]G.E. Dima, G.L. Beltramo, M.T.M. Koper, *Electrochim. Acta* **50** (2005) 4318.
[23]S. Taguchi and J.M. Feliu, *Electrochim. Acta,* **52** (2007) 6023.
[24]S. Taguchi and J.M. Feliu, *Electrochim. Acta,* (2007) 0.1016/jelectacta.2007.12. 032.
[25]G. Horanyi and E.M. Rizmayer, *J. Electroanal. Chem.,* **188** (1985) 265.
[26]H. Li, D.H. Robertson, J.Q. Chambers, and D.T. Hobbs, *J. Electrochem. Soc.,* **135** (1988) 1154.
[27]I. Bakos and G.Horanyi, *J. Electroanal. Chem.,* **370** (1994) 309.
[28]D. De, J. D. Englehardt, E.E. Kalu, *J. Electrochem. Soc.*, **147** (2000) 4224.
[29]D. De, J.D. Englehardt, E.E. Kalu, *J. Electrochem. Soc.,* **147** (2000) 4573.
[30]J.W. Peel, K.J. Reddy, B.P. Sullivan, J.M. Bowen, *Water Res.,* **37** (2003) 2512.
[31]F.V. Andrade, L.J. Deiner, H. Arela, J.F.R. de Castro, I.A. Rodrigues and F.C. Nart, *J. Electrochem. Soc.*, **154** (2007) F159.
[32]Y. Yoshinaga, T. Akita, I. Mikami, and T.Okuhara, *J. Catal.,* **207** (2002) 37.
[33]A. Pintar, *Catal. Today*, **77** (2003) 451.
[34]F. Epron, F. Gauthard, and J. Barbier, *J. Catal.,* **206** (2002) 363.
[35]K. Daub, V.K. Wunder, and R. Dittmeyer, *Catal. Today,* **67** (2001) 257.
[36]M.T. de Groot and M.T.M. Koper, *J. Electroanal. Chem.,* **562** (2004) 81.
[37]G. Horanyi and E.M. Rizmayer, *J. Electroanal Chem.,* **140** (1982) 347.
[38]M. Wasberg and G. Horanyi, *Electrochim. Acta,* **40** (1995) 615.
[39]J.F.E. Gootzen, P. Peeters, J.M.B. Dukers, L. Lefferts, W. Visscher, J.A.R. vanVeen, *J. Electroanal. Chem.*, **434** (1997) 171.
[40]G.E. Dima, V. Rosca and M.T.M. Koper, *J. Electroanal. Chem.*, **599** (2007) 167.
[41]O. Radovici, G.E. Badea, T. Badea, *Rev. Roumaine de Chimie,* **48** (2003) 591.

[42]D. Pletcher and Z. Poorabedi, *Electrochim. Acta,* **24** (1979) 1253.
[43]X. Xing and D.A. Scherson, *J. Electroanal. Chem.,* **199** (1986) 485.
[44]X. Xing and D.A. Scherson, *Anal. Chem.,* **59** (1987) 962.
[45]X. Xing, D.A. Scherson, and C. Mak, *J. Electrochem. Soc.,* **137** (1990) 2166.
[46]W.J. Albery, B.G.D. Haggett, C.P. Jones, M.J. Pritchard, and L.R. Svanberg, *J. Electroanal. Chem.,* **188** (1985) 257.
[47]W.J. Albery, P.N. Bartlett, A.E.G. Cass, D.H. Craston, and B.G.D. Haggett, *J. Chem. Soc. Faraday Trans I,* **82** (1986) 1033.
[48]A.G. Fogg, S.P. Scullion, T.E. Edmonds, and B.J. Birch, *The Analyst,* **116** (1991) 573.
[49]N.G. Carpenter and D. Pletcher, *Anal. Chim. Acta* **317** (1995) 287.
[50]J. Davis, M.J. Moorcroft, S.J. Wilkins, R.G. Compton, and M.F. Cardosi, *Analyst,* **125** (2000) 737.
[51]C.M. Welch, M.E. Hyde, C.E. Banks and R.G. Compton, *Anal. Sci.,* **21** (2005) 1421.
[52]S. Ward-Jones, C.E. Banks, A.O. Simm, L. Jiang, and R.G. Compton, *Electroanalysis,* **17** (2005) 1806.
[53]D. Reyter, M. Odziemkowski, D. Bélanger and L. Roué, *J. Electrochem. Soc.,* **154** (2007) K36.
[54]T.R.L.C. Paixao, J.L. Cardoso and M. Bertotti. *Talanta,* **71** (2007) 186.
[55]L.D. Burke, A.M. O'Connell, R. Sharna, and C.A. Buckley, *J. Applied Electrochem.,* **36** (2006) 919.
[56]L.D. Burke, A.J. Ahern, and A.P. O'Mullane, *Gold Bull.,* **35** (2002) 3.
[57]L.D. Burke, *Gold Bull.,* **37** (2004) 125.
[58]D. Reyter, G. Chamoulaud, D. Belanger and L. Roue, *J. Electroanal. Chem.,* **596** (2006) 13.
[59]G.A. Olah, R.Malhotra, and S.C. Narang, *Nitration Methods and Mechanisms,* VCH, 1989.
[60]K.J. Vetter, *Z. Phys. Chem.,* **194** (1950) 199.
[61]O.W.J. Rutten, A. Van Sandwijk and G. Van Weert, *J. Applied Electrochem.,* **29** (1999) 87.
[62]I.G. Schevbackov and D.M. Libina, *Z. Elektrochem.,* **35** (1932) 977.
[63]J.J. Kaczur, L.L. Scott and R.L. Dotson, *11th International Forum on Electrolysis in the Chemical Industry*, Clearwater Beach, November 1997.
[64]R.L. Dotson and D.Y. Hernandez, *U.S. Patent* 4,849,073.
[65]D.W. Caulfield, *U.S. Patent* 5,318, 762.
[66]R.L. Dotson, *Interface*, **3,** Issue 3 (1994) 35.
[67]L.I. Halaoui, H. Sharifian and A.J. Bard, *J. Electrochem. Soc.,* **148** (2001) E386.
[68]R.J. Davenport and D.C. Johnson, *Anal. Chem.,* **45** (1973) 1979.
[69]M.E. Bodini, and D.T. Sawyer, *Anal. Chem.,* **45** (1977) 485.
[70]D. Pletcher and Z. Poorabedi, unpublished work.
[71]J.J. Kaczur, *8th International Forum on Electrolysis in the Chemical Industry*, Lake Buena Vista, November 1994.
[72]M.J. Moorcroft, L. Nei, J. Davis and R.G. Compton, *Anal. Lett.,* **33** (2000) 3127.
[73]B.K. Simpson and D.C. Johnson, *Electroanalysis,* **16** (2004) 532.
[74]T. Ohmori, M.S. El-Deab and M. Osawa, *J. Electroanal. Chem.,* **470** (1999) 46.
[75]K. Nishimura, K. Machida and M.Enyo, *Electochim. Acta,* **36** (1991) 877
[76]S.Wasmus, E.J.Vasini, M.Krausa, H.T. Mishima and W. Vielstich, *Electrochim. Acta,* **39** (1994) 23.
[77]T.Y. Safonova and O.A. Petrii, *Russ. J. Electrochem.,* **31** (1995) 1269.

[78]M.C.P.M. da Cunha, M. Weber, F.C. Nart, *J. Electroanal. Chem.,* **414** (1996) 162.
[79]K.J. Vetter, *Z. Elektrochem.,* **63** (1959) 1183.
[80]K.J. Vetter, *Z. Elektrochem.,* **63** (1959) 1189.
[81]G. Schmid and J. Delfs, *Z. Elektrochem.,* **63** (1959) 1192.
[82]G. Schmid, *Z. Elektrochem.,* **65** (1961) 531.
[83]F. Balbaud, G. Sanchez, G. Santarini and G. Picard, *Eur. J. Inorg. Chem.*, (2000) 665.
[84]G. Horanyi and E.M. Rizmayer, *J. Electroanal. Chem.,* **140** (1982) 347.
[85]I. Bakos and G. Horanyi, *J. Electroanal. Chem.*, **370** (1994) 309.
[86]G. Horanyi and M. Wasberg, *Electrochim. Acta,* **42** (1996) 261.
[87]S. Ureta-Zanartu and C. Yanez, *Electrochim. Acta,* **42** (1997) 1725.
[88]J.F. van der Plas and E. Barendrecht, *Electrochim. Acta,* **25** (1980) 1463.
[89]T.Y. Safonova and O.A. Petrii, *J. Electroanal. Chem.,* **331** (1992) 897.
[90]T.Y. Safonova and O.A. Petrii, *J. Electroanal. Chem.,* **448** (1998) 211.
[91]K. Shimazu, R. Goto and K. Tada, *Chem Lett.,* **2** (2002) 204.
[92]K. Tada and K. Shimazu, *J. Electroanal. Chem.,* **577** (2005) 303.
[93]F. Armijo, M. Isaacs, G. Ramirez, E. Trollund, J. Canales and M.J. Aguirre, *J. Electroanal. Chem.,* **566** (2004) 315.
[94]K-W. Kim, S-M. Kim, Y-H. Kim, E-H. Lee and K-Y. Jee, *J. Electrochem. Soc.*, **154** (2007) E145.
[95]A.C.A. de Vooys, R.A. van Santen and J.A.R. van Veen, *J. Mol. Catal. A,* **154** (2000) 203.
[96]S. Hwang, J. Lee and J, Kwak, *J. Electroanal. Chem.,* **579** (2005) 143.
[97]J.F.E. Gootzen, L. Lefferts and J.A.R. van Veen, *Appl. Catal. A. Gen.* **188** (1999) 127.
[98]J. Garcia-Domenech, M.A. Climent, A. Aldaz, J.L. Vazquez and J. Clavilier, *J. Electroanal. Chem.,* **159** (1983) 223.
[99]K. Shimazu, T. Kawaguchi and K. Tada, *J. Electroanal. Chem.,* **529** (2002) 20.
[100]K. Shimazu, R. Goto, S. Piao, R. Kayama, K. Nakata and Y. Yoshinaga, *J. Electroanal. Chem.*, **601** (2007) 161.
[101]I.G. Casella and M. Contursi, *J. Electroanal. Chem.,* **588** (2006) 147.
[102] Y. Wang and J. Qu, *Water Envir. Res.,* **78** (2006) 724.
[103]R. Prasad and A. Kumar, *J. Electroanal. Chem.,* **576** (2005) 295.
[104] G.W.D. Briggs and W.F.K. Wynne-Jones, *Electrochim. Acta,* **7** (1962) 241.
[105]S.U. Falk and A.J. Salkind, *Alkaline Storage Batteries,* Wiley, 1969.
[106]D.De, E.E. Kalu, P.P. Tarjan and J.D. Englehardt, *Chem. Eng. Technol.*, **27** (2004) 1.
[107]O. Brylev, M. Sarrazin, D. Belanger and L. Roue, *Applied Catal. B,* **64** (2006) 243.
[108]O. Brylev, M. Sarrazin, D. Belanger and L. Roue, *Electrochim. Acta,* **52** (2007) 6237.
[109]C. Politades and G. Kyriacou, *J. Applied Electrochem.,* **35** (2005) 421.
[110]Y. Wang, J.H. Qu and H.J. Lui, *Chinese Chem. Lett.,* **17** (2006) 61.
[111]N. Vila, M. Van Brussel, M. D'Amours, J. Marwan, C. Buess-Herman and D. Belanger, *J. Electroanal. Chem.*, **609** (2007) 85.
[112]B.P. Dash and S. Chaudhari, *Water Research,* **39** (2005) 4065.
[113]I. Katssounaros, D. Ipsakis, C. Polatides and G. Kyriacou, *Electrochim. Acta,* **52** (2006)1329.
[114]I. Katssounaros and G. Kyriacou, *Electrochim. Acta,* **52** (2007) 6412.

[115]F. Bouamrane, A. Tadjeddine, J.E. Butler, R. Tenne and C. Levy-Clement, *J. Electroanal. Chem.*, **405** (1996) 95.
[116]C. Levy-Clement, N.A. Ndao, A. Katty, M. Bernard, A. Deneuville, C. Comninellis and A. Fujishima, *Diamond and Related Materials,* **12** (2003) 606.
[117]I. Taniguchi, N. Nakashima, K. Matsushita and K. Yasukouchi, *J. Electroanal. Chem.,* **224** (1987) 199.
[118]X. Zhang, J. Wang, Z. Wang and S. Wang, *Synthetic Metals,* **155** (2005) 95.
[119]S. Cattarin, *J. Applied Electrochem.,* **22** (1992) 1077.
[120]M. Paidur, I. Rousar and K. Bouzek, *J. Applied Electrochem.*, **29** (1999) 611.
[121]K. Bouzek, M. Paidur, S. Sadilkova and H. Bergmann, *J. Applied Electrochem.,* **31** (2001) 1185.
[122]M. Paidur, K Bouzek, and H. Bergmann, *Chem. Eng. J.*, **85** (2002) 99.
[123]I.C. Casella and G. Gatta, *J. Electroanal. Chem.,* **568** (2004) 183.
[124]C. Milhano and D.Pletcher, *J. Electroanal. Chem.*, **614** (2008) 24.
[125]L. Durivault, O. Brylev, D. Reyter, M. Sarrazin, D. Bélanger and L. Roué, *J. Alloys Cpds*, **432** (2007) 323.
[126]D. Reyter, D. Bélanger and L. Roué, *J. Electroanal. Chem.* **622** (2008) 64.
[127]L. Roué, *personal communication.*
[128]C. Lu, S. Lu, W. Qiu and Q.Liu, *Electrochim. Acta,* **44** (1999) 2193.
[129]H.L. Li, D.H. Robertson, J.Q. Chambers, D.T. Hobbs, *J. Electrochem. Soc.,* **135** (1988) 1154.
[130]H.L. Li, J.Q. Chambers and D.T. Hobbs, *J. Applied Electrochem.*, **18** (1988) 454.
[131]J.D. Genders, D. Hartsough and D.T. Hobbs, *J. Applied Electrochem.,* **26** (1996) 1.
[132]C. Milhano and D.Pletcher, *Phys.Chem.Chem.Phys.*, **7** (2005) 3545.
[133]H.L. Li, J.Q. Chambers and D.T. Hobbs, *J. Electroanal. Chem.*, **256** (1988) 447.
[134]H.L. Li, W.C. Anderson, J.Q. Chambers and D.T. Hobbs, *Inorg. Chem.,* **28** (1989) 863.
[135]N. Chebotareva and T. Nyokong, *J. Applied Electrochem.,* **27** (1997) 975.
[136]M. J. Moorcroft, J. Davis and R. G. Compton, *Talanta,* **54** (2001) 785.
[137]J.W. Collat and J.J. Lingane, *J. Am. Chem. Soc.,* **76** (1954) 4214.
[138]D.C. Johnson and G.A. Sherwood, *Anal. Chim. Acta,* **129** (1981) 87.
[139]D.C. Johnson and G.A. Sherwood, *Anal. Chim. Acta,* **129** (1981) 101.
[140]A.O. Solak and P. Cekirdek, *Anal. Lett.,* **38** (2005) 271.
[141]T.R.L.C. Paixao, J.L. Cardoso and M. Bertotti, *Talanta,* **71** (2007) 186.
[142]J. Davis, M.J. Moorcroft, S.J. Wilkins, R.G. Compton, and M.F. Cardosi, *Electroanalysis,* **17** (2000) 12.
[143]M. Shibata, K. Yoshida and N. Furuya, *J. Electrochem. Soc.,* **145** (1998) 2348.
[144]M. Nobial, O. Devos, O.R. Mattos and B. Tribollet, *J. Electroanal. Chem.,* **600** (2007) 87.
[145]J. Lee and Y. Tak, *Electrochem. Solid State Lett.,* **4** (2001) C63.
[146]M.J. Natan, D. Belanger, M.K. Carpenter and M.S. Wrighton, *J. Phys. Chem.,* **91** (1987) 1834.
[147]C. Natarajan and G. Nogami, *J. Electrochem. Soc.,* **143** (1996) 1547.
[148]M. Izaki and T. Ohmi, *J. Electrochem. Soc.,* **144** (1997) 1949.
[149]M. Izaki, *J. Electrochem. Soc.,* **146** (1999) 4517.
[150]J. Lee and Y. Tak, *Electrochem. Solid State Lett.,* **3** (2000) 69.
[151]Y. Matsumoto, T. Morikawa and H. Adachi, J. Mater. Sci. Lett., 27 (1992)1319.
[152]J. Lee and Y. Tak, *Electrochem. Commun.,* **2** (2000) 765.
[153]S.B. Abolmaali and J.B. Talbot, *J. Electrochem. Soc.,* **140** (1993) 443.

[154]D.T. Hobbs in *Electrochemistry for a Cleaner Environment*, Eds J.D. Genders and N.L. Weinberg, The Electrosynthesis Co, 1992.
[155]D.H. Coleman, R.E. White and D.T. Hobbs, *J. Electrochem. Soc.*, **142** (1995) 1152.
[156]E.E. Kalu, R.E. White and D.T. Hobbs, *J. Electrochem. Soc.*, **143** (1996) 3094.
[157]D.T. Hobbs, *Sep. Purif. Technol.*, **15** (1999) 239.
[158]J.D. Genders, D. Chai and D.T. Hobbs, *J. Applied Electrochem.*, **30** (2000) 13.
[159]J. O'M. Bockris and J. Kim, *J. Electrochem. Soc.*, **143** (1996) 3801.
[160]J. O'M. Bockris and J. Kim, *J. Applied Electrochem.*, **27** (1997) 623.
[161]Y. Hirose and T.Sawa, *9th International Forum on Electrolysis in the Chemical Industry*, Lake Clearwater Beach, November 1995.
[162]Y. Hirose and K. Itou, *10th International Forum on Electrolysis in the Chemical Industry*, Lake Clearwater Beach, November 1996.
[163]K.-W. Kim, S.-H. Kim and E.-H Lee, *J. Radioanal. Nucl. Chem.*, **260** (2004) 99.
[164]C. Mallika, B. Keerthika and U. Kamachi Mudali, *Electrochim. Acta,* **52** (2007) 6656.
[165]A. Kapoor and T. Viraraghavan, *J. Environ. Eng.*, **123** (1997) 371.
[166]L.W. Canter in *Nitrates in Groundwater,* CRC Press, 1997.
[167]J.J. Kaczur, *12th International Forum on Electrolysis in the Chemical Industry*, Clearwater Beach, November 1998.
[168]S. Otten and M. Lubin, *12th International Forum on Electrolysis in the Chemical Industry*, Clearwater Beach, November 1998.
[169]Y Wang, J. Qu, R. Wu and P. Lei, *Water Research,* **40** (2006) 1224.
[170]M.J. Waite at *5th European Electrochemical Processing Conference,* Chester, UK, April 1999.
[171]M.J. Waite, P.M. Tucker, D. Wilson and D. Foster in *Ion Exchange at the Millennium,* Ed. J.A. Greig, Imperial College Press, 2000.
[172]P.M. Tucker at *16th International Forum on Applied Electrochemistry*, Amelia Island, Florida, November 2002.
[173]G.R. Elder and M.J. Waite, *3rd European Electrochemical Processing Conference,* Toulouse, France, April 1995.
[174]http://www.ionex.co.uk/
[175]D. van Velzen and H. Langenkamp, *3rd European Electrochemical Processing Conference,* Toulouse, France, April 1995.
[176]N. Hiro, T. Koizumi, T. Rakuma, D. Takizawa and K. Takzawa, *Electrochemistry*, **70** (2002) 111.
[177]N. Hiro, J. Hirose and N. Kitayama, *Electrochemistry*, **71** (2002) 24.
[178]L. Szpyrkowicz, S. Daniele, M. Radaelli and S. Specchia, *Applied Catal. B: Eviron.*, **66** (2006) 40.
[179]H. Cheng, K. Scott and P.A. Christensen, *J. Applied Electrochem.*, **35** (2005) 551.
[180]H. Cheng, K. Scott and P.A. Christensen, *Chem. Eng. J.*, **108** (2005) 257.
[181]J. Dziewinski, at *14th International Forum on Applied Electrochemistry*, Clearwater Beach, Florida, November 2000.
[182]J. Dziewinski and S. Marczak, *Chem. Innovation*, **30** (4) (2000) 33.
[183]A. Sabzali, M. Gholami, A.R. Yazdanbakhsh, A. Khodadadi, B. Musavi and R. Mirzaee, *Iran J. Environ. Health Sci. Eng*., **3** (2006) 141.
[184]T.A. Davis, J.D. Genders and D. Pletcher, *A First Course in Ion permeable Membranes,* The Electrochemical Consultancy, 1997.
[185]B. Gillery, *12th International Forum on Electrolysis in the Chemical Industry*, Lake Clearwater Beach, November 1998.

[186]D. Bar, *15th International Forum on Electrolysis in the Chemical Industry*, Lake Clearwater Beach, November 2001.
[187]F. Hell, J. Lahnsteiner, H Frischherz and G. Baumgartner, *Desalination*, **117** (1998) 173.
[188]http://www.ameridia.com/
[189]http://www.ionics.com/applications/denitrification/index.htm
[190]T. Prato and R.G. Parent, *1993 AWWA Membrance Conference,* (1993).
[191]D. Elyanow and J. Persechino, 3rd Annual Conference of the Israel Desalination Society, Tel-Aviv, Israel, December 2000.
[192]T.Sawa, Y. Hirose and Y. Ishii, Soda to Enso, **49** (1998) 248
[193]E. Gain, S. Laborie,Ph. Viers, M. Rakib, G. Durand and D. Hartmann, *J. Applied Electrochem.,* **32** (2002) 1572.
[194]D. Bar and F. Lutin, *4th European Conference on Industrial Electrochemistry*, Barcelona, Spain, April 1997.
[195]S. Graillon, F. Persin, G. Pourcelly and C. Gavach, *Desalination,* **107** ((1996) 159.
[196]A. Ben Ali, S. Laborie, P. Viers, M. Rakib and G. Durand, *Récents Progrèsen Génie des Procédés*, **15** (86) (2001) 203.
[197]D. Reyter, M. Odziemkowski, D. Bélanger and L. Roué, *J. Electrochem. Soc.,*154 (2007) K36.

2

Electrochemistry of Room-Temperature Ionic Liquids and Melts

Tetsuya Tsuda and Charles L. Hussey

Department of Chemistry and Biochemistry, The University of Mississippi, University, Mississippi 38677-1848, USA

I. INTRODUCTION

Articles about substances designated as *ionic liquids* have begun to appear with increasing regularity in chemistry journals around the world. The recent advent of the terms *ionic liquid* or *ionic liquids*, and the publication of numerous articles promoting the unusual properties and potential uses of these materials suggest that they are new and heretofore unrecognized substances. However, these names are just a more modern way to describe *molten* or *fused salts*. Such liquid salts have been recognized since the very beginning of modern chemistry, and they form the basis for several key industrial processes, e.g., the electrolytic production of aluminum. A careful review of the literature indicates that the ionic liquids label is almost universally applied to salts that exist in the liquid state at or proximate to room temperature, leading to the useful abbreviations RTIL or RTILs. As such, these labels provide a convenient way to differentiate low-melting salts from their higher-melting cousins.

R.E. White (ed.), *Modern Aspects of Electrochemistry, No. 45*, Modern Aspects of Electrochemistry 45, DOI 10.1007/978-1-4419-0655-7_2,

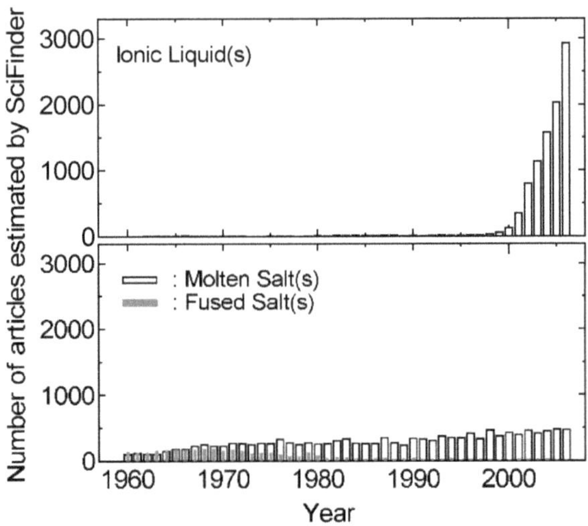

Figure 1. Number of published articles as a function of year estimated from on-line searches using *SciFinder*. "Ionic liquid(s)", "molten salt(s)", and "fused salt(s) were used as keywords."

Figure 1 illustrates the explosive growth in scientific articles about ionic liquids during the past decade compared to publications containing the descriptors *molten* or *fused salts*. These data were collected by using SciFinder. Figure 2 depicts the number of patents issued for all aspects of ionic liquids and for molten or fused salts. Although interest in the technological applications of latter materials is stronger than ever before, the increase in the number of patents devoted to ionic liquids parallels the explosive growth in the number of scientific articles on this subject. There are many factors behind the intense interest in low-melting ionic liquids. Some of these factors include high electrical conductivities that reach to nearly 100 mS cm^{-1}; large liquidus ranges, commonly 173 ~ 450 K; wide electrochemical windows extending to ~5.8 V in some cases; negligible vapor pressure at room temperature; and easily tunable physical and chemical properties.

Recently, research groups in the UK,[1] US,[2,3] and Japan[4] have attempted to present definitions for what exactly defines or constitutes an ionic liquid. We have restated some of the more universal-

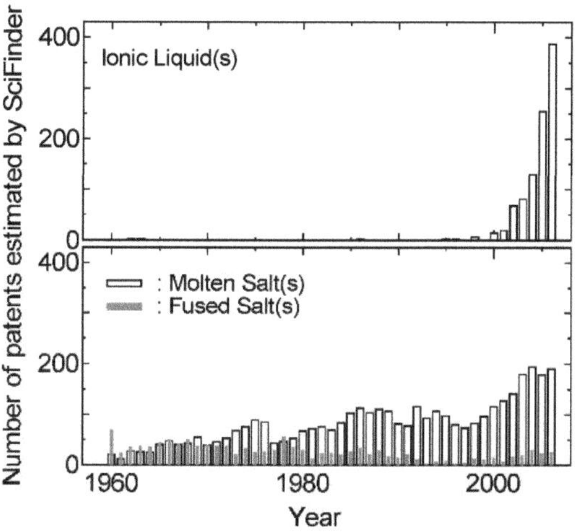

Figure 2. Number of patents as a function of year estimated from on-line searches using *SciFinder*. The keywords were the same as those used to prepare Fig. 1.

ly accepted simple characteristics of these materials as an aid to the readers of this Chapter:

a) Ionic liquids are composed entirely of cations and anions.
b) Ionic liquids contain no molecular solvent(s).
c) Although the ionic liquid label properly applies equally well to a molten or fused salt, currently it is by agreement assigned only to salts that are liquid at or below 373 K.
d) Low-melting mixtures with melting points ≤ 300 K, such as choline chloride-urea or those based on zwitter ions or acetamide, are not ionic liquids in the truest sense because they do not adhere to rules a) and b). However, they can be included with ionic liquids under the broad label *room-temperature melts*.

There are numerous short reviews or focused articles about the general properties of ionic liquids,[3,5-18] as well as applications involving analytical chemistry,[19-23] batteries,[24,25] catalysis,[26-35] electrochemistry,[36-38] inorganic materials,[39-41] organic synthesis,[42-45] supercritical fluid extraction,[46] and even the use of ionic liquids for the preparation of cellulose materials.[47] In addition to these reviews, there are numerous monographs devoted to ionic liquids[48-59] and proceedings volumes derived from meetings dedicated to ionic liquids, especially the biennial international meetings sponsored by The Electrochemical Society, Inc.[60] The preponderance of early research on ionic liquids was focused on haloaluminates, especially room-temperature chloroaluminate ionic liquids, and several detailed reviews about these materials have appeared.[6,7,9,11,36,37,39] In view of this previous work, the current article will be devoted primarily to non-haloaluminate ionic liquids. (These previous reviews are still worth reading even if you are not interested in haloaluminate ionic liquids because many of the fundamental techniques used to prepare and synthesize the other classes of ionic liquids are based on work initiated with haloaluminates, especially the chloroaluminates.)

A great many room-temperature ionic liquids, abbreviated from this point forward as RTILs (plural form) or RTIL (singular form), have been prepared. These RTILs can be conveniently subdivided on the basis of the structure of the cationic component as depicted in Fig. 3, but it is considerably more difficult to classify them on the basis of anion structure due to the many different anions that have been used to formulate these materials (Table 1). The physicochemical properties of RTILs can be tuned across a broad spectrum by choice of the cations and anions. This is perhaps the most interesting and useful feature of this class of solvents. Those low-melting mixtures that fail conditions a. and b. listed above, but form room-temperature melts (RTMs), such as those based on zwitter ions,[242-246] urea,[8,247-257] or acetamide,[255-266] also possess some unique physicochemical properties that make them suitable for electrochemistry as summarized in Table 2 and will also be discussed herein.

Although the popularity of RTILs (and some RTMs) cannot be disputed, there is often little understanding of the techniques and procedures that must be used to prepare and purify these materials. Therefore, we have compiled what we think to be the best information about these methods for each class of RTILs in the Section presented below. In other Sections of this article, we introduce fundamental electroanalytical chemistry in RTILs and discuss electrochemical technologies based on RTILs and RTMs, e.g., energy and materials science.

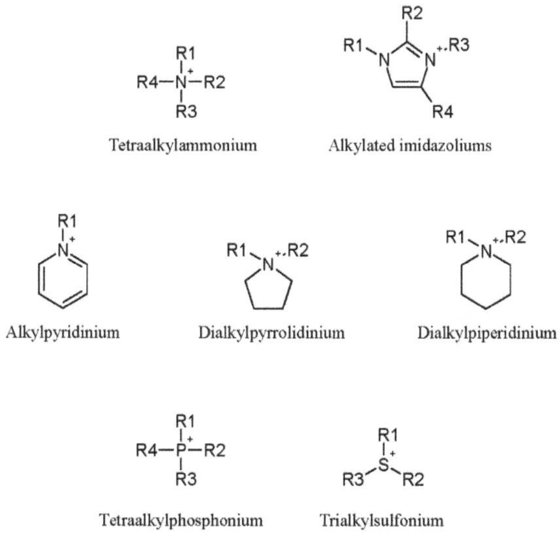

Figure 3. Common cations that have been used to prepare room-temperature ionic liquids.

Table 1
Typical Room-Temperature (RT) and Low-Temperature (LT) Ionic Liquids

Anions	Cations	Ref.
$F(HF)_n^-$	RT: Amm, Im, Pip, Py, Pyrr	61-72
	LT: Amm, Im	
Cl^-	RT: Amm, Im, P	73-80
	LT: Amm, Im, P	
ClO_4^-	RT: Amm, Im	74,81-83
	LT: Amm	
$Cl(HCl)_n^-$	RT: Im	84-86
ClI_2^-	RT: Im	87
Cl_2I^-	RT: Im	88
Br^-	RT: Amm, Im, P	73,74,78,79,
	LT: Amm, Im	81, 89
BrI_2^-	RT: Im	87,90
Br_2I^-	RT: Im	87,88,90
Br_3^-	RT: Im, Py	87,88,90
I^-	RT: Amm, Im, P, S	73,74,78,79,
	LT: Amm, Im, S	82,83,91-96

Table 1. Continuation

Anions	Cations	Ref.
I_{2n+1}^-	RT: Im	87,97
BH_4^-	RT: Amm	98
BF_4^-	RT: Amm, Im, Py, Pyrr, Others	73,77,82,83,
	LT: Amm, Im, Pip, Pyrr, Others	93,98-108
$B(CN)_4^-$	RT: Im	109
Borides (= $B_{R1R2R3R4}^-$)	RT: Amm	7,76,110-113
	LT: Amm	
$B(HSO_4)_4^-$	RT: Im	114
Tetraphenylborate (= BPh_4^-)	LT: Amm	81
Tetrakis[3,5-bis (trifluoro-methyl)phenyl]borate (= $B[Ph(CF_3)_2]_4^-$)	LT: Amm, Im, Py	115
RBF_3^-	RT: Im	116
$R_fBF_3^-$ [$R_f = C_nF_{2n+1}$]	RT: Amm, Im, Pip, Pyrr	103,107,117-
	LT: Amm, Im, Pip, Pyrr	122
$CH_3CH(BF_3)CH_2CN^-$	RT: Im	123
PF_6^-	RT: Im	77,83,98,100,
	LT: Amm, Im, Pyrr, Others	124-128
$(R_f)_3PF_3^-$ [$R_f = C_nF_{2n+1}$]	RT: Im, P	129,130
AsF_6^-	LT: Im	100,125,131
SbF_6^-	RT: Im	131-134
TaF_6^-	RT: Im	135,136
NbF_6^-	RT: Im	135,136
WF_7^-	RT: Im	131
WOF_5^-	RT: Im	137
HCO_2^-	RT: Amm	98,138
HCO_3^-	RT: Amm	98
$CH_3CO_2^-$ (= AcO^-)	RT: Amm, Im, P	80,98,99,139
$CH_3OCO_2^-$	RT(?): Im	122
$CH_3CH(OH)CO_2^-$ (= Lac^-)	RT: Im	105,140
	LT: Others	
$CH_3CH{=}CHCO_2^-$ (= $Crot^-$)	RT: Amm	98
$CF_3CO_2^-$ (= TA^-)	RT: Im, Others	105,108,139
	LT: Im, Others	
$CH_3SO_3^-$ (= MsO^-)	RT: Im	92,141
	LT: Im	
$CF_3SO_3^-$ (= TfO^-)	RT: Im	83,139,141, 142
	LT: Im	
$C_3F_7CO_2^-$ (= HB^-)	RT: Im	94,139
$C_4F_9SO_3^-$ (= NfO^-)	RT: Im	139,143
	LT: Im, Pyrr	
$(FSO_2)_2N^-$ (= FSI)	RT: Im, Pip, Py,	144,145

Table 1. Continuation

Anions	Cations	Ref.
$(CF_3SO_2)_2N^-$ (= Tf_2N^-, $TFSI^-$, $TFSA^-$)	RT: Amm, Im, P, Pip, Pyrr, S, Others LT: Amm, Im, Pip, Py, S, Others	78-80,83,87-89, 92-95,97,98, 100,101,105, 106,119,125, 127,135,139, 142,146-159
$(CH_3SO_2)_2N^-$ (= $NMes_2^-$)	RT: Amm, Im, Pyrr	160
$(CF_3SO_2)(CF_3CO)N^-$ (= $TSAC^-$)	RT: Amm, Im, Pyrr LT: Amm	161
$(C_2F_5SO_2)_2N^-$ (= $BETI^-$)	RT: Im	83,100,147
$(C_3F_7SO_2)_2N^-$	RT: Others	157
$(CF_3SO_2)_3C^-$ (= Me^-, Tf_3C^-)	RT: Im	125
NO_2^-	LT: Im	99
NO_3^-	RT: Amm, Others LT: Amm, Im, Others	73,76,98,99, 105,162-164
PO_4^{3-}	RT: Im, Others	165,166
$H_2PO_4^-$	RT: Amm, Im	98,167
SO_2^-	RT: Amm	98
SO_4^{2-}	RT: Im LT: Others	105,168
HSO_4^-	RT: Amm LT: Amm, Im	98,114,163,169
$LiCl_2^-$	RT: Im LT: Im	170
$VOCl_4^-$	LT: Im	171
$TaCl_6^-$ and Ta(V)	RT: Im LT: Im	172
W(VI)	RT: Im	173
$FeCl_4^{2-}$	LT: Im	169,174,175
$FeCl_4^{2-}$ + $FeCl_4^-$	LT: Im	169,174
$Cl(FeCl_3)_n^-$	RT: Amm, Im LT: Amm, Im	83,169,172-183
$Fe_{0.5}Ga_{0.5}Cl_4^-$	RT: Im	175,181
$FeBr_4^-$	RT: Im LT: Im	182
$CoCl_4^{2-}$	LT: Im	184
$NiCl_4^{2-}$	LT: Im	184
$Cl(CuCl)_n^-$	RT: Im, P, Pip, Py	185-189
$CuCl_{2+n}^{-1-n}$	RT: Im LT: Amm	186,188
$AgCl_4^{3-}$	LT: Im	170
$AuCl_4^-$	RT: Im LT: Im	190,191
$ZnCl_4^{2-}$	LT: Im, Py	170,192

Table 1. Continuation

Anions	Cations	Ref.
$Cl(ZnCl_2)_n^-$	RT: Amm LT: Amm, Im	178,193,194
$CdCl_4^{2-}$	LT: Im	170
$Cl(GaCl_3)_n^-$	RT: Im, Py LT: Amm, Py	83,175,181,195 -197
$InCl_6^{3-}$	LT: Im	198
$InCl_4^-$	RT: Im LT: Im	198-202
$GaCl_4^-$	RT: Im	203
$GeCl_3^-$	LT: Amm	204
$Cl(SnCl_2)_n^-$	RT: Amm, Im, Py LT: Amm, Im, Py	178,193, 204-209
$Ag(CN)_2^-$	LT: Im	210
$Au(CN)_2^-$	RT: Im LT: Im	211
$C(CN)_3^-$	RT: Im	210
$N(CN)_n^-$	RT: Amm, Im, Pyrr, S LT: Im, Pyrr	210, 212-215
SCN^-	RT: Im, Pyrr LT: Amm, Pyrr, Others	73,76,105,216- 219
$SeCN^-$	RT: Im	220
$SeO_2(OR)^-$	RT: Im	221
Chelated orthoborate	RT: Amm, Im, Py, Pyrr LT: Amm, Im	83,222
$C_6H_5CO_2^-$	RT: Im	223
Formate	RT: Im	224
Picrate $(NO_2)_3C_6H_2O^-$ (= Pic^-)	LT: Amm	216,217,225
Alkylsulfate $(RO)SO_3^-$	RT: Im LT: Im	142,226-230
$CH_3(CH_2)_{11}SO_3^-$	LT: Im	231
Bis(2-ethylhexyl) sulfosuccinate (= $BEHSS^-$)	RT: Amm	232
1,2,4-Triazolide	RT: Im	233
1,2,3,4-Tetrazolide	RT: Im	233
Amino acid based anions	RT: Im, P LT: P	234
Others	RT: Amm, Im, P, Others LT: Amm, Im, Others	79,235-240, 164,241

RT: m.p. ≤ 298 K; LT: 298 K < m.p. ≤ 373 K; Amm: Ammonium-based cation; Im: Imidazolium-based cation; P: Phosphonium-based cation; Pip: Piperidinium-based cation; Py: Pyridinium-based cation; Pyrr: Pyrrolidinium-based cation; S: Sulfonium-based cation.

Table 2
Typical Room-Temperature Melts and their Physicochemical Properties

System	Composition (mol %)	Melting point (K)	Conductivity (mS cm^{-1})	Electrochemical window (V)	Ref.
Fructose-urea	33.3-66.7	338			252
Glucose-urea-$CaCl_2$	25.0-66.0-9.0	348			252
Sorbitol-urea-NH_4Cl	42.5-36.8-20.7	340			252
Mannose-DMU	17.3-82.7	348			252
Sorbitol-DMU	24.4-75.6	350			252
α-Cyclodextrin-DMU	3.7-96.3	350			252
Citric acid-DMU	23.4-76.6	338			252
Maltose-DMU-NH_4Cl	18.5-57.7-23.8	346			252
NH_4NO_3-urea	unknown	< 317			8
NH_4NO_3-MA	unknown	< 273			8
KSCN-acetamide	unknown	< 293			8
Acetamide-urea	70.0-30.0	329			255
Acetamide-urea-$LiNO_3$	48.0-32.0-20.0		< 1 (300 K)	Cathodic limit: –0.4 (*vs*. PbO/Pb)[a] Anodic limit: +2.4 (*vs*. PbO/Pb)[a]	255
Acetamide-urea-NH_4NO_3	48.0-32.0-20.0	280.5	5 (300 K)	Cathodic limit: –0.4 (*vs*. PbO/Pb)[a] Anodic limit: +2.4 (*vs*. PbO/Pb)[a]	255
Acetamide-urea-NaBr	53.7-37.3-9.0	< 343		Cathodic limit: –1.4 (vs. SCE)[a] Anodic limit: +0.6 (vs. SCE)[a]	257
ChCl-urea	33.3-66.7	285	< 1 (300 K)		251
ChCl-N-methylurea	33.3-66.7	302			251

Table 2. Continuation

System	Composition (mol %)	Melting point (K)	Conductivity ($mS\ cm^{-1}$)	Electrochemical window (V)	Ref.
ChCl-DMU	33.3-66.7	343			251
ChCl-thiourea	33.3-66.7	342			251
ChCl-acetamide	33.3-66.7	324			251
ChCl-benzamide	33.3-66.7	365			251
NaBr-urea	20.5-79.5	336.5			248
NaBr-KBr-urea	19.5- 1.5-79.0	324			248
KI-urea	17.5-82.5	359.5			248
NaI-KI-urea	18.0- 3.0-79.0	301			248
$NaNO_3$-urea	22.5-77.5	357.5			248
$NaNO_3$-KNO_3-urea	22.3-15.0-70.7	351.5			248
$LiTf_2N$-urea	17.2-82.8	235.4			249
	21.7-78.3		0.23 (298 K)		249
	23.3-76.7			Cathodic limit: 0.8 (*vs.* Li^+/Li)[b] Anodic limit: 3.8 (*vs.* Li^+/Li)[c]	249
$LiTf_2N$-ethyleneurea	33.3-66.7	316.8			256
	20.0-80.0		0.26 (298 K)		256
$LiTf_2N$-*N*-methylurea	33.3-66.7		0.03 (298 K)		256
	25.0-75.0		0.12 (298 K)	Cathodic limit: 0.8 (*vs.* Li^+/Li)[c] Anodic limit: 3.8 (*vs.* Li^+/Li)[b]	256
	20.0-80.0		0.18 (298 K)		256
$LiTf_2N$-DMU	33.3-66.7	245	0.01 (298 K)		256
	25.0-75.0			Anodic limit: 3.8 (*vs.* Li^+/Li)[b]	256

Table 2. Continuation

System	Composition (mol %)	Melting point (K)	Conductivity (mS cm^{-1})	Electrochemical window (V)	Ref.
$LiTf_2N$-acetamide	23.3-76.7	284			256
	20.0-80.0	206	1.07 (298 K)	Cathodic limit: 0.7 (*vs.* Li^+/Li)[c] Anodic limit: 4.4 (*vs.* Li^+/Li)[d]	261
	14.3-85.7		1.21 (298 K)	Cathodic limit: 0.6 (*vs.* Li^+/Li)[a] Anodic limit: 5.3 (*vs.* Li^+/Li)[a]	256
LiBeti-acetamide	20.0-80.0	216		Cathodic limit: 0.6 (*vs.* Li^+/Li)[c] Anodic limit: 3.8 (*vs.* Li^+/Li)[b]	263
	14.3-85.7		1.27 (303 K)		263
NaI-urea	19.5-80.5	307			248
KI-urea	17.5-82.5	359.5			248
NaI-KI-urea	18.0- 3.0-79.0	301			248
$NaNO_3$-urea	22.5-77.5	357.5			248
$NaNO_3$-KNO_3-urea	22.3-15.0-70.7	351.5			248
$LiTf_2N$-urea	17.2-82.8	235.4			249
	21.7-78.3		0.23 (298 K)		249
	23.3-76.7			Cathodic limit: 0.8 (*vs.* Li^+/Li)[b] Anodic limit: 3.8 (*vs.* Li^+/Li)[c]	249
$LiTf_2N$-ethyleneurea	33.3-66.7	316.8			256
	20.0-80.0		0.26 (298 K)		256
$LiTf_2N$-*N*-methylurea	33.3-66.7		0.03 (298 K)		256
	25.0-75.0		0.12 (298 K)	Cathodic limit: 0.8 (*vs.* Li^+/Li)[c] Anodic limit: 3.8 (*vs.* Li^+/Li)[b]	256
	20.0-80.0		0.18 (298 K)		256

Table 2. Continuation

System	Composition (mol %)	Melting point (K)	Conductivity (mS cm^{-1})	Electrochemical window (V)	Ref.
$LiTf_2N$-DMU	33.3-66.7	245	0.01 (298 K)		256
	25.0-75.0			Anodic limit: 3.8 (*vs.* Li^+/Li)[b]	256
$LiTf_2N$-acetamide	23.3-76.7	284			256
	20.0-80.0	206	1.07 (298 K)	Cathodic limit: 0.7 (*vs.* Li^+/Li)[c] Anodic limit: 4.4 (*vs.* Li^+/Li)[d]	261
	14.3-85.7		1.21 (298 K)	Cathodic limit: 0.6 (*vs.* Li^+/Li)[a] Anodic limit: 5.3 (*vs.* Li^+/Li)[a]	256
LiBeti-acetamide	20.0-80.0	216		Cathodic limit: 0.6 (*vs.* Li^+/Li)[c] Anodic limit: 3.8 (*vs.* Li^+/Li)[b]	263
	14.3-85.7		1.27 (303 K)		263

DMU: *N*,*N*-dimethylurea; MA: *N*-methylacetamide; ChCl: Chorine chloride; Tf_2N: $N(CF_3SO_2)_2^-$; Beti: $N(C_2F_5SO_2)_2^-$

II. SYNTHESIS AND PURIFICATION OF ROOM-TEMPERATURE IONIC LIQUIDS

As noted above, there are many classes of non-chloroaluminate RTILs. However, not all of them are suitable for electrochemistry because they may be poorly conductive and highly viscous, or they may have a limited electrochemical windows. In this Section, we describe the general synthetic methods used to prepare and purify some selected RTILs, such as $EtMeIm^{+}F(HF)_{2.3}^{-}$, $EtMeIm^{+}MF_{x}^{-}$ [M: B (x = 4); P, As, Sb, Nb, Ta (x = 6); W (x = 7).], and n-$Bu_3MeN\ ^{+}Tf_2N^{-}$. These particular systems were chosen because they have been used as electrochemical solvents on many occasions. The various procedures that are used to prepare and purify other RTILs are far too numerous to list here, and they tend to be specific for the RTIL that is being prepared. The interested reader is referred to a recent article for more information.[267]

1. Dialkylimidazolium Chlorides

Among the various RTILs that have been used as solvents for electrochemistry, the greatest proportion is based on cations derived from 1,3-dialkylimidazolium chloride salts. The synthesis and purification of these materials were reported by Wilkes et al.[75] more than two decades ago in one of the most highly-cited articles to appear in the journal *Inorganic Chemistry*. Although several newer procedures have been reported, the technique described in this seminal article is still a reliable and simple method to make high-quality salts. Experience has shown that carefully prepared 1,3-dialkylimidazolium salts are relatively easy to purify and ultimately lead to transparent RTILs with low levels of impurities. In fact, if nontransparent RTILs are obtained, they can be readily decolorized by passage of the RTILs dissolved in dichloromethane through a suitable chromatography column.[268]

A typical procedure for preparing one of the more popular salts, 1-ethyl-3-methylimidazolium chloride (EtMeImCl), is as follows: 1-ethyl chloride (600 mL) is condensed into a 1 L glass pressure vessel containing distilled 1-methylimidazole (300 mL) and dry acetonitrile (100 mL). The mixture is heated at 333 K with continuous stirring until the solution exhibits a milky-white color if the stirring is halted. The bulk of the unreacted ethyl chloride is removed by sparging the reaction mixture with a stream of dry N_2. The remaining traces of ethyl chloride are removed by evacuation. At this point, a large quantity of white crystals should be evident in the reaction vessel. These crystals are crude EtMeImCl. The crystals can be purified by precipitation from acetonitrile with ethyl acetate. Because this salt is very hygroscopic, a good quality vacuum line and an assortment of Kontes Airless-ware® flasks (Kjeldahl-shaped) and filter funnels with coarse porosity filters will greatly simplify this procedure. To carry out this purification step, the resulting crude EtMeImCl is first dissolved in warm (~333 K) acetonitrile. Ethyl acetate is then added to this solution. Upon cooling, fluffy white crystals of the salt precipitate from the solution. It may be necessary to scratch the inside of the flask with a glass rod or to shake the flask vigorously in order to induce precipitation. The salt is collected in an Airless-ware® filter funnel and dried under vacuum. Experience dictates that this procedure must be repeated at least three times in order to obtain a high-quality product. The final purification step involves melting the salt under vacuum (1×10^{-3} torr) at 373 K to remove the last traces of solvent.

2. Dialkylimidazolium Salts with Fluorohydrogenate Anions

Fluorohydrogenate RTILs, such as 1-ethyl-3-methylimidazolium fluorohydrogenate, $EtMeIm^+F(HF)_{2.3}^-$, display low viscosity and high conductivity at room temperature. If properly prepared, these RTILs exhibit negligible release of HF gas and will not etch Pyrex® glassware.[61-63] Because of these properties, they are excellent, if unappreciated, electrochemical solvents. Only modest skills with the safe handling of fluorine and/or HF are required to synthesize high purity $EtMeIm^+F(HF)_{2.3}^-$. The starting material for this RTIL is EtMeImCl. Figure 4 shows a schematic of a vacuum line recommended for the synthesis of fluorohydrogenates. Because anhydrous HF reacts rapidly with borosilicate glassware, the vacuum line and reaction cell are made of fluorine-resistant materials such as SUS-316, FEP (fluoroethylene-propylene copolymer), PFA (perfluoroalkoxide polymer), or Teflon. Anhydrous HF (AHF) is obtained by contact with K_2NiF_6 overnight in a sealed cell. The volatile AHF is then transferred from the HF–K_2NiF_6 container to

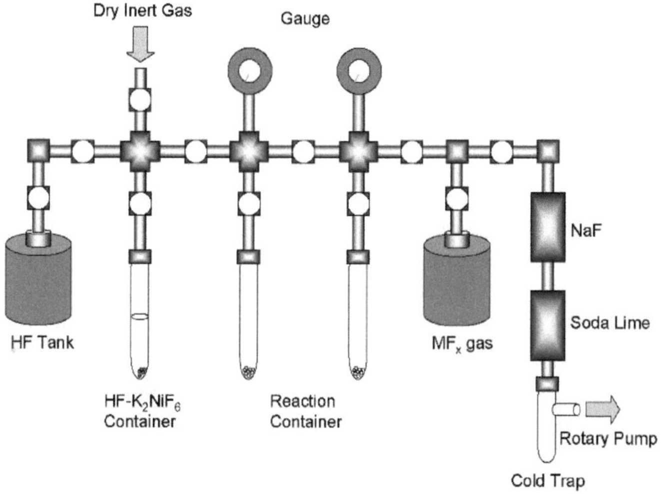

Figure 4. Diagram of a vacuum line used for the synthesis of fluorohydrogenate-based RTILs.

the FEP reaction tube containing EtMeImCl and a Teflon-coated stir bar by cooling the reaction tube with liquid N_2. After melting the icy AHF in the reaction tube, the AHF (added in excess) reacts with the EtMeImCl and generates HCl gas. The resulting HCl can be eliminated by sparging the RTIL with dry nitrogen gas. This procedure should be repeated several times so as to ensure that no unreacted Cl^- remains. Unreacted anhydrous HF is eliminated from the $EtMeIm^+F(HF)_x^-$ under vacuum by reducing the pressure to 7.5×10^{-3} torr (1 Pa) or less.

3. Dialkylimidazolium Salts with Fluorocomplex Anions

RTILs based on dialkylimidazolium salts with polyfluoroanions, such as MF_x^-, where M: B ($x = 4$); P, As, Nb, Ta ($x = 6$); W ($x = 7$), are normally prepared by using the corresponding aqueous acids or metal salts. This pathway leads to RTILs that are contaminated with water and/or metal chlorides. Such RTILs are difficult to purify. Recently, Matsumoto et al.[131] have reported a novel procedure to make $EtMeIm^+MF_x^-$ salts by using $EtMeIm^+F(HF)_{2.3}^-$.

The reaction between $EtMeIm^+F(HF)_{2.3}^-$ and the corresponding Lewis fluoroacid, $MF_{x\text{-}1}$, is conducted in the reaction vessel shown in Fig. 4. There do not appear to be any significant side reactions, and the HF produced during the reaction can be readily eliminated from the resulting RTILs under vacuum. If $EtMeIm^+F(HF)_{2.3}^-$ is available, this is a quick and reliable method for preparing these salts.

4. Tetraalkylammonium Salts with Bis[(trifluoromethyl)-sulfonylimide Anions

Tetraalkylammonium salts with bis((trifluoromethyl)sulfonyl) imide (Tf_2N^-) anions, such as tri-*n*-butylmethylammonium bis ((trifluoromethyl)sulfonyl)imide, n-$Bu_3MeN^+Tf_2N^-$, are highly viscous RTILs.[146] However, n-$Bu_3MeN^+Tf_2N^-$ (386 cP at 303 K) in particular shows greater chemical and electrochemical stability than bis((trifluoromethyl)sulfonyl)imide salts based on dialkylimidazolium cations.[150] Furthermore, it is very hydrophobic. This suggests that this RTIL might be a promising solvent for the chemistry of radioactive materials, radical chemistry, and high-energy density batteries.

The n-$Bu_3MeN^+Tf_2N^-$ RTIL is prepared by mixing exactly equal molar amounts of n-Bu_3MeNCl and $LiTf_2N$ in ultrapure water. (The former salt was purified by a procedure similar to that described above for EtMeImCl.) This solution is agitated at room temperature for 24 hours, and the resulting hydrophobic ionic liquid is extracted with ultrapure dichloromethane. This solution is washed with several portions of purified water until the wash water contains no chloride as determined by the addition of a drop or two of a dilute solution of silver nitrate. If necessary, the chloride content can also be evaluated with ion chromatography. Finally, the dichloromethane is removed by evacuating the solution at 1×10^{-3} torr for 24 h while it is heated to 373 K. The resulting RTIL is usually colorless and very dry ($H_2O < 3$ ppm). All hydrophobic Tf_2N^--based RTILs can be prepared by using this same basic method.

III. FUNDAMENTAL PROPERTIES OF ROOM-TEMPERATURE IONIC LIQUIDS

As a class of solvents, RTILs display many interesting and useful physical and chemical properties. However, newcomers to this field need to know that there are also limitations and special requirements associated with these ionic solvents that need to be understood before employing them as solvents for electrochemistry.

1. Thermal Stability

The use of RTILs at elevated temperatures requires information about their thermal stabilities. The thermal stability of a RTIL is usually defined through experiments carried out with differential scanning calorimetry (DSC), thermogravimetric analysis (TGA), or differential thermal analysis (DTA). Although these methods are easy to use and show good reproducibility, the results obtained with these techniques, especially fast TGA scans (10 ~ 20 K min^{-1}), do not always provide an accurate assessment of the thermal stabilities of the RTILs.[35,269-274] Table 3 shows thermal decomposition rates at different temperatures for some popular RTILs compared to the decomposition onset temperature as measured by TGA-DTA. Obviously, significant decomposition occurs at temperatures considerably lower than the onset of the decomposition process. Therefore, electrochemical experiments should not be done at temperatures close to the decomposition temperature as determined by TGA/DTA, especially when the experiments are expected to last for a considerable period of time. Furthermore, several research groups have compared the thermal stabilities of selected RTILs in moist air and dry N_2[139,272,275] and found notable differences.[272,275] These differences are directly related to the reactivity of the RTILs with atmospheric moisture. In addition to this factor, sample quantity,[272] scan rate,[272] and the material used to make the sample crucible[271,272,276] can also affect the results.

Table 3
Thermal Decomposition Rates for Imidazolium-Based RTILs

Temperature (K)	Decomposition rate (mass % min^{-1})			
	$EtMeIm^+BF_4^-$ [269]	$BuMeIm^+BF_4^-$ [269]	1-Bu-2,3-$Me_2Im^+PF_6^-$ [270]	1,2-Me_2-3-$PrIm^+Tf_2N^-$ [269]
473	–	–	0.01	
523	0.0130	0.0228	0.03	0.0030
573	0.0260	0.0398	0.25	0.0320
623	0.4699	0.240	2.18	0.0765
673	1.637	1.507	–	0.5204
723	7.193	8.333	–	9.941
Onset decomposition temperature (K)	718	697	693	730

Limited investigations have been carried out on the pyrolysis reactions associated with decomposition of the organic cations. The main products of the pyrolysis process seem to be related to the nucleophilicity of the anions and the basic skeleton of the cations. Two general decomposition models have been proposed for tetraalkylammonium cations: the well-known Hofmann elimination with the production of an alkene (Eq. 1) and the reverse Menschutkin reaction, leading to a tertiary amine and alkyl halide (Eq. 2):[20,277]

$$R_4N^+X^- \rightarrow R_3NH^+X^- + CH_2{=}CH_2 \qquad (1)$$

$$R_4N^+X^- \rightarrow R_3N + RX \qquad (2)$$

Similar investigations using phosphonium-based RTILs have been attempted, but the decomposition mechanisms are more complicated than for the tetraalkylammonium cations and often depend on the anion. The original literature should be consulted for more information.[278,279]

In the case of those RTILs based on dialkylimidazolium cations, the decomposition pathway most likely proceeds via S_N1 (Eq. 3) or S_N2 (Eq. 4) reactions:[74,275]

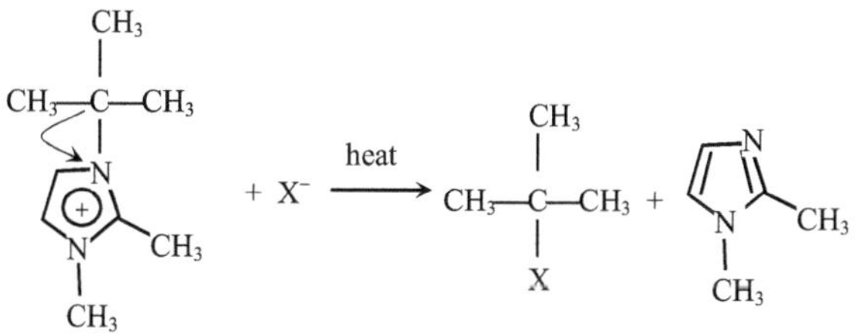

(3)

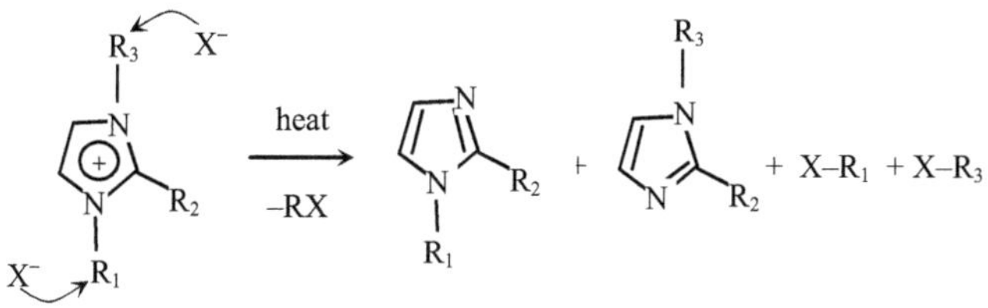

(4)

The reaction pathway depends on the nucleophilicity of the anions and the structure of the alkyl group on the nitrogen atom of the imidazolium ring. Thus, if the anions are halides having strong nucleophilicity, and the cations have linear alkyl chains, the reaction will likely follow the S_N2 pathway.[74,275]

Several groups have listed the thermal stabilities of the imidazolium cation-based RTILs. For example, Ngo et al.[276] indicates $PF_6^- > Beti^-$ (bis(perfluoroethylsulfonyl)imide) $> Tf_2N^- \approx BF_4^- > Tf_3C^- \approx AsF_6^- > I^-$, Br^-, and Cl^-. Awad, et al.[275] gives the following order: $PF_6^- > Tf_2N^- > BF_4^- > Br^-$, and Cl^-, whereas Fredlake et al.[280] proposes $Tf_2N^- > Tf_3C^- > TfO^- > BF_4^- > N(CN)_2^- > Br^-$. There is little doubt that the anion plays a crucial rule in the thermal stability. Overall, imidazolium-based cations are reported to be more stable than ammonium-based cations.[276] The reason is unclear, but the aromatic properties of the imidazolium ring may contribute to the increased stability.

The effect of atmospheric oxygen on the decomposition of imidazolium salts was also investigated by Awad, et al.[275] They reported that the tendency to undergo oxidative decomposition increased as the length of the alkyl chain bound to the nitrogen of the imidazolium ring was increased. However, this was not the case for the halide salts, leading them to propose that the activation energy for oxidative decomposition was higher than the activation energy for thermal decomposition. There is also evidence to indicate that the thermal stabilities of dialkylimidazolium salts declines considerably after the addition of strong nucleophilies.[281] The details of the pyrolysis reactions of the various classes of RTILs are just emerging, but more will be known as further investigations are conducted in this field.

2. Water Contamination

The contamination of RTILs by water is a long standing problem. The discovery that some RTILs possess hydrophobic properties has probably led to inattention during the purification of these materials in some cases because researchers assumed that the water content of these materials would be diminutive. Only recently has it come to be appreciated that hydrophobic RTILs may contain appreciable amounts of water and that this water contamination may affect the physicochemical properties of these liquids. In fact, it is now established that adventitious water can result in measurements that lead to lower densities,[76,282-285] lower viscosities,[76,282-287] and higher conductivities[284,287] than neat RTILs. In addition, *wet* RTILs may exhibit a smaller electrochemical window because the reduction of water to produce hydrogen gas is likely to become the limiting cathodic process.[284,288,289] In the latter case, it is important to note that the electrochemical window is highly dependent on the material used for the working electrode.[288,290] For example, glassy carbon may give the illusion that the RTIL under study has a large electrochemical window because it displays a large hydrogen overpotential. Platinum, which exhibits a relatively low hydrogen overpotential compared to glassy carbon, may give an entirely different result.

Many different methods have been used for the detection of water in RTILs, including NMR spectroscopy, infrared spectroscopy, Karl-Fischer (K-F) titration methods, and even voltammetry at platinum electrodes. The general consensus is that the spectroscopic techniques are only adequate for detecting relatively large amounts of water, whereas K-F titration and voltammetric methods are better suited for the detection of trace contamination. Of these two techniques, the voltammetric method is more universally applicable because some RTILs may react with the K-F reagents. The voltammetric method is easy to apply, provided that it has been established beforehand that the current for the reduction of water or H^+ varies linearly or in some predictable way with the concentration of these impurities so that a standard curve can be prepared. Then the observed current recorded in a contaminated RTIL can be compared to the standard curve to estimate the water or H^+ content. If the current versus concentration response is linear, then it is also possible to employ a simple standard addition method. However, the application of platinum electrodes for the detection of trace water or H^+ contamination is not without complication; it may be necessary to activate the platinum surface by using one of the many literature recipes in order to obtain the lowest overpotential and greatest sensitivity and reproducibility.

As discussed above, the electrochemical window of most wet RTILs is expected to be smaller than that of their dry counterparts. However, in some cases, wet RTILs display a potential window comparable to dry ones. This phenomenon appears to be related to the formation of symmetric hydrogen bonds between the water and RTIL anions, e.g., BF_4^-, PF_6^-, and NTf_2^-.[291-294] This hydrogen bonding apparently results in highly associated water that is less reactive than water in the free state. Water also plays a more insidious role in RTILs. Swatloski, et al.[293] report that 1-butyl-3-methylimidazolium hexafluorophosphate is readily hydrolyzed by adventitious water to produce HF among other products.

It should be realized that most RTILs readily absorb water if exposed to the atmosphere, even those considered to be hydrophobic.[282,284,295] Therefore, it is often necessary to handle RTILs under an inert atmosphere to obtain accurate data, regardless of their supposed *inertness*. The procedure for removing water from ionic liquids is straightforward, if the material under investigation is water stable, i.e., the anion does not readily hydrolyze. Most water can be removed with a liquid N_2-trapped vacuum line ($< 1 \times 10^{-3}$ torr) using conventional procedures, under moderate heating and vigorous stirring. Also, water can be removed by sparging the RTIL with a dry inert gas such as N_2 or Ar. When ultrapure RTILs are needed, further treatment should be done by using controlled-potential electrolysis at an applied potential sufficiently cathodic so as to reduce any water or H^+ without reductive decomposition of the RTIL. To attain maximum benefit, this procedure should be carried out under an inert atmosphere in a divided cell. For RTILs containing fluoroanions such as MF_x^-, where M: B ($x = 4$); P, As, Nb, Ta ($x = 6$); W ($x = 7$), care should be taken to avoid any HF that may result from the hydrolysis of these anions. In particular, the contamination of wet RTILs by protons, chloride, and hard acid impurities requires special attention, because, under such conditions, these impurities will catalyze anion hydrolysis.[296-301]

3. Chloride Ion Contamination

Most RTILs are produced from a halide salt intermediate, usually a chloride salt, because such salts are easy to prepare and purify. However, RTILs produced in this way often contain residual chloride. In a careful study, Seddon et al.[282] found that some physicochemical properties are highly dependent on the amount of residual chloride in the RTIL. Generally speaking, chloride ion, because of its strong nucleophilicity, increases the viscosity and decrease the density of the ionic liquid. In addition, because chloride is relatively easy to oxidize, RTILs containing chloride ion impurities exhibit a smaller electrochemical window.[302] Chloride contaminated RTILs often show decreased thermal stabilities as mentioned above (see Section III.1). Needless to say, if RTILs containing chloride contaminants are employed as reaction media, the results may be much different from those that would be obtained in the neat RTILs. Therefore it is very important to purify RTILs prepared from halide salts. Recommended procedures for the purification of RTILs are described in a recent monograph.[303] After the RTILs have been purified, the Cl^- concentration should be examined in some way. A simple approach is to add a few drops of a dilute silver nitrate solution and watch for the formation of a precipitate of silver chloride. However, quantitative analysis of the chloride concentration is best done with ion chromatography.[134,298,304]

4. Oxygen Contamination

Oxygen does not react with most RTILs. However, the electrochemical reduction of dissolved O_2 in dry RTILs is similar to that observed in anhydrous aprotic organic solvents such as acetonitrile, dimethylformamide, dimethylsulfoxide, and similar solvents under anhydrous conditions and results in the formation of superoxide ion, O_2^-:[305-313]

$$O_2 + e^- \leftrightarrows O_2^- \tag{5}$$

Unfortunately, O_2^- is a moderately powerful reductant, leading to the decomposition of the organic cations in imidazolium- and phosphonium-based RTILs.[305,306] Therefore, if these RTILs are used for electrochemistry, it is imperative to avoid contamination by oxygen. If both protons (water) and O_2 are present, the results can be even more complicated. Numerous investigations of the reduction of O_2 have been undertaken in RTILs. The data resulting from these investigations are summarized in Table 4. As found in most aprotic solvents, the O_2/O_2^- electrode reaction is quasi-reversible. However, the heterogeneous kinetic data and diffusion coefficients resulting from the many prior investigations do show some differences. The reason for these variations is not clear, but it is probable that the concentration of dissolved O_2 has not been established accurately and/or that the electrogenerated O_2^- reacts with impurities such as protons[313] or water.[305,309] It may also attack the organic cation.[305,306] These differences could also result from

Table 4
Electrochemical Parameters for O_2 Reduction in RTILs

RTILs	Electrode	Temp. (K)	ΔE_p	C_{O2} (mmol L^{-1})	$D_{O2} \times 10^5$ ($cm^2\ s^{-1}$)	$k^0 \times 10^3$ (cm s^{-1})	α	Ref.
$EtMeIm^+BF_4^-$	Pt	298 ± 1	0.26	1.1 ± 0.2		0.94 ± 0.13	0.46 ± 0.02	309
	Au	298 ± 1	0.18	1.1 ± 0.2	1.7 ± 0.2	2.2 ± 0.1	0.44 ± 0.01	309
	GC	298 ± 1	0.09	1.1 ± 0.2		6.4 ± 0.1	0.47 ± 0.01	309
	PG	298 ± 1				4.3 ± 0.1		312
	PG+CNT	298 ± 1				8.3 ± 0.1		312
$EtMeIm^+Tf_2N^-$	Au	298	0.22					305
	Au	293 ± 3		3.9	0.73			308
	Au	RT		10.8 ± 0.6	0.25			310
$MePrIm^+BF_4^-$	Pt	298 ± 1	0.26	0.97 ± 0.05		1.5 ± 0.4	0.36 ± 0.03	309
	Au	298 ± 1	0.20	0.97 ± 0.05	1.3 ± 0.2	3.3 ± 0.5	0.35 ± 0.01	309
	GC	298 ± 1	0.08	0.97 ± 0.05		7.3 ± 1.6	0.42 ± 0.06	309
	PG	298 ± 1				2.9 ± 0.1		312
	PG+CNT	298 ± 1				10.4 ± 0.1		312
$BuMeIm^+BF_4^-$	Pt	298 ± 1	0.29	1.1 ± 0.1		0.83 ± 0.20	0.36 ± 0.02	309
	Au	298 ± 1	0.18	1.1 ± 0.1	1.2 ± 0.1	1.2 ± 0.2	0.37 ± 0.02	309
	GC	298 ± 1	0.11	1.1 ± 0.1		7.5 ± 0.1	0.39 ± 0.01	309
	PG	298 ± 1				2.3 ± 0.1		312
	PG+CNT	298 ± 1				4.2 ± 0.1		312
$BuMeIm^+PF_6^-$	GC	RT	0.32	3.6	0.22		0.42	307
$1,2\text{-}Me\text{-}3\text{-}PrIm^+Tf_2N^-$	Au	298	0.13					305
$1\text{-}Bu\text{-}2,3\text{-}MeIm^+Tf_2N^-$	Au	RT		7.2 ± 0.8	0.21			310
$MePenIm^+FAP^-$	Au	RT		9.7 ± 0.2	0.35			310

Table 4. Continuation

RTILs	Electrode	Temp. (K)	ΔE_p	C_{O2} (mmol L^{-1})	$D_{O2} \times 10^5$ (cm^2 s^{-1})	$k^0 \times 10^3$ (cm s^{-1})	α	Ref.
$MeOctIm^+Tf_2N^-$	Au	RT		8.3 ± 0.8	0.28			310
$DecMeIm^+Tf_2N^-$	Au	RT		13.0 ± 0.9	0.25			310
$HexMe_3N^+Tf_2N^-$	Au	298	0.21					305
$Et_3HexN^+Tf_2N^-$	Au	293 ± 3		11.6	0.148			308
	Au	308 ± 1		8.9 ± 0.8	0.32 ± 0.03	5.0	0.35	306
	Pt	308 ± 1		8.9 ± 0.8	0.32 ± 0.03	3.0	0.33	306
$BuMePyrr^+Tf_2N^-$	Au	298	0.13					305
	Au	308 ± 1		6.1 ± 0.8	0.52 ± 0.04	3.5	0.38	306
	Au	298		13.6 ± 0.8	0.18 ± 0.02			311
	Pt	308 ± 1		6.1 ± 0.8	0.52 ± 0.04	0.8	0.35	306
$P_{14.666}{}^+Tf_2N^-$	Au	308 ± 1		6.0 ± 0.5	0.75 ± 0.06			306
$P_{14.666}{}^+FAP^-$	Au	308 ± 1		7.8 ± 1.5	0.61 ± 0.11			306

inconsistent efforts to account for the uncompensated resistance during these measurements. Overall, the presence of dissolved O_2 can complicate electrochemical experiments in RTILs. Fortunately dissolved O_2 can be easily removed under vacuum or by sparging with dry N_2.

5. Conductance Measurements at High Frequencies

More than a decade ago, the equipment needed to carry out high-frequency AC impedance measurements was forbiddingly expensive. Therefore only a few electrochemists were well versed in the theory and experimental practice of AC impedance methods. However, the prospects for carrying out such studies have increased enormously because software-controlled equipment is now readily available at reasonable cost. AC impedance methods are widely used for measuring the conductivity of RTILs, but there is considerable variation among the data measured by different workers who have studied the same systems. Impurities aside, these deviations are likely caused by insufficient electrochemical knowledge. It is well-known that there are very strong interactions between anions and cations in most RTILs. Under such conditions, the Debye-Falkenhagen effect must be taken into account. Unfortunately, in many cases it has been ignored.

Conductivity cells are typically calibrated with aqueous KCl solutions prepared from highly purified water according to IUPAC recommendations.[314] Depending on the cell geometry, the measured cell resistances may vary with frequency, ω, over the range from 100 Hz to 100 kHz. However, the cell resistance will usually show a linear dependence on the reciprocal of the square root of the frequency, $\omega^{-1/2}$. The frequency-independent resistances of the calibration solutions and the RTILs under investigation are obtained by extrapolating the cell resistance to infinite frequency.[315] If such plots do not exhibit the requisite linear dependence, an empirical equation can be employed to obtain the frequency-independent cell resistance.[315] Normally the data resulting from such measurements exhibit high precision.

6. Other Considerations

The self-diffusion coefficients of cations and anions in neat RTILs have been measured with pulsed-field gradient spin–echo NMR (PGSE–NMR).[70,101,316-319] Such data can also be used to estimate transport numbers and ion mobilities, and it can be combined with other physicochemical properties to gain insight into the dynamics of ionic motion in RTILs.

A recent article by Earle et al.[320] reported on the distillation and volatility of several common RTILs. It is generally held that most RTILs do not exhibit significant vapor pressure at room temperature. In practice this is true, but contrary to popular belief, many ionic liquids can be distilled under high vacuum with little or no decomposition in the temperature range of 473-573 K. The rate of distillation is roughly proportional to the molecular weight of the RTIL.

IV. GENERAL ELECTROCHEMICAL TECHNIQUES

In 1914, Walden published an article describing the electrical conductivity of some molten salts or ionic liquids.[162] Thus, it would not be an exaggeration to say that electrochemistry in ionic liquids has a very long history. Since the publication of that article, the field of electrochemistry in ionic liquids has become a mature science, and almost every electrochemical technique that has been applied to aqueous electrolyte solutions or solutions prepared from aprotic organic solvents and organic salts has been applied to solutions derived from ionic liquids as well. In the case of reactive or high-temperature ionic liquids, it has been necessary to overcome numerous experimental obstacles. The purpose of this Section is to review a few of the more common electrochemical methods that have been commonly applied to RTILs. For more complete information, the reader is advised to consult one of the many excellent comprehensive texts on electrochemistry such as the classic text coauthored by Allen Bard and Larry Faulkner.[321]

1. Electrodes and Experimental Considerations

Electrochemical experiments with RTILs are normally carried out in three-electrode cells. Working electrodes (WE) that have been used for such experiments in RTILs include polycrystalline metals such as gold, platinum, and tungsten, and non-metals, principally glassy (vitreous) carbon and pyrolytic graphite. Various single crystal metals have also been used for specialized surface studies and will not be discussed here.[322]

It is important that the surface of the WE be carefully prepared before use by polishing with aqueous slurries containing successively finer grades of alumina, electrochemical polishing under appropriate conditions that depend on electrode material, and finally, degreasing with ethanol or acetone. The procedures for carrying out such polishing are well established. However, because such polishing involves exposure of the electrode surface to water, the electrode should be carefully dried under vacuum before use in dry RTILs. In some cases, it may be advantageous to electrochemically precondition the WE before use. For example, the electrode might be poised at a potential close to the positive limit of the solvent to remove surface impurities such as adsorbed water. However, all preconditioning methods must be carefully evaluated to ensure that they produce reproducible results and do not diminish the active surface area of the electrode. Cyclic voltammetry is a good technique for assessing such preconditioning methods. In some cases, it may be necessary to employ more drastic cleaning procedures that would seldom be used during electrochemical experiments in aqueous solutions such as cleaning the WE surface with an abrasive tissue after each electrochemical experiment. However, if electrochemical experiments performed in RTILs result in intractable surface films, there is often no other recourse except to re-polish the electrode surface.

At the present time, there is no way to determine the active surface areas of working electrodes used in RTILs. Most investigators simple refer all quantitative data, e.g., current densities, to the geometrical area of the electrode. This matter is further complicated by the tendency of organic salt-based ionic liquids to form films at their anodic and/or cathodic potential limits. It is doubtful that surface area measurements derived from experiments carried out in aqueous solutions are applicable to electrodes used in

RTILs. Furthermore, it is often assumed that the electrode surface area does not vary with the type of ionic liquid under investigation; this may be a false assumption. At the present time, there are few answers for these questions. The best strategy is to polish the electrode until the maximum voltammetric current is obtained for a test redox couple such as ferrocenium/ferrocene (Fc^+/Fc). By referring to the original scan at the fresh electrode, it is possible to track any electrode surface area changes that occur during the course of a series of experiments.

A useful procedure for determining the electrochemical window of a RTIL is to record two cyclic voltammograms at a WE; one is initiated from the rest potential toward cathodic potentials, whereas the second is initiated from the rest potential toward anodic potentials or vice versa. Because sampling the potential limit of the RTIL is sure to produce some type of surface film, care should be taken to refresh or clean the electrode surface between scans. However, one must be mindful that the concept of an electrochemical window is somewhat subjective and depends on the level of the background current ascribed to the potential limit of the ionic liquid.

Most electrochemical texts provide little or no information about the desirable characteristics of the counter electrode (CE). When current is passed through an electrochemical cell, the function of the CE is to support the current that flows through the WE. The electrochemical reaction occurring at the CE must in no way affect the overall cell response. Therefore, the surface area of the CE should be considerably larger than that of the WE to ensure that all current limitations in the cell are due to the WE response. In some cases, it may be necessary to take into account the geometry and placement of the CE in order to provide uniform current distribution at the WE surface. Finally, some thought must be given to the electrode reaction that takes place at the CE. Will the products that are produced at the CE affect the cell reaction? In RTILs, the CE reaction is almost always the oxidative or reductive decomposition of the solvent. Thus, it is necessary to isolate the CE in a compartment that is separated from the bulk solution by a glass frit in order to limit the mixing of the contents of this compartment with the bulk solution. In addition, the volume of the CE compartment should be sufficient to contain enough of the RTIL to support the total charge passed during the experiment, or the CE

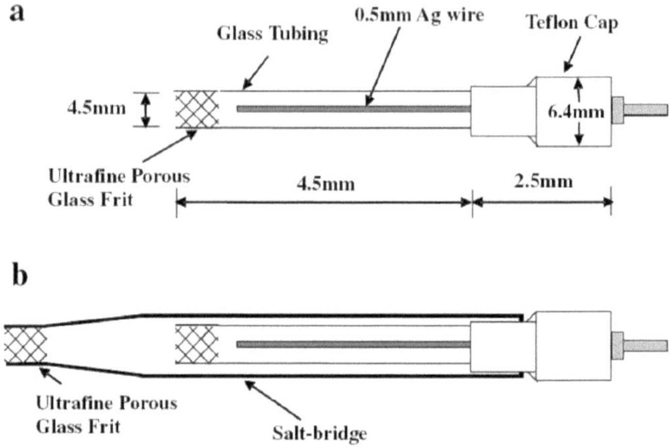

Figure 5. Diagrams of (a): a Ag^+/Ag reference electrode for use with RTILs, and (b): this reference electrode inserted into a salt-bridge compartment. Reproduced from reference 323 with permission of Elsevier B.V.

reaction will limit the cell response. Typically, the CE is fashioned from a coil of Pt or W wire or a large surface area coupon or flag of the same materials. One of the authors has observed the successful use of a large stainless-steel spatula as the CE while visiting the laboratory of a prominent organic electrochemist!

At the present time, there is no universal reference electrode (RE) that can be used in all RTILs. However, Ag^+/Ag couples prepared by immersing a clean Ag wire in a RTIL solution prepared from a soluble silver salt or by electrogenerating Ag^+ show the most promise. Like the CE, the RE must be isolated from the bulk ionic liquid with a fritted glass barrier such as a glass frit or porous Vycor plug (Fig. 5a).[323] If the leakage of Ag^+ into the bulk solution is problematic, then it may be necessary to employ a salt bridge so that the hydraulic flow of the Ag^+ reference electrode solution into the test solution is protected by two barriers. One such sample is depicted in Fig. 5(b).[323] Another approach is to employ a simple quasireference electrode such as a Ag, Au, or Pt wire immersed in the RTIL and isolated from the bulk solution with a porous barrier. The potential of the quasi-reference electrode can be related to the one of the internal standards recommended by the IUPAC such as

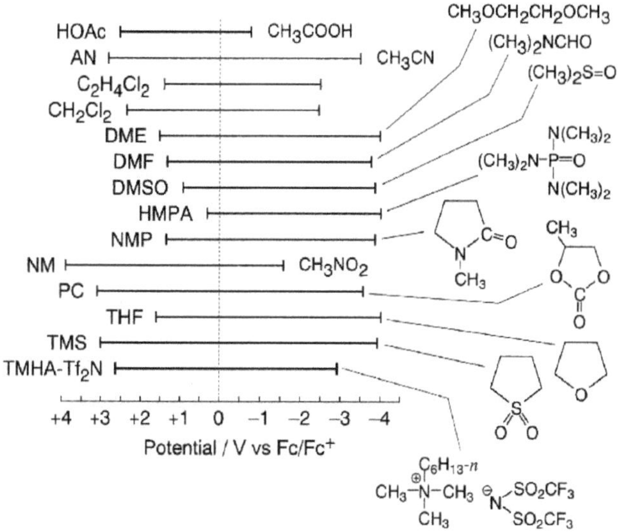

Figure 6. Comparison of electrochemical window of n-$HexMe_3N^+Tf_2N^-$ (TMHA-Tf_2N) and some common organic solvents that have been used for electrochemistry. All potentials are referenced to the standard electrode potential for the Fc^+/Fc. Reproduced from Ref. 328 with permission of Kluwer Academic Publishers.

the Fc^+/Fc couple, bis(biphenyl)chromium couple (BCr^+/BCr), or other metallocenes.[324] The number of research groups using this approach seems to be increasing. But unfortunately, Fc^+ may react with some RTILs[38,325,326] and with traditional organic solvents (DMF and DMSO).[327] If the decomposition of one of the reference compounds takes place during the electrochemical measurements, standard electrode reactions such as the aforementioned Ag^+/Ag couple or the I^-/I^{3-} couple can be used as an alternative to Fc^+/Fc. Figure 6 shows the electrochemical windows of n-$HexMe_3N^+Tf_2N^-$ and conventional organic solvents calibrated relative to the Fc^+/Fc standard electrode potential.[328]

As electrochemical solvents, most RTILs are not especially conductive. In fact, at room temperature, one of the very best salts, e.g., $EtMeIm^+F(HF)_{2.3}^-$, exhibits a specific conductivity akin to that of a 1.0 mol L^{-1} aqueous KCl solution. Unfortunately, there are a

Table 5
Physicochemical Data for Selected Electrochemically-Viable RTILs Having Suitable Viscosity (≤ 50 cP) and Conductivity (5.0 mS cm^{-1} ≤)

RTIL		Density	Viscosity	Conductivity	Ref.
Cation	Anion	(g cm^{-3})	(cP)	(mS cm^{-1})	
$DMeIm^+$	$CF_3BF_3^-$	1.40 (298 K)	27 (298 K)	15.5 (298 K)	118,120
$DMeIm^+$	$C_2F_5BF_3^-$	1.47 (298 K)	33 (298 K)	11.7 (298 K)	118
$DMeIm^+$	n-$C_3F_7BF_3^-$	1.55 (298 K)	47 (298 K)	7.3 (298 K)	118
$DMeIm^+$	$N(SO_2CF_3)_2^-$	1.559 (295 K)	44 (293 K)	8.4 (293 K)	139
$DMeIm^+$	$F(HF)_{2.3}^-$	1.17 (298 K)	5.1(298 K)	110 (298 K)	66
$EtMeIm^+$	BF_4^-		43 (299 K)	13 (299 K)	100
$EtMeIm^+$	BF_4^-	1.2853 (298 K)	38 (298 K)	13.6 (298 K)	116
$EtMeIm^+$	BF_4^-	1.279 (298 K)	32 (298 K)	13.6 (298 K)	101
$EtMeIm^+$	$CH_3BF_3^-$	1.1536 (298 K)	47 (298 K)	9.0 (298 K)	116
$EtMeIm^+$	$CH_2CHBF_3^-$	1.1614 (298 K)	41 (298 K)	10.5 (298 K)	116
$EtMeIm^+$	$CF_3BF_3^-$	1.35 (298 K)	26 (298 K)	14.8 (298 K)	118
$EtMeIm^+$	$CF_3BF_3^-$		26 (298 K)	14.6 (298 K)	120
$EtMeIm^+$	$C_2F_5BF_3^-$	1.42 (298 K)	27 (298 K)	12 (298 K)	118
$EtMeIm^+$	n-$C_3F_7BF_3^-$	1.49 (298 K)	32 (298 K)	8.6 (298 K)	118
$EtMeIm^+$	n-$C_4F_9BF_3^-$	1.55 (298 K)	38 (298 K)	5.2 (298 K)	118
$EtMeIm^+$	$CF_3CO_2^-$	1.285 (295 K)	35 (293 K)	9.6 (293 K)	139
$EtMeIm^+$	$CF_3SO_3^-$	1.390 (295 K)	45 (293 K)	8.6 (293 K)	139
$EtMeIm^+$	$N(CN)_2^-$	1.08 (293 K)	17 (295 K)	27 (293 K)	210
$EtMeIm^+$	$N(SO_2CF_3)_2^-$	1.51 (298 K)	34 (298 K)	9.2 (298 K)	329
$EtMeIm^+$	$N(SO_2CF_3)_2^-$	1.518 (298 K)	31 (298 K)	5.7 (298 K)	101
$EtMeIm^+$	$N(SO_2CF_3)_2^-$		28 (299 K)	8.4 (299 K)	100

Table 5. Continuation

RTIL		Density	Viscosity	Conductivity	Ref.
Cation	Anion	($g\ cm^{-3}$)	(cP)	($mS\ cm^{-1}$)	
$EtMeIm^+$	$N(SO_2CF_3)_2^-$	1.52 (298 K)	33 (298 K)	8.7 (298 K)	330
$EtMeIm^+$	$N(SO_2CF_3)_2^-$		45.9 (298 K)	8.4 (298 K)	145
$EtMeIm^+$	$N(SO_2CF_3)_2^-$		33 (298 K)	8.3 (298 K)	144
$EtMeIm^+$	$N(SO_2CF_3)_2^-$	1.520 (295 K)	34 (293 K)	8.8 (293 K)	139
$EtMeIm^+$	$N(COCF_3)(SO_2CF_3)^-$	1.46 (298 K)	25 (298 K)	9.9 (298 K)	330
$EtMeIm^+$	$N(SO_2F)_2^-$		24.5 (298 K)	16.5 (298 K)	145
$EtMeIm^+$	$N(SO_2F)_2^-$		18 (298 K)	15.4 (298 K)	144
$EtMeIm^+$	$C(CN)_3^-$	1.11 (293 K)	18 (295 K)	18 (293 K)	210
$EtMeIm^+$	$SeCN^-$		25 (294 K)	14.1 (294 K)	220
$EtMeIm^+$	$CuCl_2^-$		45.9 (298 K)	11.6 (298 K)	188
$EtMeIm^+$	$AlCl_4^-$	1.2943(298 K)	17.8 (298 K)	22.9 (298 K)	331
$EtMeIm^+$	$FeCl_4^-$	1.42 (293 K)	14 (303 K)	18 (293 K)	175
$EtMeIm^+$	$Fe_{0.5}Ga_{0.5}Cl_4^-$	1.46 (293 K)	12 (303 K)	20 (293 K)	175
$EtMeIm^+$	$GaCl_4^-$	1.53 (293 K)	13 (303 K)	20 (293 K)	175
$EtMeIm^+$	NbF_6^-	1.67 (298 K)	49 (298 K)	8.5 (298 K)	136
$EtMeIm^+$	$F(HF)_{2.3}^-$	1.13 (298 K)	4.9 (298 K)	100 (298 K)	66
$EtMeIm^+$	1,2,3,4-tetrazolide		42.5 (298 K)	8.9 (298 K)	233
$MePrIm^+$	$CF_3BF_3^-$	1.31 (298 K)	43 (298 K)	8.5 (298 K)	118
$MePrIm^+$	$CF_3BF_3^-$		41 (298 K)	8.5 (298 K)	120
$MePrIm^+$	$C_2F_5BF_3^-$	1.38 (298 K)	35 (298 K)	7.5 (298 K)	118
$MePrIm^+$	$n\text{-}C_3F_7BF_3^-$	1.44 (298 K)	44 (298 K)	5.3 (298 K)	118
$MePrIm^+$	$F(HF)_{2.3}^-$	1.11 (298 K)	7.0 (298 K)	61 (298 K)	66
$BuMeIm^+$	$CF_3BF_3^-$		49 (298 K)	5.9 (298 K)	120

Table 5. Continuation

RTIL		Density ($g\ cm^{-3}$)	Viscosity (cP)	Conductivity ($mS\ cm^{-1}$)	Ref.
Cation	Anion				
$BuMeIm^+$	$C_2F_5BF_3^-$	1.34 (298 K)	41 (298 K)	5.5 (298 K)	118,120
$BuMeIm^+$	ICl_2^-	1.78 (RT)	35 (303 K)	6.7 (303 K)	88
$BuMeIm^+$	IBr_2^-	1.547 (RT)	45 (303 K)	6.7 (303 K)	88
$BuMeIm^+$	$F(HF)_{2.3}^-$	1.08 (298 K)	19.6 (298 K)	33 (298 K)	66
$MePenIm^+$	$F(HF)_{2.3}^-$	1.05 (298 K)	26.7 (298 K)	27 (298 K)	66
$HexMeIm^+$	$F(HF)_{2.3}^-$	1.00 (298 K)	25.8 (298 K)	16 (298 K)	66
$Me(MeOCH_2)Im^+$	$C_2F_5BF_3^-$	1.46 (298 K)	47 (298 K)	6.0 (298 K)	118
$Me(MeO(CH_2)_2)Im^+$	$CF_3BF_3^-$	1.36 (298 K)	43 (298 K)	6.9 (298 K)	118
$Me(MeO(CH_2)_2)Im^+$	$C_2F_5BF_3^-$	1.42 (298 K)	38 (298 K)	6.1 (298 K)	118
$1\text{-}Et\text{-}2,3\text{-}DMeIm^+$	$F(HF)_{2.3}^-$	1.10 (298 K)	6.8 (298 K)	56.9 (298 K)	69
$1\text{-}Et\text{-}3,5\text{-}DMeIm^+$	$N(SO_2CF_3)_2^-$	1.47 (293 K)	37 (293 K)	6.6 (293 K)	332
$1\text{-}Et\text{-}3,5\text{-}DMeIm^+$	$N(SO_2CF_3)_2^-$	1.470 (295 K)	37 (293 K)	6.6 (293 K)	139
$DEtIm^+$	$N(SO_2CF_3)_2^-$	1.45 (293 K)	35 (293 K)	8.5 (293 K)	332
$DEtIm^+$	$N(SO_2CF_3)_2^-$	1.452 (294 K)	35 (293 K)	8.5 (293 K)	139
$DEtIm^+$	$CF_3CO_2^-$	1.250 (295 K)	43 (293 K)	7.4 (293 K)	139
$1,3\text{-}DEt\text{-}4\text{-}MeIm^+$	$N(SO_2CF_3)_2^-$	1.432 (296 K)	36 (293 K)	6.2 (293 K)	139
$1,3\text{-}DEt\text{-}4\text{-}MeIm^+$	$N(SO_2CF_3)_2^-$	1.43 (293 K)	36 (293 K)	6.2 (293 K)	332
$BuPy^+$	$F(HF)_{2.3}^-$	1.084 (298 K)	24.9 (298 K)	36.9 (298 K)	62
$EtMePyrr^+$	$F(HF)_{2.3}^-$	1.07 (298 K)	11.5 (298 K)	74.6 (298 K)	67
$MePrPyrr^+$	$N(SO_2F)_2^-$		40 (298 K)	8.2 (298 K)	144
$MePrPyrr^+$	$F(HF)_{2.3}^-$	1.05 (298 K)	13.0 (298 K)	58.1 (298 K)	67
$BuMePyrr^+$	$F(HF)_{2.3}^-$	1.04 (298 K)	6.0 (298 K)	35.9 (298 K)	67
$EtMePip^+$	$F(HF)_{2.3}^-$	1.07 (298 K)	21.9 (298 K)	37.2 (298 K)	67

Table 5. Continuation

RTIL		Density ($g\ cm^{-3}$)	Viscosity (cP)	Conductivity ($mS\ cm^{-1}$)	Ref.
Cation	Anion				
$MePrPip^+$	$F(HF)_{2.3}^-$	1.06 (298 K)	27.4 (298 K)	23.9 (298 K)	67
$BuMePip^+$	$F(HF)_{2.3}^-$	1.04 (298 K)	34.4 (298 K)	12.3 (298 K)	67
$TMeS^+$	$N(CN)_2^-$		27.2 (293 K)	20.8 (293 K)	215
$DEtMeS^+$	$N(CN)_2^-$		22.9 (293 K)	26.8 (293 K)	215
$MeDPrS^+$	$N(CN)_2^-$		29.5 (293 K)	15.0 (293 K)	215
$EtDMeS^+$	$N(CN)_2^-$		25.3 (293 K)	22.1 (293 K)	215
$TEtS^+$	$N(CN)_2^-$		20.9 (293 K)	22.4 (293 K)	215
$TEtS^+$	$N(SO_2CF_3)_2^-$	1.46 (298 K)	30 (298 K)	7.1 (298 K)	333
$TEtS^+$	$N(SO_2CF_3)_2^-$	1.49 (298 K)	33 (298 K)	8.5 (298 K)	330
$TEtS^+$	$N(COCF_3)(SO_2CF_3)^-$	1.42 (298 K)	28 (298 K)	9.0 (298 K)	330
$EtDPrS^+$	$N(CN)_2^-$		29.4 (293 K)	14.7 (293 K)	215

$DMeIm^+$: 1,3-dimethylimidazolium$^+$; $EtMeIm^+$: 1-ethyl-3-methylimidazolium$^+$; $MePrIm^+$: 1-methyl-3-*n*-propylimidazolium$^+$; $BuMeIm^+$: 1-*n*-butyl-3-methylimidazolium$^+$; $MePenIm^+$: 1-methyl-3-*n*-pentylimidazolium$^+$; $HexMeIm^+$: 1-*n*-hexyl-3-methylimidazolium$^+$; $Me(MeOCH_2)Im^+$: 1-methyl-3-metoxymethylimidazolium$^+$; $Me(MeO(CH_2)_2)Im^+$: 1-methyl-3-metoxyethylimidazolium$^+$; 1-Et-2,3-$DMeIm^+$: 1-ethyl-2,3-dimethylimidazolium$^+$; 1-Et-3,5-$DMeIm^+$: 1-ethyl-3,5-dimethylimidazolium$^+$; $DEtIm^+$: 1,3-Diethylimidazolium$^+$; 1,3-DEt-4-$MeIm^+$: 1,3-Diethyl-4-methylimidazolium$^+$; $BuPy^+$: 1-*n*-butylpyridinium$^+$; $EtMePyrr^+$: 1-ethyl-1-methylpyrrolidinium$^+$; $MePrPyrr^+$: 1-methyl-1-*n*-propylpyrrolidinium$^+$; $BuMePyrr^+$: 1-*n*-butyl-1-methylpyrrolidinium$^+$; $EtMePip^+$: 1-ethyl-1-methylpiperidinium$^+$; $MePrPip^+$: 1-methyl-1-*n*-propylpiperidinium$^+$; $BuMePip^+$: 1-*n*-butyl-1-methylpiperidinium$^+$; $TMeS^+$: trimethylsulfonium$^+$; $DEtMeS^+$: diethylmethylsulfonium$^+$; $MeDPrS^+$: methyl-di-*n*-propylsulfonium$^+$; $EtDMeS^+$: ethyl-dimethylsulfonium$^+$; $TEtS^+$: trimethylsulfonium$^+$; $EtDPrS^+$: ethyl-di-*n*-propylsulfonium$^+$.

number of electrochemically favorable RTILs having useful conductivities (~ 5.0 mS cm^{-1}) and low viscosities (~ 50 cP); some examples are given in Table 5. Even when employing the salts in this table, a substantial potential error can be expected during controlled potential measurements due to the effects of uncompensated resistance, R_u, between the WE and RE, despite the most careful electrochemical cell design. (This is rarely a concern at small macroelectrodes in conventional high-melting inorganic molten salts or ionic liquids, which tend to be very conductive.). As pointed out in a recent monograph,[288] this effect is most noticeable in RTILs under conditions of high current density or large electrode area. In aqueous solutions, the potential error or iR_u drop is often minimized by placing the RE in a compartment with a Luggin capillary. The tip of this capillary is placed close to the WE surface. In principle, there is no reason why such an approach cannot be employed in RTILs, but the high viscosity of these ionic solvents makes the use of Luggin capillaries difficult in practice. Although careful attention to the cell design can eliminate some of the R_u, most of the better quality commercial potentiostats/galvanostats designed for use at moderate currents with macroelectrodes have provisions for electronic resistance compensation. That is, they have circuitry that provides positive feedback to the control amplifier circuit (albeit at the expense of potentiostatic stability) that artificially removes the potential error. In almost every case examined by the authors, the use of electronic resistance compensation in RTILs is sufficient to reduce the potential error to negligible levels.

As described in Section III, because invisible impurities derived from air such as water and oxygen affect electrochemical reactions in RTILs, electrochemical experiments with these solvents, even those thought to be hydrophobic, should be conducted under a dry, inert gas atmosphere. Electrochemical reactions at a relatively positive potential, e.g., Al deposition, can be safely carried out under dry nitrogen. It is no secret that the N_2/N^{3-} electrode reaction has been observed in inorganic molten salts or ionic liquids. For example, in molten LiCl-KCl, the standard electrode potential for this redox couple, $E^0_{N_2/N^{3-}}$, is estimated to be 0.382 V *vs.* Li^+/Li at 723 K.[334] Thus, when alkali metal deposition/stripping reactions or other electrode processes are investigated at very negative potentials, it is wise to use dry argon as an inert gas since nitrogen may be reduced to nitride ion, N^{3-}, in RTILs under these conditions.

2. An Overview of the Techniques Used for Electrochemical Analysis in Ionic Liquids

The various electrochemical techniques that have been used to probe electrochemical reactions are discussed in detail in many excellent comprehensive texts and monographs. This Section is intended as an aid for those researchers who are interested in doing electrochemistry in RTILs, but do not have research experience or academic training in electrochemistry. Brief descriptions of some of the more popular techniques are described here along with references to examples of studies in ionic liquids where these techniques were effectively employed.

(*i*) *Cyclic Voltammetry*

Cyclic voltammetry (CV) is widely employed as a survey technique, i.e., it is the first technique that is commonly applied during the preliminary analysis of an electrochemical system. With the demise of analog signal generators, true analog sweep cyclic voltammetry has been largely supplanted by cyclic staircase voltammetry (CSV). This technique varies from analog sweep voltammetry in that the potential is changed in small steps. As long as the step size is sufficiently small, the theories developed for CV can be applied to CSV.[321,335] The correct analysis of cyclic voltammetric data is somewhat of an art, and there are many examples of investigations where such CV data have been grossly over-interpreted or misinterpreted altogether.

If the electrochemical reaction (oxidation or reduction) is a simple diffusion-controlled process, i.e., an electrochemically reversible or Nernstian system in which charge transfer is very fast so that the reaction rate is limited by the rate of diffusion of the soluble electroactive species to the electrode surface in the test solution, then, the absolute magnitude of the voltammetric peak current, i_p, is given by the well known Randles-Sevick equation:[321]

$$i_p = 0.4463 \frac{n^{3/2} F^{3/2}}{R^{1/2} T^{1/2}} A D^{1/2} C^* v^{1/2} \tag{6}$$

where A is the electrode area (cm^2), C^* is the bulk solution concentration (mol cm^{-3}), D is the diffusion coefficient of the electroactive species (cm^2 s^{-1}), n is the number of electrons involved in the reaction, v is the potential scan rate (V s^{-1}), and all of the other symbols have their usual meaning. Linear plots of i_p as a function of $v^{1/2}$ that pass through the origin suggest that the system under study is diffusion controlled. However, this is not a definitive test because the same behavior is also observed for an irreversible electrode reaction.[336] Fortunately, there are other criteria that can be examined in order to verify the reversibility of the system under study. For example, the peak potential for the oxidation or reduction wave in question is independent of scan rate in the case of a reversible system, but varies with scan rate for a quasi-reversible or irreversible systems. Likewise, the nonlinearity of a plot of i_p versus $v^{1/2}$ does not necessarily indicate a slow charge transfer process. It may simply indicate that the electrode reaction rate is governed by the rate of a chemical reaction preceding the charge-transfer step.[336] If n, C^*, A, and T are known and the system in question has been proven to be electrochemically reversible, then D can be estimated from the slope of a plot of i_p versus $v^{1/2}$ by using Eq. (6).

For a reversible (Nernstian) reaction, the following criteria are valid

$$\left| E_{pa} - E_{pc} \right| = \frac{2.3RT}{nF} \tag{7}$$

$$\left| E_p - E_{p/2} \right| = \frac{2.2RT}{nF} \tag{8}$$

where E_{pa}, E_{pc}, and $E_{p/2}$ are the anodic peak potential, cathodic peak potential, and the appropriate half-peak potential, respectively. $E_{p/2}$ is simply the potential at 50% of i_p. When the reaction is known to be reversible, the number of electrons involved in the reaction can be estimated from Eqs. (7) and (8).[321] Note, however, that Eq. (7) varies slightly with changes in the voltammetric switching potential. Thus, Eq. (8) is probably superior to Eq. (7) for estimating n. For a reversible electrode reaction, the voltammetric half-wave potential, $E_{1/2}$, can be determined by estimating the potential at which 85.17% of the peak current is observed.[336] $E_{1/2}$ can also be estimated from the relationship given in Eq. (9):

$$E_{1/2} = \frac{E_{\text{pa}} + E_{\text{pc}}}{2} \tag{9}$$

$E_{1/2}$ is related to the formal potential, $E^{0'}$, of the redox reaction by the following expression

$$E_{1/2} = E^{0'} + \frac{2.303RT}{nF}\log\left(\frac{D_{\text{R}}}{D_{\text{O}}}\right)^{1/2} \tag{10}$$

where D_{R} and D_{O} are the diffusion coefficients for the reduced and oxidized species, respectively.

The chemical stability of the reduction or oxidation product produced during the initial voltammetric scan and/or the presence of homogeneous reactions coupled to the charge transfer process can often be determined by examining the appropriate peak-current ratio as a function of scan rate.[336] For a cathodic process, the ratio of interest is $i_{\text{pa}} / i_{\text{pc}}$, where i_{pa} and i_{pc} are the absolute magnitudes of the anodic and cathodic peak currents, respectively. For an anodic reaction, the ratio $i_{\text{pc}} / i_{\text{pa}}$ provides the desired information. Because the peak current obtained on the reverse scan, whether i_{pa} or i_{pc}, depends strongly on the switching potential for the initial scan, E_{λ}, there is no simple graphical procedure that can be used to obtain this current. However, as noted by Nicholson,[337] the peak current ratio for a *reversible reaction* can be expressed a function of three parameters in the following empirical relationship for a cathodic reaction

$$\frac{i_{\text{pa}}}{i_{\text{pc}}} = \frac{(i_{\text{pa}})_0}{i_{\text{pc}}} + \frac{0.485(i_{\text{sp}})_0}{i_{\text{pc}}} + 0.086 \tag{11}$$

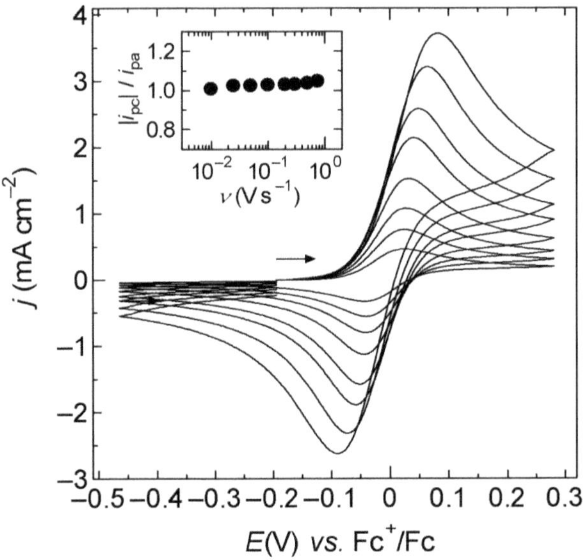

Figure 7. Cyclic staircase voltammograms recorded at different sweep rates at a platinum disk electrode in $EtMeIm^+F(HF)_{2.3}^-$ containing 11.83 mmol L^{-1} ferrocene.[326] (inset) ratio of the cathodic to anodic peak currents as a function of sweep rate for the voltammograms shown in this figure.

where $(i_{pa})_0$ and $(i_{sp})_0$ are the magnitudes of the anodic current and the current at E_λ, respectively, measured with respect to the baseline for the cathodic current. The corresponding equation for i_{pc}/i_{pa} is then

$$\frac{i_{pc}}{i_{pa}} = \frac{(i_{pc})_0}{i_{pa}} + \frac{0.485(i_{sp})_0}{i_{pa}} + 0.086 \qquad (12)$$

where $(i_{pc})_0$ is the absolute magnitude of the cathodic current measured with respect to the baseline for the anodic current. Figure 7 shows a series of CSVs recorded for the oxidation of ferrocene $[Fe(cp)_2]$ in $EtMeIm^+F^-(HF)_{2.3}^-$ at different scan rates and the corresponding plots of i_{pc}/i_{pa} as a function of scan rate, which was estimated by using Eq. (12).[326] The oxidation of $Fe(cp)_2$ in this

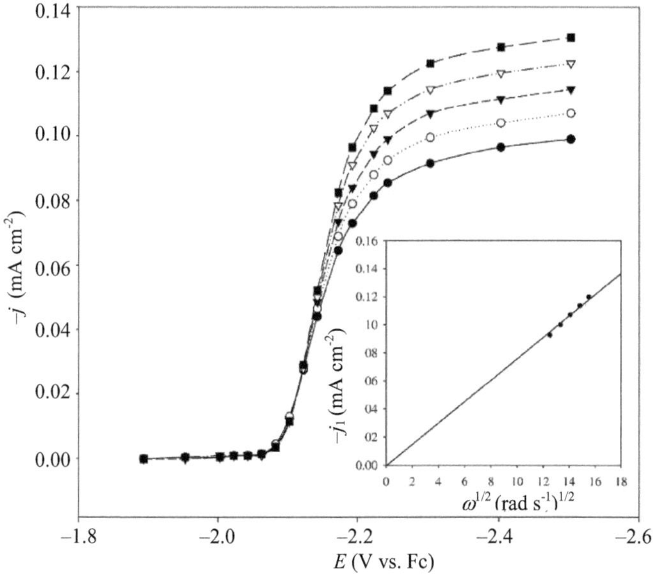

Figure 8. Voltammograms recorded as a function of angular velocity, ω, at a mercury film rotating disk electrode in n-$Bu_3MeN^+Tf_2N^-$ containing 27.8 mmol L^{-1} Cs(I). (inset) relationship between the limiting current density, j_l, and $\omega^{1/2}$. Reproduced from Ref. 150 with permission of Elsevier Ltd.

solvent is known to be electrochemically and chemically reversible, and as expected, i_{pc}/i_{pa} is close to unity and essentially independent of the scan rate. For a complete discussion of the effects of coupled homogeneous chemistry on the peak current ratio, the original article by Nicholson and Shain[336] should be consulted. It is the experience of the authors that Eqs. (11) and (12) can also be qualitatively applied to most quasi-reversible systems with reasonably good results.

(ii) Hydrodynamic Voltammetry

The most common hydrodynamic technique employed for voltammetric measurements is that based on rotating electrodes, mainly rotating disk electrodes (RDE) and rotating ring-disk electrodes (RRDE). The main benefit of hydrodynamic or convective-diffusion voltammetry is that the rate of mass transport of the electroactive species to the electrode surface can be precisely controlled by adjusting the rotation rate of the electrode. Because the electrode potential is scanned very slowly, there is no measurable current contribution from double-layer charging. Changing the electrode rotation rate in RDE voltammetry is analogous to changing the scan rate in stationary electrode voltammetry. Figure 8 shows a voltammogram recorded as a function of angular velocity at a mercury film RDE in n-$Bu_3MeN^+Tf_2N^-$ containing Cs^+.[150] Under mass-transport-limited conditions, the absolute magnitude of the limiting current, i_l, should always vary linearly with the square root of the electrode rotation rate, ω, according to the Levich equation:

$$i_l = 0.62nFAC^*D^{2/3}\upsilon^{-1/6}\omega^{1/2} \tag{13}$$

where υ is the kinematic viscosity of the solution ($cm^2\ s^{-1}$) and all other symbols have their usual meaning. An example of a *Levich plot* is depicted in the inset of Fig. 8. In the absence of complications, this plot should be a straight line passing through the origin. The slope of this plot can be used to estimate D_O or D_R, depending whether the oxidized or reduced form of the solute, respectively, is present in the solution at the beginning of the experiment. Following the U.S. sign convention for current, the relationship between the potential and current for a RDE voltammetric wave in the case of a reversible electrode reaction, where both the oxidized and reduced species are present in the solution, is given by:[321]

$$E = E_{1/2} + \frac{2.303RT}{nF}\log\frac{i_{lc} - i}{i - i_{la}} \tag{14}$$

where i_{lc} and i_{la} are the cathodic and anodic limiting currents, respectively. If only the oxidized or reduced species is initially present in the solution, this equation simplifies accordingly. The number of electrons involved in the electrode reaction under investigation and $E_{1/2}$ can be estimated from plots of $\log[(i_{lc} - i)/(i - i_{la})]$ versus E constructed from the rising portion of the voltammetric wave. In the case of RDE voltammetry, $E_{1/2}$ and $E^{0'}$ are related through an expression similar to Eq. (10):

$$E_{1/2} = E^{0'} + \frac{2.303RT}{nF} \log\left(\frac{D_{\mathrm{R}}}{D_{\mathrm{O}}}\right)^{2/3} \qquad (15)$$

RDE voltammetric measurements are a useful practical tool for determining the completeness of bulk electrolysis experiments. For example, consider a solution containing only the oxidized species. In the case of a chemically and electrochemically reversible reaction, a RDE voltammogram of the solution containing only the oxidized species will exhibit the usual S-shaped reduction wave such as that shown in Fig. 8. As the electrolysis experiment progresses, the wave will be displaced from the zero current axis in proportion to the ratio of the oxidized to reduced species in solution. Eventually, when electrolysis is complete, only an anodic wave will remain.

RDE voltammetry is perhaps the preeminent technique for studying quasi-reversible electrode reactions because of the ease with which kinetic information can be unequivocally separated from mass transport effects. To carry out such an analysis, RDE voltammograms are recorded as a function of rotation rate. If the system under investigation is quasi-reversible, then plots of i versus $\omega^{1/2}$ constructed from data at the same potential on the rising portion of each wave will be non-linear. In fact, the current in plots of i versus $\omega^{1/2}$ constructed with data taken at potentials near the foot of the RDE wave may be nearly independent of ω, whereas those plots prepared at potentials approaching that where the limiting current is observed may be nearly linear. Such behavior is clear indication of a quasi-reversible electrode reaction because in the case of a reversible reaction plots of i versus $\omega^{1/2}$ are linear at all potentials from the foot of the wave to the limiting current. In order to extract kinetic data from the quasi-reversible plots of i versus $\omega^{1/2}$, the data are plotted as $1/i$ versus $1/\omega^{1/2}$ according to the Koutecký-Levich equation,

$$\frac{1}{i} = \frac{1}{i_{\mathrm{k}}} + \frac{1}{0.62nFAC^{*}D^{2/3}\upsilon^{-1/6}\omega^{1/2}} \qquad (16)$$

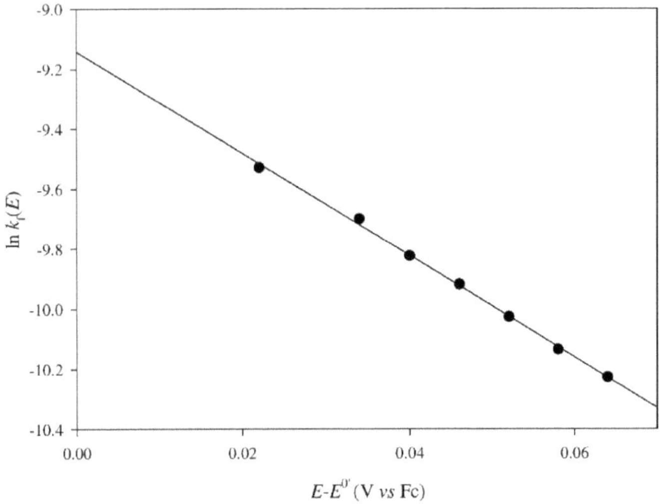

Figure 9. Dependence of the logarithm of the heterogeneous rate constant for the reduction of Cs(I) at a mercury film rotating disk electrode on the applied potential. Reproduced from Ref. 150 with permission of Elsevier Ltd.

provided that the reverse reaction can be ignored.

The kinetic current, i_k, can be estimated as a function of potential from the intercept of each plot of $1/i$ versus $1/\omega^{1/2}$. This current leads to the potential dependent heterogeneous rate constant for the forward reaction, $k_f(E)$, via the following expression

$$i_k = nFAk_f(E)C^* \tag{17}$$

Furthermore, k^0 and αn can then be determined from the intercept and the slope of a plot of ln $k_f(E)$ *vs*. $E - E^{0'}$, according to Eq. (18):

$$\ln k_f(E) = \ln k^0 + \frac{\alpha nF(E - E^{0'})}{RT} \tag{18}$$

Such a plot is illustrated in Fig. 9 for the data in Fig. 8. The formal potential, $E^{0'}$, can be estimated by the use of Eq. (15). But normally the diffusion coefficients for the oxidized and reduced forms of the solute are nearly the same, making $E^{0'}$ approximately equal to $E_{1/2}$. From the data in Fig. 9, $E^{0'}$ was found to be -2.273 V, and k^0 and αn were estimated to be $9.8 \pm 0.8 \times 10^{-5}$ cm s^{-1} and 0.36 ± 0.04, respectively.[150]

RRDE voltammetry has also been employed for electrochemical experiments in ionic liquids. It has been of great value in the analysis of plating reactions in chloroaluminate ionic liquids involving the underpotential deposition of aluminum-transition metal alloys.[39,338] For the complete details of this procedure the reader should consult the original literature.[339]

(*iii*) *Chronoamperometry*

Chronoamperometry is the simplest controlled-potential technique available to the electrochemist and traces its roots to the seminal work of Frederick Gardner Cottrell in 1903.[340] In spite of this simplicity, chronoamperometric techniques have been applied to a host of electrochemical problems in conventional molecular solvents and ionic liquids. Some of these uses include the measurement of diffusion coefficients, studies of metal nucleation, and the investigation of coupled homogeneous chemistry. This technique also forms the basis of a pulse voltammetric method for recording current-potential curves (see below).

The absolute magnitude of current obtained following a potential step of sufficient magnitude to deplete the electroactive species at the electrode surface in an unstirred solution, i.e., to produce a diffusion-controlled current, is given by the Cottrell equation,[321]

$$i(t) = \frac{nFAD^{1/2}C^*}{\pi^{1/2}t^{1/2}} \tag{19}$$

where t is the time (s) and all other symbols have their usual meaning. For a potentiostatic current-time transient resulting from simple, uncomplicated, diffusion-controlled reaction, i is proportional to $t^{-1/2}$ or $it^{1/2}$ is constant with t within the time limitations associated with the convective movement of the test solution caused by vibrations or thermal gradients. (The latter can be significant in RTILs at elevated temperatures in poorly designed cells.) Thus, knowledge of $it^{1/2}$ measured under diffusion-controlled conditions can be used along with Eq. (19) to calculate D.

The chronoamperometric technique has been employed extensively in RTILs and their higher temperature cousins to investigate nucleation/crystallization processes during the electrodeposition of metals on foreign substrates, e.g., substrates composed of materials different from the depositing metal. The presence of a nucleation overpotential can be readily determined by examining a cyclic voltammogram of the reaction of interest. For a nucleation-

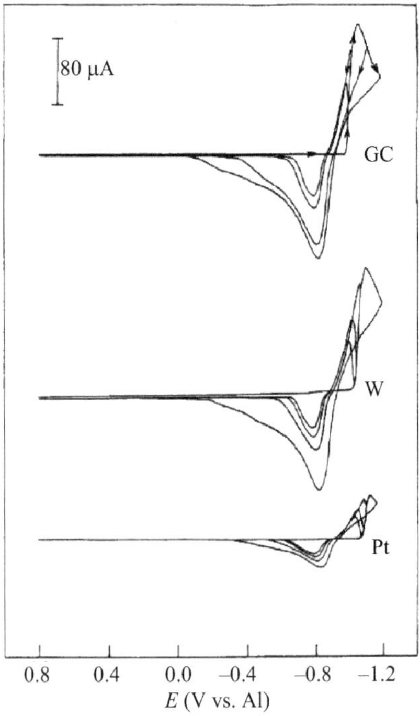

Figure 10. Cyclic staircase voltammograms recorded at different electrodes in mixtures of EtMeImCl + $EtMeIm^+BF_4^-$ containing 37.3 mmol L^{-1} Cd(II). The sweep rates were 50 mV s^{-1}. Reproduced from Ref. 341 with permission of Elsevier Ltd.

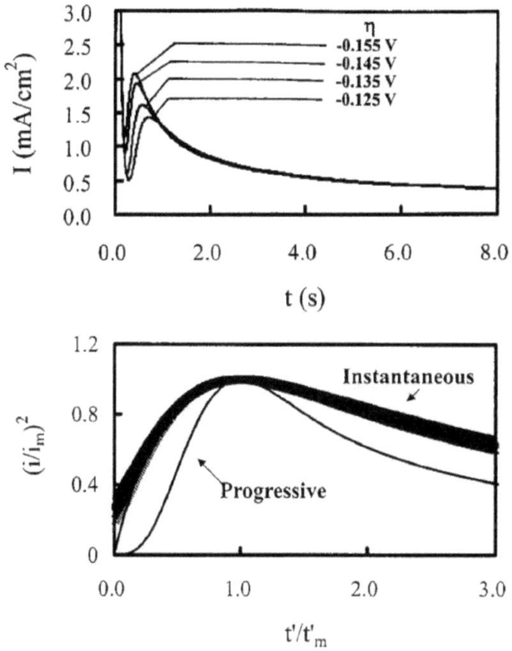

Figure 11. Top: Current-time transients resulting from potential step experiments recorded at a Pt electrode in the 50.0-50.0 mol % mixture of $ZnCl_2$ + EtMeImCl containing 30 mmol L^{-1} Cd(II). Bottom: Plots of t'/t_m' versus $(i/i_m)^2$ constructed from the current-time transients. Reproduced from Ref. 342 by permission of ECS—The Electrochemical Society.

controlled reaction, an anomalous loop will be observed in the voltammogram (Fig. 10).[341] This loop occurs because of the overpotential required to initiate the first stages of the metal deposition process on the forward or reductive scan. When the scan is reversed, metal continues to deposit at more positive potentials because the electrode surface has changed from a foreign or hostile surface to one that is covered with the depositing metal. Metal deposition finally ceases when the potential is scanned positive of E_{eq}.

Nucleation followed by crystal growth also gives rise to current-potential curves with maxima such as those shown in Fig. 11 (top). These examples were recorded at a Pt disk electrode in 50.0 mol% $ZnCl_2$–EtMeImCl containing 30 mmol L^{-1} Cd(II).[342] The shape of these transients is complex and consists of four main elements:

(1) a current due to charging of the electrode double layer,
(2) a potential-dependent time delay relevant to nucleation,
(3) a rising current due to nucleation and crystal growth, and
(4) a region following the current maximum where the current decays with $t^{-1/2}$, according to Eq. (19).

The maximum in the magnitude of the current density, j_m, is observed at t_m where the spherical diffusion zones of the developing nuclei coalesce to form a planar diffusion zone. The value of t_m depends on the applied potential, i.e., t_m becomes shorter with an increase in the overpotential, $\eta = |E - E_{eq}|$.

The theory of chronoamperometry has been developed for several different three-dimensional nucleation/growth mechanisms. These mechanisms have been discussed in detail.[343] Simple analysis of the rising portion of the experimental chronoamperometric current-time transients can usually lead to identification of the nucleation model. For example, instantaneous nucleation on a fixed number of active sites is indicated if the current grows linearly with $t^{1/2}$, whereas progressive nucleation on an infinite number of active sites is indicated when the current grows linearly with $t^{3/2}$.[344] However, for a more unequivocal test, the appropriate theoretical models can be compared to the entire experimental current-time transients. In order to simplify this process, it is useful to convert the experimental data into dimensionless form by dividing the current density, j, by the maximum current density, j_m, and the time by t_m. Thus, the dimensionless models for instantaneous and progressive nucleation are given by Eqs. (20) and (21), respectively:[344]

$$\left(\frac{j}{j_m}\right)^2 = \frac{1.9542 t_m}{t}\left[1 - \exp\left(\frac{-1.2564 t}{t_m}\right)\right]^2 \tag{20}$$

$$\left(\frac{j}{j_m}\right)^2 = \frac{1.2254t_m}{t}\left[1-\exp\left(\frac{-2.3367t}{t_m}\right)^2\right]^2 \tag{21}$$

However, before making this comparison, the data must be refined by taking into account the nucleation delay time, $t' = t - t_0$. Fortunately, t_0 can be obtained from the intercepts of plots of $(j/j_m)^2$ or $(j/j_m)^{2/3}$ vs. t, whichever is appropriate, or plots of j *vs.* $t^{1/2}$ or $t^{3/2}$, but the former method generally gives better results.[345] As an example of the application of this analysis to a metal deposition reaction in a RTIL, the dimensionless plots of $(j/j_m)^2$ vs. (t'/t_m') in Fig. 11 (top) along with theoretical curves from Eqs. (20) and (21) are given in Fig. 11 (bottom).[342] In this case, the experimental results are in good agreement with the instantaneous three-dimensional diffusion controlled nucleation process described by Eq. (20).

When one of the theoretical dimensionless current-time transients described above agrees well with the experimental data it is often possible to extract useful information about the deposition process such as the nucleation rate constant, the number density of active sites on the electrode surface, and the saturation density of nuclei by using the appropriate mathematical expressions.[341-350]

Those cases where the experimental current-time transients appear to fall between the theoretical transients for the limiting cases of progressive and instantaneous nucleation have also been considered,[345,351-355] and a theoretical transient has been developed for this situation as well:[356,357]

$$\left(\frac{j}{j_m}\right)^2 = \frac{t_m}{t}\frac{\left\{1-\exp\left[-xt/t_m+\alpha\left(1-\exp^{-xt/\alpha t_m}\right)\right]\right\}^2}{\left\{1-\exp\left[-x+\alpha\left(1-\exp^{-x/\alpha}\right)\right]\right\}^2} \tag{22}$$

In this equation, α and x are adjustable parameters that contain information about the number density of active sites and the potential dependent nucleation rate per active site. For the limiting cases of instantaneous and progressive nucleation, α approaches 0 with $x \approx 1.2564$ and ∞ with $x \approx 2.3367$, respectively. Figure 12 shows the dimensionless experimental data derived from the current-time transients for copper deposition on glassy carbon in a solution of

Cu(I) in the 66.7 m/o urea–choline chloride room-temperature melt.[354] Equation (22) was fit to the experimental data to give values of α and x of 0.422 and 2.140, respectively, with a correlation coefficient of 0.9922. In this case, α is a relatively large, suggesting that the deposition reaction is initiated mainly by progressive nucleation. A more comprehensive discussion of electrochemical crystallization can be found in Ref. 358.

(*iv*) *Sampled Current Voltammetry*

Several kinds of pulse voltammetric techniques have been proposed with the aim of avoiding charging current contributions to the overall current, thereby increasing the analytical detection limit of voltammetric measurements. These techniques, which are usually employed in conjunction with a dropping mercury electrode (DME), have been discussed at length in several prominent

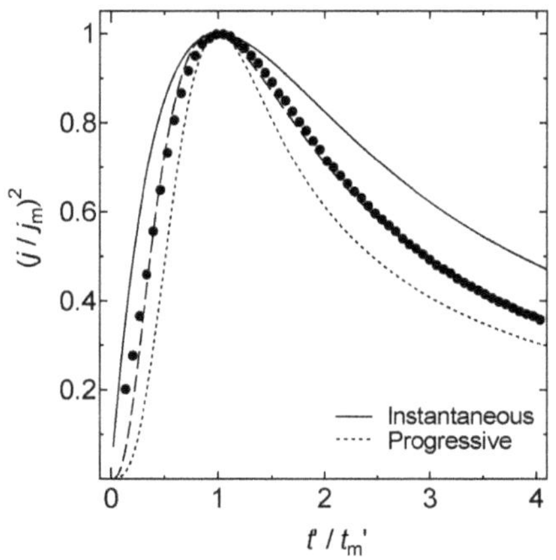

Figure 12. Examples of plots of t'/t_m' *vs.* $(j/j_m)^2$ constructed from the current-time transients resulting from potential step experiments recorded at a GC electrode in a 66.7-33.3 mol % mixture of urea + choline chloride containing 20.1 mmol L^{-1} Cu(I). The theoretical transient was fitted to the experimental data by using the adjustable parameters, α and x, in Eq. (22).[354]

texts cited in Ref. 321. However, analytical techniques involving DMEs are seldom important in ionic liquid research, particularly if the RTIL is to be heated, but such pulse techniques can be advantageous at solid electrodes in RTILs. The most commonly applied pulse technique, sampled current voltammetry (SCV), is carried out by applying potential steps of increasingly negative (reduction) or positive (oxidation) amplitude to a stationary solid electrode to produce a series of potentiostatic current-time transients. The current for each transient is sampled at a designated time after the application of each step and then the potential is returned to the initial value potential while the solution is stirred. The advantage of the SCV technique is that the electrode diffusion layer and the electrode surface are renewed between each potential pulse. When the sampled currents are plotted as a function of the applied potential for a freely diffusing, i.e., kinetically uncomplicated system, the result is a current-potential curve identical to that obtained at a RDE that can be analyzed by using the current-potential expressions usually applied to polarographic and RDE waves. Because the electrode surface is renewed between data points, this technique can sometimes be used to produce well-defined current-potential curves for systems that are intractable and do not produce useful voltammograms with conventional scanning methods at stationary and rotating electrodes. Several investigations in which this technique has been used to advantage in RTILs have been published by our research group.[39,338,359]

V. ELECTROCHEMICAL APPLICATIONS

Electrochemical applications of RTILs include their use as solvents for the study of redox reactions and metal deposition and as electrolytes for a variety of technological applications, including surface finishing, electromechanical actuators, dye-sensitized photoelectrochemical cells, electrochromic devices, fuel cells, double-layer capacitors, and nonaqueous batteries. Many of these applications were featured in a recent article.[18] Electrochemical data for a variety of organic, organometallic, and inorganic solutes that have been studied in RTILs are given in Table 6. By the far, the most extensively studied solute in RTILs is oxygen, and so we have placed this data in a separate table (Table 4). From these tables, it

Table 6
Summary of Electrochemical Data Obtained in RTILs and RTMs

	Electrode reaction	Reference electrode	T (K)	$10^7 D$ ($cm^2\ s^{-1}$)	$E^{0'}$ (V)	$E_{1/2}$ (V)	$10^5 k^0$ ($cm\ s^{-1}$)	α	Ref.
			RTILs						
$EtMeIm^+BF_4^-$	Fc^+/Fc	Al(III)/Al	295	5.1 [h,r]		+0.491 [h,r]			360
$EtMeIm^+BF_4^-$	Fc^+/Fc	Ag^+/Ag	RT			–0.64			361
$EtMeIm^+BF_4^-$	Fc^+/Fc	Hg/Hg_2SO_4	RT			–0.24			361
$BuMeIm^+PF_6^-$	Fc^+/Fc	NHE	RT		+0.37				362
$BuMeIm^+Tf_2N^-$	Fc^+/Fc	Ag^+/Ag	298			–0.387 [i,m]			363
$BuMeIm^+Tf_2N^-$	Fc^+/Fc	Ag^+/Ag	298			–0.383 [i,m]			364
$HexMeIm^+FAP^-$	Fc^+/Fc	Pt-quasi	RT		+0.55 [h,m]				365
$HexMe_3N^+Tf_2N^-$	Fc^+/Fc	I^-/I_3^-	323		+0.16 [i,m]				328
$HexMe_3N^+Tf_2N^-$	Fc^+/Fc	I^-/I_3^-	298	1.15 [i,m]			20000 [i,m]		366
$Bu_3MeN^+Tf_2N^-$	Fc^+/Fc	Ag^+/Ag	298			–0.340 [i,m]			363
$BuMePyrr^+Tf_2N^-$	Fc^+/Fc	Fc^+/Fc	298	3.3 [h]	0.00 [h]				367
$EtMeIm^+TSAC^-$	Li(I)/Li	Fc^+/Fc	298		–3.36 [u,m]				330
$BuMeIm^+PF_6^-$	Li(I)/Li	Fc^+/Fc	RT	0.19 [e,k]	–2.448				362
$Me_3PrN^+Tf_2N^-$	Li(I)/Li	Fc^+/Fc	298		–3.16 [u,m]				330
$Et_4N^+TSAC^-$	Li(I)/Li	Fc^+/Fc	298		–3.40 [u,m]				330
$BuMePyrr^+C_2F_5BF_3^-$	Li(I)/Li	Fc^+/Fc	298		–3.2 [u,m]				107
$Me(CH_3OCH_2)Pip^+Tf_2N^-$	Li(I)/Li	Fc^+/Fc	298		–3.20 [u,m]				330

Table 6. Continuation

	Electrode reaction	Reference electrode	T (K)	$10^7 D$ ($cm^2\ s^{-1}$)	$E^{0'}$ (V)	$E_{1/2}$ (V)	$10^5 k^0$ ($cm\ s^{-1}$)	α	Ref.
			RTILs						
$Me(CH_3OC_2H_4)Pip^+Tf_2N^-$	Li(I)/Li	Fc^+/Fc	298		–3.22[u,m]				330
$Me(CH_3(C_2H_4O)_2)Pip^+Tf_2N^-$	Li(I)/Li	Fc^+/Fc	298		–3.37[u,m]				330
$BuMeIm^+PF_6^-$	Na(I)/Na	Fc^+/Fc	RT	0.14 [e,k]	–2.959	–2.02 [e,s]			362
$BuMeIm^+PF_6^-$	K(I)/K	Fc^+/Fc	RT	0.45 [e,k]	–3.354	–2.24 [e,s]			362
$Bu_3MeN^+Tf_2N^-$	Cs(I)/Cs(Hg)	Fc^+/Fc	303	0.104[f,l]	–2.273[f,l]		9.8[f,l]	0.36[f,l]	150
$Bu_3MeN^+Tf_2N^-$	Cs(I)[a]/Cs	Fc^+/Fc	303	0.027[f,k]					368
$Bu_3MeN^+Tf_2N^-$	Sr(II)[b]/Sr	Fc^+/Fc	303	0.021[f,l]					368
EtMeImCl-$TaCl_5$	Ta(V)/Ta(IV)	Al(III)/Al	373			–0.19 [h,s]			369
EtMeImCl-$TaCl_5$	Ta(IV)/Ta(II)	Al(III)/Al	373			–1.33 [h,s]			369
$BuMeIm^+PF_6^-$	$Mo_6O_{19}^{2-}$/ $Mo_6O_{19}^{3-}$	Fc^+/Fc	293	0.14 [g,m]	–0.419[g,m]				370
$BuMeIm^+PF_6^-$	α-$SiMo_{12}O_{40}^{4-}$/ α-$SiMo_{12}O_{40}^{5-}$	Fc^+/Fc	293	0.07 [g,m]	–0.274[g,m]				370
$BuMeIm^+PF_6^-$	α-$S_2Mo_{18}O_{62}^{4-}$/ α-$S_2Mo_{13}O_{62}^{5-}$	Fc^+/Fc	293	0.06 [g,m]	+0.390[g,m]				370
$BuMeIm^+PF_6^-$	$W_6O_{19}^{2-}$/ $W_6O_{19}^{3-}$	Fc^+/Fc	293	0.13 [g,m]	–0.989[g,m]				370
$BuMeIm^+PF_6^-$	α-$SiW_{12}O_{40}^{4-}$/ α-$SiW_{12}O_{40}^{5-}$	Fc^+/Fc	293	0.08 [g,m]	–0.720[g,m]				370
$BuMeIm^+PF_6^-$	α-$S_2W_{18}O_{62}^{4-}$/ α-$S_2W_{18}O_{62}^{5-}$	Fc^+/Fc	293	0.06 [g,m]	+0.020[g,m]				370

Table 6. Continuation

	Electrode reaction	Reference electrode	T (K)	10^7D (cm2 s-1)	$E^{0'}$ (V)	$E_{1/2}$ (V)	10^5k^0 (cm s-1)	α	Ref.
RTILs									
$BuMeIm^+PF_6^-$	α-$S_2W_{18}O_{62}^{5-}$/ α-$S_2W_{18}O_{62}^{6-}$	Fc^+/Fc	293		$-0.257^{g,m}$				370
$BuMeIm^+PF_6^-$	α-$S_2W_{18}O_{62}^{6-}$/ α-$S_2W_{18}O_{62}^{7-}$	Fc^+/Fc	293		$-0.720^{g,m}$				370
$BuMeIm^+PF_6^-$	α-$S_2W_{18}O_{62}^{7-}$/ α-$S_2W_{18}O_{62}^{8-}$	Fc^+/Fc	293		$-1.005^{g,m}$				370
$BuMeIm^+PF_6^-$	α-$S_2W_{18}O_{62}^{8-}$/ α-$S_2W_{18}O_{62}^{9-}$	Fc^+/Fc	293		$-1.373^{g,m}$				370
$BuMeIm^+PF_6^-$	α-$S_2W_{18}O_{62}^{9-}$/ α-$S_2W_{18}O_{62}^{10-}$	Fc^+/Fc	293		$-1.628^{g,m}$				370
$BuMePyrr^+Tf_2N^-$	Fe(III)/Fe(II)	Fc^+/Fc	298	$1.1^{h,k,n}$	$-0.22^{h,m}$				367
$BuMePyrr^+Tf_2N^-$	Fe(III)/Fe(II)	Fc^+/Fc	298	$0.97^{h,k,n}$	$0.02^{h,m}$				367
$BuMePyrr^+Tf_2N^-$	Fe(III)/Fe(II)	Fc^+/Fc	298	$0.53^{h,k,n}$	$-1.05^{h,m}$				367
$BuMePyrr^+Tf_2N^-$	Fe(II)/Fe(III)	Ag^+/Ag	298	$0.97^{h,k}$	$1.068^{h,m}$				371
$Bu_3MeN^+Tf_2N^-$	$Ru(bpy)_3^{2+}$	Ag^+/Ag	RT	$0.012^{h,k}$					372
$BuMeIm^+Tf_2N^-$	$Ru(bpy)_3^{2+}$	Ag^+/Ag	RT	$0.048^{h,k}$					372
$BuMePyrr^+Tf_2N^-$	Co(II)/Co	Ag^+/Ag	301	$1.0^{h,n}$					373
$BuMeIm^+BF_4^-$	Ni(II)/Ni(I)^c	$Cd^{2+}/CdHg$	298	$0.18^{i,m}$					374

Table 6. Continuation

	Electrode reaction	Reference electrode	T (K)	$10^7 D$ (cm2 s-1)	$E^{0'}$ (V)	$E_{1/2}$ (V)	$10^5 k^0$ (cm s-1)	α	Ref.
			RTILs						
$EtMeIm^+Cl^-BF_4^-$	Pd(II)/Pd	Al(III)/Al	393	$5.25^{i,k}$					375
$EtMeIm^+Cl^-BF_4^-$	Pd(II)/Pd	Al(III)/Al	308	$2.03^{i,k}$					376
$EtMeIm^+Cl^-BF_4^-$	Cu(I)/Cu(II)	Al(III)/Al	303		$0.192^{i,m}$	$0.196^{i,m}$		$0.48^{i,m}$	377
					$0.202^{j,m}$	$0.193^{j,m}$		$0.56^{j,m}$	
					$0.192^{h,m}$	$0.192^{h,m}$		$0.50^{h,m}$	
$EtMeIm^+Cl^-BF_4^-$	Cu(I)/Cu(II)	Al(III)/Al	303	2.3(Cu(I))			$10.5^{i,l}$	0.48^{i}	377
				1.5(Cu(II))			$45^{j,l}$	0.58^{j}	
							$18^{h,l}$	$0.45^{h,l}$	
$EtMeIm^+BF_4^-$	Ag(I)/Ag	Ag^+/Ag	298	$6^{h,m}$			$1^{h,m}$		378
$BuMePyrr^+Tf_2N^-$	Ag(I)/Ag	Fc^+/Fc	RT		+0.44				379
$EtMeIm^+Cl^-BF_4^-$	Ag(I)/Ag	Al(III)/Al	308	$2.64^{i,k}$					376
$EtMeIm^+Cl^-BF_4^-$	Au(I)/Au	Al(III)/Al	353	$22.4^{i,k}$					380
$EtMeIm^+Cl^-BF_4^-$	Cd(II)/Cd	Al(III)/Al	303	$2.4\pm0.1^{j,l}$					341
$EtMeIm^+Cl^-BF_4^-$	In(III)/In	Al(III)/Al	393	$4.92^{i,k}$					375
$BuPy^+Cl^-–SnCl_2$	Sn(II)/Sn	Sn(II)/Sn	403		$-0.65^{h,m,o}$				206
					$+0.07^{h,m,p}$				
$EtMeIm^+Cl^-BF_4^-$	Sb(III)/Sb(V)	Al(III)/Al	303	$2.98^{i,l}$	$+0.60^{i}$		$6.2^{i,l}$	$0.22^{i,l}$	381
							$9.8^{j,l}$	$0.37^{j,l}$	
$EtMeIm^+Cl^-BF_4^-$	Sb(III)/Sb	Al(III)/Al	303		-0.27^{i}				381

Table 6. Continuation

	Electrode reaction	Reference electrode	T (K)	$10^7 D$ ($cm^2\ s^{-1}$)	$E^{0'}$ (V)	$E_{1/2}$ (V)	$10^5 k^0$ ($cm\ s^{-1}$)	α	Ref.
RTILs									
$BuMeIm^+Tf_2N^-$ + 0.06 M $BuMeIm^+Cl^-$	Cl^-/Cl	Ag^+/Ag	298			+0.584[i,l,v]			364
$BuMeIm^+BF_4^-$	I^-/I_3^-	Ag-quasi	RT?	2.0 [h,k]	0.500 ~ 0.518[h,m,q]	0.503 ~ 0.517 [h,m]			382
$BuMeIm^+BF_4^-$	I_3^-/I_2	Ag-quasi	RT?		0.944[h,m,q]	0.928[h,m]			382
$BuMePyrr^+Tf_2N^-$	Sm(III)/Sm(II)	Ag^+/Ag	298	0.31 [i,k] 0.27 [i,n]					383
$BuMePyrr^+Tf_2N^-$	Eu(III)/Eu(II)	Ag^+/Ag	298	0.31 [i,k] 0.28 [i,n]	–0.269[i]				383
$EtMeIm^+Tf_2N^-$	Eu(III)/Eu(II)	Ag^+/Ag	298	0.78 [i,k] 0.66 [i,n]					383
$BuMeIm^+Tf_2N^-$	Eu(III)/Eu(II)	AgCl/Ag	298	2.32 [h,i,m]	0.160[h,i,m]		8.6 [h,i,m]	0.55[h,i,m]	384
$DEtMeMetN^+Tf_2N^-$	Eu(III)/Eu(II)	AgCl/Ag	298	1.61 [h,i,m]	0.354[h,i,m]		7.6 [h,i,m]	0.31[h,i,m]	384
$BuMePyrr^+Tf_2N^-$	Yb(III)/Yb(II)	Ag^+/Ag	298	0.36 [i,k] 0.34 [i,n]	–0.933[i]				383
$BuMe_3N^+Tf_2N^-$	Th(IV)/Th	SHE Fc^+/Fc	298		–1.80[i,m] –2.20[i,m]				385
$BuMeIm^+Tf_2N^-$	UCl_6^{2-}/UCl_6^-	Fc^+/Fc	298		+0.266[i,m,q]				363
$BuMeIm^+Tf_2N^-$	UCl_6^{2-}/UCl_6^{3-}	Fc^+/Fc	298		–1.96[i,m,q]				363

Table 6. Continuation

	Electrode reaction	Reference electrode	T (K)	10^7D ($cm^2\ s^{-1}$)	$E^{0'}$ (V)	$E_{1/2}$ (V)	10^5k^0 ($cm\ s^{-1}$)	α	Ref.
			RTILs						
$Bu_3MeN^+Tf_2N^-$	UCl_6^{2-}/UCl_6^-	Fc^+/Fc	298		+0.189[i,m,q]				363
$Bu_3MeN^+Tf_2N^-$	UCl_6^{2-}/UCl_6^{3-}	Fc^+/Fc	298		–2.2[i,m,q]				363
$BuMeIm^+Cl^-$	$UO_2(II)/UO_2$	Pd-quasi	343	0.0664[i,m]					386
$BuMeIm^+Cl^-$	$UO_2(II)/UO_2$	Pd-quasi	373	0.169 [i,m]					386
$BuMeIm^+Tf_2N^-$ + 0.04 M $BuMeIm^+Cl^-$	Np(IV)/Np(III)	Ag^+/Ag	298			–0.97 [i,l,v]			364
$BuMeIm^+Tf_2N^-$ + 0.06 M $BuMeIm^+Cl^-$	Np(IV)/Np(III)	Ag^+/Ag	298			–0.82 [i,l,v]			364
$BuMeIm^+Tf_2N^-$ + 0.06 M $BuMeIm^+Cl^-$	Pu(IV)/Pu(III)	Ag^+/Ag	298			–1.01 [i,l,v]			364
$EtMeIm^+BF_4^-$	TTF^d/TTF^+	AgCl/Ag	295	4.2 [h,r]		+0.401[h,r]			360
	TTF^+/TTF^{2+}			–		+0.707[h,r]			
$Hex_4N^+C_6H_5CO_2^-$	O_2	AgCl/Ag	296			–0.37 [e,t]			223
$Hex_4N^+C_6H_5CO_2^-$	Fumaric acid	AgCl/Ag	296			–0.50 [e,t]			223
$Hex_4N^+C_6H_5CO_2^-$	Benzophenone	AgCl/Ag	296			–1.42 [e,t]			223
$Hex_4N^+C_6H_5CO_2^-$	Anthracene	AgCl/Ag	296			–1.7 [e,t]			223
$Hex_4N^+C_6H_5CO_2^-$	β-naphthol	AgCl/Ag	296			–2.3 [e,t]			223
$BuMePyrr^+Tf_2N^-$	benzaldehyde	Fc^+/Fc	RT			–1.73 [h,m]			387

Table 6. Continuation

	Electrode reaction	Reference electrode	T (K)	$10^7 D$ ($cm^2\ s^{-1}$)	$E^{0'}$ (V)	$E_{1/2}$ (V)	$10^5 k^0$ ($cm\ s^{-1}$)	α	Ref.
			RTMs						
Urea-NaBr	Fe(II)/Fe	Ag-quasi	373	27.4 [h,k]					250
Urea-Acetamide-NaBr	Co(II)/Co	SCE	373	6.4 [h,k]					257
Urea-Acetamide-NaBr-KBr	Fe(II)/Fe	SCE	343	9.53 [h,k]		−1.07 ~ -1.12 [h,m]		0.31 [h,m]	388

$EtMeIm^+$: 1-ethyl-3-methylimidazolium$^+$; $BuMeIm^+$: 1-*n*-butyl-3-methylimidazolium$^+$; $HexMeIm^+$: 1-*n*-hexyl-3-methylimidazolium$^+$; $HexMe_3N^+$: *n*-hexyltrimethylammonium$^+$; Bu_3MeN^+: *n*-tributylmethylammonium$^+$; Me_3PrN^+: trimethyl-*n*-propyl ammonium$^+$; Et_4N^+: tetraethylammonium$^+$; $BuMePyrr^+$: 1-*n*-butyl-1-methylpyrrolidinium$^+$; $BuPy^+$: 1-*n*-butylpyridinium$^+$; $BuMe_3N^+$: *n*-butyl-trimethylammonium$^+$; Hex_4N^+: tetra-*n*-hexylammonium$^+$; $DEtMeMetN^+$: *N*,*N*-diethyl-*N*-methyl-*N*-(2-methoxyethyl)ammonium$^+$; FAP^-: trifluorotris(pentafluoroethyl)phosphate$^-$; $TSAC^-$: $(CF_3SO_2)(CF_3CO)N^-$; Tf_2N^-: $(CF_3SO_2)_2N^-$.

[a]BOBCalixC6·2Cs(I); [b]DCH18C6·Sr(II); [c]Ni(II) salen; [d]Tetrathiafulvalene; [e]Hg drop electrode; [f]Hg film electrode; [g]GC or Au electrode; [h]Pt electrode; [i]GC electrode; [j]W electrode; [k]Chronoamperometry; [l]Rotating disk electrode voltammetry; [m]Cyclic voltammetry; [n]Chronopotentiometry; [o]33.3 mol % $SnCl_2$; [p]66.7 mol % $SnCl_2$; [q]$(E_{pa} + E_{pc}) / 2$; [r]Square wave voltammetry; [s]Normal pulse voltammetry; [t]Polarogram; [u]Ni Electrode; [v]500 rpm.

is obvious that the number of electrochemical investigations of solutes in non-chloroaluminate RTILs is relatively small, much smaller than in chloroaluminate RTILs. Although chloroaluminate RTILs have been studied for a much longer period of time, there are now more investigators who are interested in RTILs than in the past. The real reason for the paucity of electrochemical data may be simple: non-chloroaluminate RTILs are relatively inert materials and do not solvate the wide variety of materials that dissolve in the corresponding chloroaluminates. Recently, however, Chen and Hussey[368] reported that ionophores can be used effectively to enhance the solubility of otherwise intractable materials, in this case, Cs(I) and Sr(II) so that they can be subjected to electrochemical studies. If this proves to be an effective method to increase the solubility of other solutes, then perhaps more electrochemical data will accrue in the near future. In this Section, we will introduce recent electrochemical research results derived from RTILs.

1. Surface Finishing

The application of RTILs, in particular the room-temperature haloaluminates, to surface finishing technology has a long history that spans more than half a century. In 1948, Hurley and Wier[389,390] successfully used mixtures of ethylpyridinium chloride and $AlBr_3$ as a bath for electroplating Al. At that time, the electrochemical reaction leading to the electroplating of Al had not been clarified, but it is now known to be

$$4\,Al_2X_7^- + 3\,e^- \leftrightarrows Al + 7\,AlX_4^- \tag{23}$$

where X is either Br or Cl. Most research is now carried out with chloroaluminate RTILs, rather than the mixed chloro-bromoaluminate systems, because those containing Br tend to undergo photodecomposition. However, in this Section, we review the progress attained to date in surface finishing technology with RTILs and RTMs. Tables 7 and 8 summarize the data about the electrodeposition of metal and alloy films from RTILs and RTMs that were available at the time this chapter was prepared. For convenience, we have classified the surface finishing techniques into three categories according to the type of solvent.

Table 7
Metal Deposition from Non-Chloroaluminate RTILs and RTMs

	Solute	Metal deposited	T (K)	Additives	Ref.
		RTILs			
$EtMeIm^+TSAC^-$	$LiTf_2N$	Li	298		330
$BuMeIm^+PF_6^-$	$LiPF_6$	LiHg	RT		362
$HexMe_3N^+Tf_2N^-$	$LiTf_2N$	Li	RT		391
$Me_3PrN^+Tf_2N^-$	$LiTf_2N$	Li	298		330
$Et_4N^+TSAC^-$	$LiTf_2N$	Li	298		330
$Me_3(CH_2CN)N^+Tf_2N^-$	$LiTf_2N$	Li	298		392
$C_nH_{2n+1}Me_2(CH_2CN)N^+Tf_2N^-$	$LiTf_2N$	Li	298	$EtMeIm^+Tf_2N^-$	393
$BuMePyrr^+C_2F_5BF_3^-$	$LiBF_4$	Li	298		107
$BuMePyrr^+Tf_2N^-$	$LiTf_2N$	Li	RT		379
$BuMePyrr^+Tf_2N^-$	$LiTf_2N$	Li	RT	Ethylene carbonate	379
$BuMePyrr^+Tf_2N^-$	$LiTf_2N$	Li	RT		394
$MePrPyrr^+Tf_2N^-$	$LiTf_2N$	Li	RT		395
$MePrPyrr^+Tf_2N^-$	$LiTf_2N$	Li	RT	$Pyrr^+MeBuSO_3^-$	395
$BuMePip^+Tf_2N^-$	$LiTf_2N$	Li	298		396
$MePrPip^+Tf_2N^-$	$LiTf_2N$	Li	293		397
$MePrPip^+Tf_2N^-$	$LiTf_2N$	Li/LiSi	298		398
$Me(CH_3OCH_2)Pip^+Tf_2N^-$	$LiTf_2N$	Li	298		330
$Me(CH_3OC_2H_4)Pip^+Tf_2N^-$	$LiTf_2N$	Li	298		330
$Me(CH_3(C_2H_4O)_2)Pip^+Tf_2N^-$	$LiTf_2N$	Li	298		330
$C_2dabco^+Tf_2N^-$	$LiTf_2N$	Li	353		153
$Li^+TFA\text{-}n^-$		Li	353		240

Table 7. Continuation

	Solute	Metal deposited	T (K)	Additives	Ref.
RTILs					
$BuMeIm^+PF_6^-$	$NaPF_6$	NaHg	RT		362
$BuMeIm^+PF_6^-$	KPF_6	KHg	RT		362
$Bu_3MeN^+Tf_2N^-$	Cs(I)[a]	CsHg	303		150
$Bu_3MeN^+Tf_2N^-$	BOBCalixC6·2Cs(I)	CsHg	303		368
$Bu_3MeN^+Tf_2N^-$	BOBCalixC6·2Cs(I)	CsHg	303	Toluene	368
$Bu_3MeN^+Tf_2N^-$	DCH18C6·Sr(II)	SrHg	303~323		368
$Bu_3MeN^+Tf_2N^-$	DCH18C6·Sr(II)	SrHg	303	Toluene	368
$BuMeIm^+BF_4^-$	$MgTf_2$	Mg	RT		399
$MePrPip^+Tf_2N^-$	$MgTf_2$	Mg	323		400
$BuMeIm^+Tf_2N^-$	$TiCl_4$	Ti	RT		401
$EtMeImCl\text{-}TaCl_5$		Ta	373		369
$BuMePyrr^+Tf_2N^-$	TaF_5	Ta	473		402
$BuMePyrr^+Tf_2N^-$	TaF_5	Ta	473	LiF	402
$BuMePyrr^+Tf_2N^-$	$Fe(Tf_2N^-)_2$	Fe	298		371
$EtMeImCl\text{-}ZnCl_2$	$FeCl_2$	Fe	363		403
$BuPyCl\text{-}ZnCl_2$	$CoCl_2$	Co	403		404,405
$EtMeImCl\text{-}ZnCl_2$	$CoCl_2$	Co	353		406
$EtMeImCl\text{-}ZnCl_2$	$CoCl_2$	Co	353		407
$BuMePyrr^+Tf_2N^-$	$Co(Tf_2N^-)_2$	Co	300		373
$EtMeIm^+Cl^-BF_4^-$	$PdCl_2$	Pd	393		375
$EtMeIm^+Cl^-BF_4^-$	$PdCl_2$	Pd	308		376
$EtMeIm^+Cl^-BF_4^-$	$PdCl_2$	Pd	303		380

Table 7. Continuation

	Solute	Metal deposited	T (K)	Additives	Ref.
		RTILs			
$Bu_3MeN^+Tf_2N^-$	Mn(II)	Mn	353		355
$BuMePyrr^+Tf_2N^-$	Mn(II)	Mn	323		408
BuPyCl-CuCl		Cu	363		189,409
BuPyCl-CuCl		Cu	313	Acetonitrile	189,409
$EtMeIm^+BF_4^-$	Cu(I)[a]	Cu	303		377
$EtMeImCl\text{-}ZnCl_2$	CuCl	Cu	323~373		410
$HexMe_3N^+Tf_2N^-$	$Cu(Tf_2N)_2$	Cu	323		328
$BuMePyrr^+Tf_2N^-$	Cu(I)	Cu	298~423		411
$EtMeIm^+Cl^-BF_4^-$	AgCl	Ag	308		376
$EtMeIm^+BF_4^-$	$AgBF_4$	Ag	298		378
$Ag(H_2N\text{-}C_nH_{2n+1})_2{}^+Tf_2N^-$		Ag	298		158
$EtMeIm^+Cl^-BF_4^-$	Au(I)[a]	Au	303		380
$BuMeIm^+Tf_2N^-$	$NaAlCl_4$	Au	RT		412
$Zn(H_2N\text{-}C_nH_{2n+1})_4{}^{2+}2Tf_2N^-$		Zn	343		158
$(HOC_2H_4)Me_3NCl\text{-}ZnCl_2$		Zn	333		413
$EtMeImCl\text{-}ZnCl_2$		Zn	323~353		414
$EtMeImCl\text{-}ZnCl_2$		Zn	303~353	Propylene carbonate	414
$EtMeImCl\text{-}ZnCl_2$		Zn	343~403		415
$EtMeImCl\text{-}ZnCl_2$		Zn/ZnPt	393		416
$EtMeImCl\text{-}ZnCl_2$		Zn/ZnAg	363~423		417
$EtMeImCl\text{-}ZnCl_2$		Zn/ZnCu	393		418
$EtMeImBr\text{-}ZnBr_2$		Zn	393	Ethylene glycol	419

Table 7. Continuation

	Solute	Metal deposited	T (K)	Additives	Ref.
		RTILs			
$EtMeImBr\text{-}ZnBr_2$		Zn/ZnCu	393		420
$EtMeImBr\text{-}ZnBr_2$		Zn/ZnCu	393	Ethylene glycol	420
$EtMeImBr\text{-}ZnBr_2$		Zn/ZnCu	393	1,2-propanediol	420
$EtMeImBr\text{-}ZnBr_2$		Zn/ZnCu	393	1,2-butanediol	420
$EtMeImBr\text{-}ZnBr_2$		Zn/ZnCu	393	1,3-butanediol	420
$BuMeIm^+BF_4^-$	$ZnCl_2$	Zn	RT		421
$Me_3PrN^+Tf_2N^-$	$Zn(Tf_2N)_2$	Zn	393		422
$Bu_3MeN^+Tf_2N^-$	Zn(II)	Zn	353		355
$EtMeIm^+Cl^-BF_4^-$	$CdCl_2$	Cd	303		341
$EtMeImCl\text{-}ZnCl_2$	$CdCl_2$	Cd	363		342
$BuMePyrr^+Tf_2N^-$	$AlCl_3$	Al	373		423
$BuPyCl\text{-}GaCl_3$		Ga	313		195
$EtMeImCl\text{-}GaCl_3$		Ga	295~303		195,196
$EtMeIm^+Cl^-BF_4^-$	$InCl_3$	In	393		375
$EtMeIm^+Cl^-BF_4^-$	$InCl_3$	In	303		424
$EtMeIm^+Cl^-BF_4^-$	$InCl_3$	In	353		425
$BuMePyrr^+Tf_2N^-$	$InCl_3$	In	298~423		411
$(EtMeIm^+)_2SiF_6^{2-}$		Si	363		426
$EtMeIm^+Tf_2N^-$	$(EtMeIm)_2SiF_6$	Si	298		426
$BuMePyrr^+Tf_2N^-$	$SiCl_4$	Si	RT		427
$BuMePyrr^+Tf_2N^-$	$SiCl_4$	Si	296±1		428
$BuMeIm^+PF_6^-$	GeI_4	Ge	RT		429

Table 7. Continuation

	Solute	Metal deposited	T (K)	Additives	Ref.
		RTILs			
$BuMeIm^+PF_6^-$	GeI_4	Ge	297±1		430
$BuMeIm^+PF_6^-$	$GeCl_4$	Ge	RT		431
$BuMeIm^+PF_6^-$	$GeCl_4$, $GeBr_4$	Ge	RT		432
$BuMeIm^+PF_6^-$	$GeBr_4$	Ge	297±0.5		433
$BuMeIm^+PF_6^-$	$GeCl_4$	Ge	297±0.5		434
$BuMeIm^+PF_6^-$	$GeCl_4$, $GeBr_4$	Ge	297±0.5		435
$BuPyCl\text{-}SnCl_2$		Sn	403		206
$EtMeIm^+Cl^-BF_4^-$	$SnCl_2$	Sn	303		436
$EtMeIm^+Cl^-BF_4^-$	$SnCl_2$	Sn	353		425
$EtMeIm^+Cl^-BF_4^-$	$SbCl_3$	Sb	303~393		381
$EtMeIm^+Cl^-BF_4^-$	$SbCl_3$	Sb	303		424
$BuMePyrr^+Tf_2N^-$	$SeCl_4$	Se	298~423		411
$EtMeImCl\text{-}ZnCl_2$	$TeCl_4$	Te	353		437
$BuMeIm^+Cl^-$	$UO_2(NO_3)_2$	UO_2	343		386
$BuMe_3N^+Tf_2N^-$	$[Th(Tf_2N)_4(HTf_2N)]\cdot 2H_2O$	Th	298		385
		RTMs			
Urea-NaBr	$FeCl_2$	Fe	373		250
Urea-Acetamide-NaBr-KBr	$FeCl_2$	Fe	343		388
Urea-Acetamide-NaBr	$CoCl_2$	Co	373		257
Urea-$(HOC_2H_4)Me_3NCl$	NiO	Ni	333		253
Urea-$(HOC_2H_4)Me_3NCl$	CuO	Cu	333		253
Urea-$(HOC_2H_4)Me_3NCl$	Cu_2O	Cu	323		354

Table 7. Continuation

	Solute	Metal deposited	T (K)	Additives	Ref.
		RTMs			
Urea-(HOC_2H_4)Me_3NCl	$CuCl_2$	Cu	363		438
Urea-(HOC_2H_4)Me_3NCl	Ag_2O	Ag	333		253
Urea-(HOC_2H_4)Me_3NCl	ZnO	Zn	333		253
Urea-(HOC_2H_4)Me_3NCl	$ZnCl_2$	Zn	373?		439
Urea-(HOC_2H_4)Me_3NCl	$CdCl_2$	Cd	338~373		439
Urea-(HOC_2H_4)Me_3NCl	$GaCl_3$	Ga	338		438
Urea-(HOC_2H_4)Me_3NCl	$InCl_3$	In	363		438
Urea-(HOC_2H_4)Me_3NCl	PbO_2	Pb	333		253
Urea-(HOC_2H_4)Me_3NCl	$SeCl_4$	Se	363		438

$EtMeIm^+$: 1-ethyl-3-methylimidazolium$^+$; $BuMeIm^+$: 1-*n*-butyl-3-methylimidazolium$^+$; Bu_3MeN^+: *n*-tributylmethylammonium$^+$; $HexMe_3N^+$: *n*-hexyltrimethylammonium$^+$; Me_3PrN^+: trimethyl-*n*-propylammonium$^+$; Et_4N^+: tetraethylammonium$^+$; $BuMePyrr^+$: 1-*n*-butyl-1-methylpyrrolidinium$^+$; $BuPy^+$:1-*n*-butylpyridinium$^+$; $BuMe_3N^+$: *n*-butyl-trimethylammonium$^+$; $MePrPyrr^+$: 1-methyl-1-*n*-propylpyrrolidinium; $BuMePip^+$: 1-*n*-buthyl-1-methylpiperidinium$^+$; $MePrPip^+$: 1-methyl-1-*n*-propylpiperidinium$^+$; C_2dabco^+: 1-ethyl-4-aza-1-azoniabicyclo[2.2.2]octane; $TSAC^-$: $(CF_3SO_2)(CF_3CO)N^-$; Tf_2N^-: $(CF_3SO_2)_2N^-$; TFA-n^-: $(CH_3(OCH_2CH_2)_nO)_2B(O_2CCF_3)_2^-$. [a]Anodic dissolution of pure metal.

Table 8
Alloy Deposition from Non-Chloroaluminate RTILs and RTMs

	Solute	Alloy deposition	*T* (K)	Additives	Ref.
		RTILs			
$MePrPip^+Tf_2N^-$	$LiTf_2N$	Li-Si	298		398
$EtMeImBr-ZnBr_2$	$MgBr_2$	Zn-Mg	393	Ethylene glycol	440
$EtMeImBr-ZnBr_2$	$MgBr_2$	Zn-Mg	413	Glycerin	441
$EtMeImCl-ZnCl_2$	$FeCl_2$	Zn-Fe	363		403
$BuPyCl-ZnCl_2$	$CoCl_2$	Zn-Co	403		404,405
$BuPyCl-ZnCl_2$	$CoCl_2$	Zn-Co	298	Ethanol	442,443
$BuPyCl-ZnCl_2$	$CoCl_2$	Zn-Co	298	Propylene carbonate	442
$BuPyCl-ZnCl_2$	$CoCl_2$	Zn-Co	298	Dimethylformamide	442
$EtMeImCl-ZnCl_2$	$CoCl_2$	Zn-Co	353		406
$EtMeImCl-ZnCl_2$	$CoCl_2$	Zn-Co	353		407
$(HOC_2H_4)Me_3NCl-ZnCl_2$	$CoCl_2$	Zn-Co	333		413
$EtMeImCl-ZnCl_2$	$NiCl_2$	Zn-Ni	313	Ethanol	444,445
$EtMeImCl-ZnCl_2$	$NiCl_2$	Zn-Ni	353~373		445
$EtMeImCl-ZnCl_2$	$PdCl_2$	Zn-Pd	353		446
$EtMeImCl-ZnCl_2$	–	Zn-Pt[a]	393		416
$EtMeImCl-ZnCl_2$	–	Zn-Pt[a]	363		447
$EtMeImCl-ZnCl_2$	$PtCl_2$	Zn-Pt	363		447
$EtMeImCl-ZnCl_2$	CuCl	Zn-Cu	323~373		410
$EtMeImCl-ZnCl_2$	–	Zn-Cu[a]	393		418
$EtMeImBr-ZnBr_2$	–	Zn-Cu[a]	393		420
$EtMeImCl-ZnCl_2$	–	Zn-Ag[a]	363~423		417

Table 8. Continuation

	Solute	Alloy deposition	T (K)	Additives	Ref.
		RTILs			
$Bu_3MeN^+Tf_2N^-$	Mn(II), Zn(II)	Zn-Mn	353		355
$EtMeImCl-ZnCl_2$	$CdCl_2$	Zn-Cd	363		342
$EtMeImCl-ZnCl_2$	$TeCl_4$	Zn-Te	313	Oxine + propylene carbonate	437
$EtMeImCl-ZnCl_2$	$CoCl_2$, $DyCl_3$	Zn-Co-Dy	353~373		448
$(HOC_2H_4)Me_3NCl-ZnCl_2$	$SnCl_2$	Zn-Sn	333		413
$HexMe_3N^+Tf_2N^-$	$Sn(Tf_2N)_2$	Sn-Cu[a]	373		449
$EtMeIm^+Cl^-BF_4^-$	$InCl_3$, $SnCl_2$	Sn-In	353		425
$BuPyCl-NbCl_5-SnCl_2$	–	Sn-Nb	393~403		450,451
$EtMeImCl-NbCl_5-SnCl_2$	–	Sn-Nb	403~473		207
$EtMeImCl-NbCl_5-SnCl_2$	–	Sn-Nb	433		452
$EtMeIm^+Cl^-BF_4^-$	$PdCl_2$, AgCl	Pd-Ag	308~393		376
$EtMeIm^+Cl^-BF_4^-$	$PdCl_2$, Au(I)[b]	Pd-Au	303~393		380
$EtMeIm^+Cl^-BF_4^-$	$PdCl_2$, $InCl_3$	Pd-In	393		375
$EtMeImCl-GaCl_3$	$AsCl_3$	Ga-As	RT		197
$EtMeImCl-InCl_3$	$SbCl_3$	In-Sb	318		199,200
$EtMeIm^+Cl^-BF_4^-$	$InCl_3$, $SbCl_3$	In-Sb	393		424
		RTMs			
Urea-Acetamide-NaBr-KBr	$FeCl_2$, $NdCl_3$	Fe-Nd	343		388
Urea-NaBr	$FeCl_2$, $SmCl_3$	Fe-Sm	373		250
Urea-Acetamide-NaBr	$CoCl_2$, $SmCl_3$	Co-Sm	373		257
$Urea-(HOC_2H_4)Me_3NCl$	$CuCl_2$, $InCl_3$	Cu-In	363		438
$Urea-(HOC_2H_4)Me_3NCl$	$CuCl_2$, $InCl_3$, $SeCl_4$	Cu-In-Se	363		438

Table 8. Continuation

	Solute	Alloy deposition	T (K)	Additives	Ref.
		RTMs			
Urea-$(HOC_2H_4)Me_3NCl$	$CuCl_2$, $InCl_3$, $SeCl_4$, $GaCl_4$	Cu-In-Ga-Se	338		438
Urea-$(HOC_2H_4)Me_3NCl$	ZnO, PbO_2	Zn-Pb	333		253
Urea-$(HOC_2H_4)Me_3NCl$	$ZnCl_2$, Na_2S	Zn-S	373?		439
Urea-$(HOC_2H_4)Me_3NCl$	$CdCl_2$, Na_2S	Cd-S	373		439
Urea-$(HOC_2H_4)Me_3NCl$	$CdCl_2$, Na_2SeSO_3	Cd-Se	373?		439

$EtMeIm^+$: 1-ethyl-3-methylimidazolium$^+$; $BuMeIm^+$: 1-*n*-butyl-3-methylimidazolium$^+$; $HexMe_3N^+$: *n*-hexyltrimethylammonium$^+$; $BuPy^+$:1-*n*-butylpyridinium$^+$; Bu_3MeN^+: *n*-tributylmethylammonium$^+$; $MePrPip^+$: 1-methyl-1-*n*-propylpiperidinium$^+$; Tf_2N^-: $(CF_3SO_2)_2N^-$. [a]Interdiffusion reaction between deposited metal and the substrate material; [b]Anodic dissolution of pure metal.

Table 9
Ionic Speciation and Electrodeposition in Metal Chloride (MCl_x)–Organic Chloride Salt (R^+Cl^-) RTILs

MCl_x	Ionic species			Electrodeposition			Ref.
	$33 < X^a < 50$	$X^a = 50$	$50 < X^a < 67$	$33 < X^a < 50$	$X^a = 50$	$50 < X^a < 67$	
CuCl	R^+, $CuCl_2^-$, Cl^-	R^+, $CuCl_2^-$	R^+, $Cl(CuCl)_n^-$	○	○	○	189,453
$ZnCl_2$	R^+, $ZnCl_3^-$, Cl^-	R^+, $ZnCl_3^-$	R^+, $Cl(ZnCl_2)_n^-$	○	○	○	178,193,194,415
$SnCl_2$	R^+, $SnCl_3^-$, Cl^-	R^+, $SnCl_3^-$	R^+, $Cl(SnCl_2)_n^-$	○	○	○	178,193,206,209
$AlCl_3$	R^+, $AlCl_4^-$, Cl^-	R^+, $AlCl_4^-$	R^+, $Cl(AlCl_3)^-_n$	×	×	○	39,454
$GaCl_3$	R^+, $AlCl_4^-$, Cl^-	R^+, $GaCl_4^-$	R^+, $Cl(GaCl_3)^-_n$	×	×	○	195,196

[a]Percent mole fraction of MCl_x in the RTILs
○: Electrodeposition
×: No electrodeposition

(*i*) *Metal Halide–Organic Halide Salt Binary Systems*

RTILs based on the combination of a metal halide with an organic halide salt often display dramatic changes in their physicochemical properties based on the molar ratio of the components. This behavior simply reflects the fact that the anionic speciation in these ionic liquids varies with the component ratio. The most well known and thoroughly studied examples of such RTILs are chloroaluminate salts based on $AlCl_3$ mixed with aromatic quaternary ammonium chloride salts such as 1-*n*-butylpyridinium chloride or 1-ethyl-3-methylimidazolium chloride. Table 9 lists a number of similar non-haloaluminate systems along with the corresponding composition-dependent anionic species. The prospects for electrodeposition/surface finishing in these RTILs are also indicated in the table. Fortunately, as a result of the development of commercially available computer modeling software, it is often possible to predict the structures of the anionic species that might exist in these RTILs. The results from these modeling programs show excellent agreement with the data obtained from spectroscopic methods. Note that all of the anions (or more accurately, complex anions) listed in Table 9 can be reduced to the corresponding metal, except those with the T_d structure: $AlCl_4^-$ and $GaCl_4^-$. Although $AlCl_4^-$ can be reduced to Al in high temperature inorganic chloroaluminate ionic liquids or molten salts,[39] it cannot be reduced to Al within the electrochemical window of chloroaluminate RTILs.

As can be seen from Tables 7 and 8, additives are often employed to improve of the surface morphology of electrodeposits. As reported for chloroaluminate RTILs,[13,455-459] the electrodeposit surface quality varies greatly with the type of additive used. But in this particular case, the main role of the additive is to reduce the viscosity of the RTIL.

A number of metal and alloy coatings prepared in RTILs have been shown to have interesting properties. For example, Iwagishi, et al.[440,441] electrodeposited Zn-Mg coatings on automotive steel sheet from $ZnBr_2$–EtMeImBr–ethylene glycol (or glycerin) mixtures. For pure Zn plated steel, rust appeared after 168 hrs of immersion in a 5 mass % aqueous NaCl solution at 308 K. But if the $Zn_{97.5}Mg_{2.5}$ alloy was immersed in this same solution under the same conditions, 2184 h were required to initiate rust development.[440] Not surprisingly, $Zn_{80}Mg_{20}$ alloy coatings exhibit a rust delay that exceeds 2440 hrs.[441] Clearly the corrosion resistance imparted to sheet steel by Zn-Mg coatings is improved as the Mg content of the alloy is increased.

Koura et al.[207,450-452] have prepared Nb_3Sb superconductors in $NbCl_5$–$SnCl_2$–EtMeImCl (or BuPyCl). These alloys showed superconducting behavior at a temperature of 16 K (onset) and 19 K (zero).[452] Electrochemical alloying/dealloying methods have been used in the RTILs to prepare porous metal surfaces.[416-418,447] For example, porous Ag has been produced by the dealloying of the Zn-Ag alloy formed on a Ag substrate in the $ZnCl_2$–EtMeImCl RTIL.[417] The surface is covered with a uniform porous structure (Fig. 13), and the surface morphology depends on the quantity of the deposited Zn. The advantage of this method is that the bath can be reused completely without deterioration if the counter electrode is made from Zn.

(*ii*) *RTILs with Fluoroanions*

RTILs based on fluoroanions usually exhibit outstanding electrochemical stability with electrochemical windows as large as 5.8 V. With such large electrochemical windows, it is not surprising that many attempts have been made to use these solvents to reversibly deposit active metals, such as lithium. This work is driven by the current interest in lithium batteries, and the number of reports about this work increases every year (see Section V.3). As to the surface finishing in these RTILs, several problems remain. The most significant problems are the low solubility of inorganic salts used as solutes in these ionic solvents and the difficulty of preparing thick coatings (> 1 μm) that are smooth. We think that the anodic dissolution of pure metals[150,355,377,380,408] and the addition of solubility-enhancing ionophores to the RTILs are promising strategies for overcoming these limitations.[289,295,368]

Nano-ordered deposits of pure metals, such as Mg,[399,400] Ti,[401] and Si,[426-428] have been produced in these RTILs systems. These reports are noteworthy because these metals cannot normally be produced in metal halide–organic halide RTILs. However, additional investigations will be required to achieve the practical application of this technology. Katase, et al.[449] have investigated Cu-Sn formation using the interdiffusion reaction between a Cu substrate and Sn deposited in *n*-$HexMe_3N^+Tf_2N^-$. The rate-determining step of this process was the interdiffusion reaction, and the activation energy for this reaction was estimated to be 58 kJ mol^{-1}.

(*iii*) *Room-Temperature Melts (RTMs)*

As a general rule, these systems are low cost compared to RTILs because they are formulated from off-the-shelf components that require only modest purification. For example, urea–chorine chloride mixtures are molten at room-temperature and display good environmental suitability.[251] Both urea and choline chloride are inexpensive and available in bulk quantities. Urea-based RTMs containing the appropriate metal salts have been used successfully to prepare metal and alloy coatings on a variety of substrates. Most metal salts display good solubilities in these melts. There do not seem to be many difficulties associated with the preparation of solutions containing the large amounts of dissolved metal ions needed to deposit thick coatings. Some metal oxides also appear to dissolve readily in urea-based RTMs.[253] However, the RTMs such as urea–chorine chloride mixtures typically show high viscosity.[251] Thus, from a practical standpoint, these types of solvents must necessarily be employed at elevated temperatures.

Generally speaking, the electrochemical windows of most RTMs are better than those of most aqueous solutions, but inferior to that of non-chloroaluminate RTILs. Therefore, it is difficult to deposit base metals such as the rare earths and alkali metals from these solvents. In spite of these generalizations, some workers have reported the deposition of magnetic and semiconductive materials such as Sm-Fe,[250] Sm-Co,[257] and Cu-In-Ga-Se[438] from the urea-based RTMs. Unfortunately, in many cases, these electrodeposits do not show magnetic and semiconductor properties unless they are annealed. Because most RTMs are usually very hygroscopic, care should be taken to handle them only under dry conditions.

2. Fuel Cells

The electrochemistry related to fuel cells based on RTIL electrolytes has been investigated since the early 1990's.[460] More recently, anhydrous protic RTILs and RTMs, which consist of HTf_2N (or

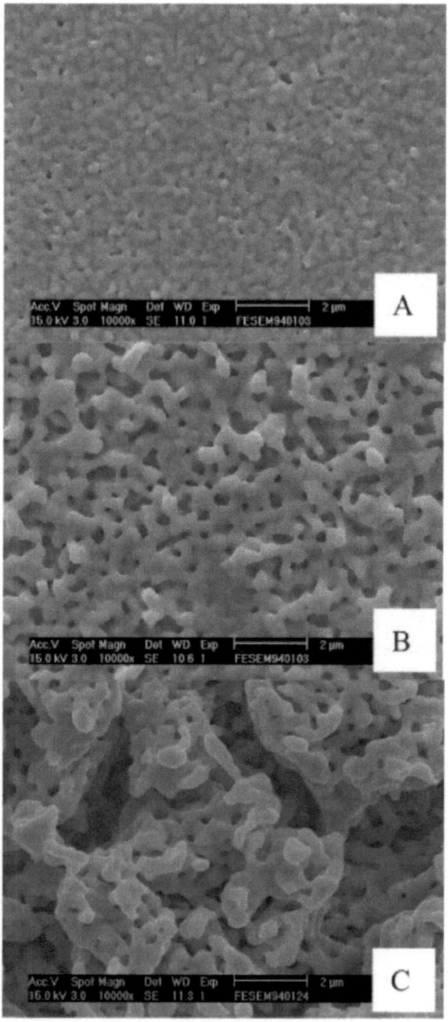

Figure 13. SEM images of Zn-Ag samples dealloyed after Zn electrodeposition on a Ag substrate. The sample areas covered by each micrograph are: (A) 0.64, (B) 5.12, and (C) 20.48 C cm^{-2}. Reprinted with permission from Ref. 417. Copyright (2006) American Chemical Society.

HTf) and imidazole-based tertiary amines, have been proposed as solvents for PEM fuel cells.[461] (Technically speaking, these solutions would only be a RTIL if the two components are mixed in equal proportions, otherwise they would simply be a RTM.) Most conventional PEM fuel cell systems suffer decreased performance at higher temperatures because of water evaporation. Non-aqueous proton conductive solvents have been proposed as an answer to this problem. There are many efforts to replace the conventional solvents used in PEM fuel cells with proton conductive RTILs (and RTMs). In addition, several research groups have focused on the advantages of RTILs in these applications and have proposed novel fuel cell systems based on the reactions that are inherent in RTILs such as imidazolium salts,[460] $AlCl_3$–EtMeImCl,[462] and EtMeImF$(HF)_n$.[463,464] In this Section, we describe some of these electrode reactions and the operation of fuel cells based on RTILs.

(*i*) *Hydrogen Electrode Reaction*

There are many reports providing electrochemical data and reaction mechanisms for the hydrogen electrode reaction in high-temperature ionic liquids.[465] Compared to aqueous solvents, the mechanism in these high-temperature systems appears to be complex, but this may be a perception that is derived from a lack of knowledge and understanding. Based on the limited information that is available, the hydrogen electrode reaction is also complex in RTILs. In the basic imidazole–HTf_2N system under oxygen-free conditions, the redox reaction at 0 V *vs*. RHE at 353 K is:[466]

$$2\,HIm^+ + 2\,e^- \leftrightarrows 2\,Im + H_2 \tag{24}$$

But, under an O_2 atmosphere, the reaction becomes:[466,467]

$$O_2 + 4\,HIm^+ + 4\,e^- \rightarrow 2\,H_2O + 4\,Im \tag{25}$$

The diffusion coefficient for HIm^+ in these mixtures depends on the imidazole–HTf_2N composition ratio, and the value was estimated by pulse-gradient spin-echo NMR spectroscopy to be ca. 1 ~ 6 × 10^{-7} $cm^2 s^{-1}$.[466]

(*ii*) *Oxygen Electrode Reaction*

It is very important to know about the mechanism of the oxygen electrode reaction in RTILs because as mentioned above oxygen is easily reduced to O_2^- in anhydrous aprotic solvents (Eq. 5). Although the electrochemistry of oxygen in RTILs has been investigated by several scientists (see Section III.4), most of these experiments were carried out in very dry RTILs under an inert atmosphere with O_2 and H_2O < 3 ppm. There is not much information about the O_2 electrode reaction in RTILs containing H^+[313] and/or H_2O,[305,309] which is the environment that is required for the operation of fuel cells. It is well known that in the presence of H^+ and H_2O, the reduction of O_2 takes a different mechanism from Eq. (5):

$$O_2 + 4\,H^+ + 4\,e^- \rightarrow 2\,H_2O \qquad (26)$$

Thus, studies that are designed to probe and understand the O_2 electrode reaction in RTILs containing H^+/H_2O are required before these ionic solvents can be exploited as electrolytes for use in low-temperature fuel cells. Recently, an investigation of the O_2 electrode reaction in proton conductive RTILs, was reported.[468,469] The oxygen reduction reaction rate in RTILs containing HTf exhibited a higher reaction rate than with phosphoric acid and neutral RTILs.

(*iii*) *Fuel Cell Systems Based on RTILs*

The first reported fuel cell system based on a RTIL was of the thermally regenerative type.[460] This fuel cell employs mixtures of $MeEtIm^+Cl^-$ + HCl with or without added $AlCl_3$. The pertinent cell reactions are:

Anode:

$$2\,EtMeIm^+Cl^- + H_2 + 2\,Cl^- \rightarrow 2\,EtMeIm^+HCl_2^- + 2\,e^- \qquad (27)$$

Cathode:

$$2\,HCl + 2\,e^- \rightarrow H_2 + 2\,Cl^- \qquad (28)$$

Thus, the net reaction is:

$$2\,EtMeIm^+Cl^- + 2\,HCl \rightarrow 2\,EtMeIm^+HCl_2^- \qquad (29)$$

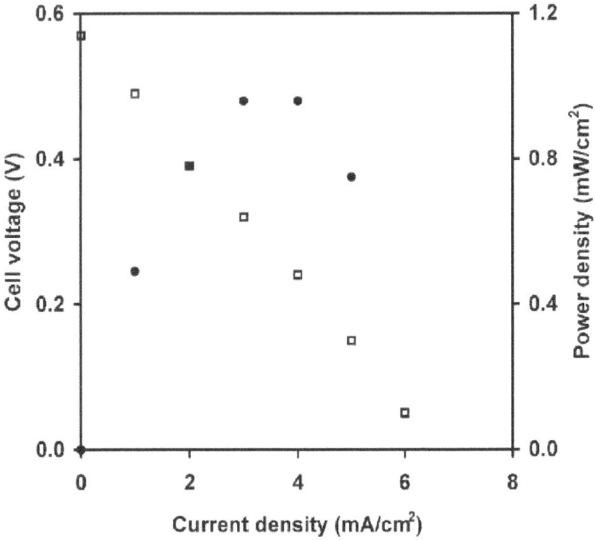

Figure 14. Variation of cell voltage (□) and power density (•) with current density for a single fuel cell using dry H_2 and O_2. The electrodes were Pt/C with a Pt loading of 0.926 mg cm^{-2}. The cell was operated at 373 K. Reproduced from Ref. 475. Copyright (2006) with permission of The Royal Society of Chemistry.

The regeneration of a sufficient amount of $EtMeIm^+Cl^-$ and HCl from the resulting $EtMeIm^+HCl_2^-$ proceeds readily when the cell is heated to more than 453 K. The actual peak open circuit voltage is 0.3 V, and the maximum power density is 0.69 mW cm^{-2} if the cell is heated to 383 K. Aluminum chloride is added to the electrolyte in order to lower the operating temperature.

Fuel cells employing proton conductive RTILs and RTMs are the most common systems that have been investigated because the prospective electrochemical reactions in these cells are basically the same as those found in conventional PEM fuel cells.[467,470-476] In addition, these anhydrous systems can operate at temperatures of more than 373 K. Unfortunately, very little comprehensive cell performance data has been produced for comparison to conventional fuel cell systems. Polymer electrolyte membranes that consist of the same quantities of 2,3-dimethyl-1-octylimidazolium triflate, and polyvinylidenefluoride-*co*-hexafluoropropylene (PVdF-HFP), and HTf (0.5 M) have been examined in a single fuel cell having a 10 cm^2 active area with 0.926 mg cm^{-2} of a Pt catalyst at 373 K.[475] The maximum power density of this cell was ca. 1.0 mW cm^{-2} (Fig. 14). Susan et al.[472] reported an unoptimized single cell prepared from a PVdF-(1,2,4-triazole/HTf_2N = 5/5) membrane with a 1.6 mg cm^{-2} Pt catalyst that generated a maximum power of 0.32 mW cm^{-2} at 0.970 mA cm^{-2}.

A non-proton transport type fuel cell system has also been proposed. This cell is based on RTILs derived from 1,3-dialkylimidazolium cations and fluorohydrogenate anions denoted by $(HF)_nF^-$.[463,464] This fuel cell system is relatively simple as illustrated in Fig. 15. What is remarkable is that this fuel cell does not involve the direct transfer of H^+. The reactions that take place in this fuel cell are shown below:

For $n = 1.3$

Anode: $$H_2 + 6\,(HF)F^- \rightarrow 4\,(HF)_2F^- + 2e^- \quad (30)$$

Cathode: $$4\,(HF)_2F^- + \tfrac{1}{2}\,O_2 + 2\,e^- \rightarrow H_2O + 6\,(HF)F^- \quad (31)$$

For $n = 2.3$

Anode: $$H_2 + 8\,(HF)_2F^- \rightarrow 6\,(HF)_3F^- + 2e^- \quad (32)$$

Cathode: $$6\,(HF)_3F^- + \tfrac{1}{2}\,O_2 + 2e^- \rightarrow H_2O + 8\,(HF)_2F^- \quad (33)$$

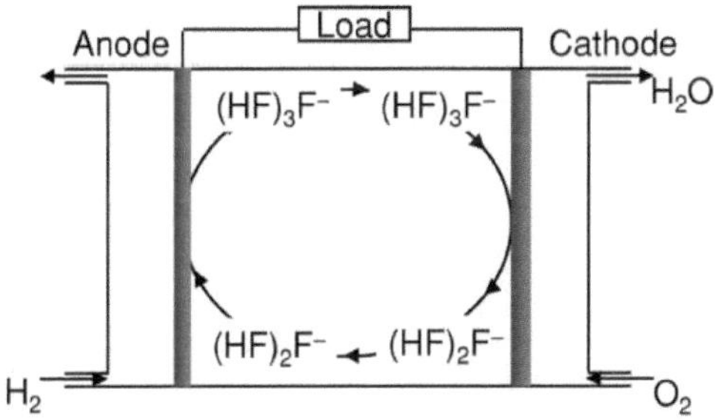

Figure 15. Operating principles of a fuel cell based on fluorohydrogenate ion conduction in $EtMeIm^+F(HF)_{2.3}^-$. Reproduced from Ref. 464 by permission of ECS—The Electrochemical Society.

The overall reaction for $n = 1.3$ or 2.3 is, thus,

$$H_2 + \frac{1}{2} O_2 \rightarrow H_2O \quad (34)$$

Figure 16 shows the open circuit voltage (OCV) of the fuel cell system illustrated in Fig. 15. The OCV was stable at ca. 1.1 V during experiments conducted over a period of 18 h and was almost independent of the n value.[464] Surprisingly, under wet condition the cell performance was maintained without HF generation. A composite electrolyte consisting of poly-2-hydroxyethyl methacrylate and $EtMeIm^+(HF)_{2.3}F^-$, which exhibits high conductivity, has great promise for use in the fabrication of a novel PEM fuel cell system.[477]

Very recently, an innovative carbonate fuel cell system was proposed by a research group at the Georgia Institute of Technology, USA.[478] This cell uses a carbonate anion exchange membrane as the electrolyte, and the cell can be operated with hydrogen or methanol as the anode gas. The proposed anodic reactions in this system are:

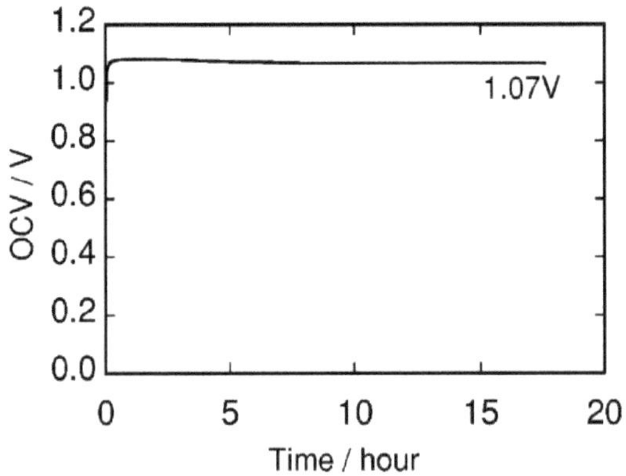

Figure 16. Open circuit voltage as a function of time for the fuel cell in Fig. 15. Reproduced from Ref. 464 by permission of ECS—The Electrochemical Society.

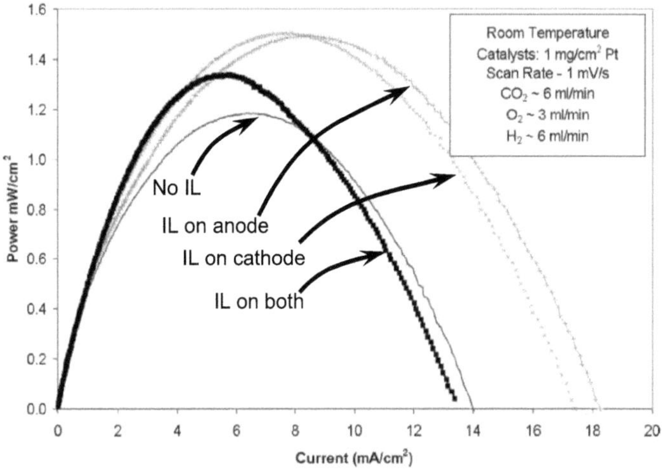

Figure 17. Power curve for room-temperature fuel cell modified with n-$BuMeIm^+BF_4^-$ operating on hydrogen. Reproduced from Ref. 478 by permission of ECS—The Electrochemical Society.

for hydrogen

$$H_2 + CO_3^{2-} \rightarrow H_2O + CO_2 + 2e^- \tag{35}$$

for methanol

$$CH_3OH + 3\,CO_3^{2-} \rightarrow 2\,H_2O + 4\,CO_2 + 6\,e^- \tag{36}$$

The cathodic reaction is:

$$2\,CO_2 + O_2 + 4\,e^- \rightarrow 2\,CO_3^{2-} \tag{37}$$

Interestingly, applying n-$BuMeIm^+BF_4^-$ to this system improve the power density if hydrogen is used as the anode gas (Fig. 17). However, the reason why this RTIL enhances the cell performance is unclear, and there is no report that this system modified with the RTIL can be operated with methanol as the fuel. This room-temperature carbonate fuel cell is a notable energy conversion system with great promise.

3. Lithium and Lithium-Ion Batteries

High energy density batteries are widely used for electronics devices such as laptop computers and cell-phones. In particular, the Li-ion battery has been the focus of constant attention.[24,479,480] The technology associated with Li-ion batteries is one of the most thoroughly studied topics in electrochemistry as a result of its significant economic impact. Unfortunately, the margin of safety for these particular batteries is less than that for many other battery systems. Because Li-ion batteries contain very reactive materials, they have a tendency for *thermal runaway*, leading to ignition and blowout. A safer nonflammable electrolyte that cannot contribute to thermal runaway is greatly desired. In addition, conventional secondary (rechargeable) cells based on Li metal do not reach their optimum performance level, owing to the instability of the electrolytes used in these cells toward Li metal. Not surprisingly, there is great interest in employing nonflammable and electrochemically stable RTILs and RTMs as electrolytes in these cells in an effort push the limits of current Li-ion battery technology.

Until several years ago, chloroaluminate RTILs were the major target for the Li battery electrolytes,[24,25,481-484] but interest in the use of non-chloroaluminate RTILs has gradually increased in recent years. A large number of RTILs have been proposed for Li battery electrolytes, and actual Li rechargeable batteries have been described.[144,148,156,261,330,397,485-496] Recent research involving Li and Li-ion rechargeable batteries based on RTIL electrolytes is described in Table 10. Those systems in which organic solvents were added to the RTILs are not included in this table. It is well known that adding organic solvents to RTILs often gives desirable results,[391,397,499,503,504] but such mixed solvent systems are outside the scope of this article.

Although Li metal can be deposited from RTILs as indicated in Table 7, almost all RTILs react with Li metal to form a solid electrolyte interphase film (SEI) layer on the Li metal surface. The mechanisms by which these passivating layers are formed and the stability of these SEI layers vary with the types of RTIL, experimental temperature, and the presence or absence of organic solvents.[330,391,499-503,505-507] Unfortunately, it seems that most RTILs are unstable during the cell discharge-charge process if the appropriate SEI layer does not form on the electrode. That is, the SEI

Table 10
Performance of Rechargeable Lithium and Lithium-Ion Cells Using RTILs and RTMs

Cell system	Discharge capacity (mAh g^{-1}) / discharge-charge efficiency (%) in first cycle	Discharge capacity (mAh g^{-1}) and discharge/charge efficiency (%)	Experimental conditions	Ref.
Li / $MePrPip^+Tf_2N^-$-$LiTf_2N$ (0.12[a]) / $LiCoO_2$	– / –	87 (50 cycles) / *ca.* 100 (25 cycles)	*C*/2[b] 3.2 ~ 4.2 V[c] RT[c]	148
Li / $MePrPip^+Tf_2N^-$-$LiTf_2N$ (0.4 M) / $LiCoO_2$	152 / 91.6	150 (~ 5 cycles) / 99.8 % (~ 5 cycles)	0.5 mA cm^{-2} [b] 3.2 ~ 4.5 V [c] RT[d]	397
$Li[Li_{1/3}Ti_{5/3}]O_4$ / $EtMeIm^+BF_4^-$-$LiBF_4$ (1 M) / $LiCoO_2$	– / 74.7	121 (5 cycles) / 92.4 (5 cycles)	0.025 mA cm^{-2} [b] 1.20 ~ 2.60 V [c] 298 K[d]	485
$Li[Li_{1/3}Ti_{5/3}]O_4$ / G($EtMeIm^+BF_4^-$-$Li^+BF_4^-$ (1 M)) / $LiCoO_2$	– / 66.0	68 (5 cycles) / 97.9 (5 cycles)	0.025 mA cm^{-2} [b] 1.20 ~ 2.60 V [c] 298 K[d]	485
Li / $HexMe_3N^+Tf_2N^-$-$LiTf_2N$ (1 M) / $LiMnO_4$	108.2 / 91.4	*ca.* 110 (at 10 cycles) / –	15 mA g^{-1} [b] 3.30 ~ 4.30 V [c] 303 K[d]	486
Li / $DMFP^+BF_4^-$-$LiBF_4$ (0.37 M) / $LiMnO_4$	– / –	– / 42~68	15 μA [b] 3.14±0.15~ 4.38±0.10 V[c] 353 K[d]	487

Table 10. Continuation

Cell system	Discharge capacity (mAh g^{-1}) / discharge-charge efficiency (%) in first cycle	Discharge capacity (mAh g^{-1}) and discharge/charge efficiency (%)	Experimental conditions	Ref.
Li / $DMFP^+BF_4^-$-$LiAsF_6$ (0.80 M) / $LiMnO_4$	– / –	– / *ca.* 100 (~ 25 cycles)	20 ~ 30 μA [b] 3.0 ~ 4.6 V [c] RT[d]	487
Li / $EMP^+BF_4^-$-$LiAsF_6$ (0.80 M) / $LiMnO_4$	– / –	– / 93 ~ 96 (after 11 cycles)	20 ~ 50 μA [b] 3.0 ~ 4.45 ± 0.15 V [c] RT [c]	488
Li / $DEME^+Tf_2N^-$-$LiTf_2N$ (0.32 mol kg^{-1}) / ZrO_2-coated $LiCoO_2$	180 / –	150 (30 cycles) / 99 (30 cycles)	0.05 mA cm^{-2} [b] 3.0 ~ 4.6 V [c] 298 K [c]	489
Li / $DEME^+Tf_2N^-$-$LiTf_2N$ (0.32 mol kg^{-1}) / $LiCoO_2$	180 / –	85 (30 cycles) / 96 (30 cycles)	0.05 mA cm^{-2} [b] 3.0 ~ 4.6 V [c] 298 K [d]	489
Li / $DEME^+Tf_2N^-$-$LiTf_2N$ (0.32 mol kg^{-1}) / $LiCoO_2$	145 / –	118 (100 cycles) / 99 (~40 cycles)	0.05 mA cm^{-2} [b] 3.0 ~ 4.2 V [c] RT [d]	490
$Li_4Ti_5O_{12}$ / $EtMeIm^+Tf_2N^-$-$LiTf_2N$ (1 M) / $LiCoO_2$	– / –	106 (200 cycles) / –	*C* [b] 298 K [d]	491
$Li_4Ti_5O_{12}$ / $DEMPyr123^+Tf_2N^-$-$LiTf_2N$ (10 mol %) / $LiFePO_4$	127 / 88	– / –	*C*/12 [b] 1.2 ~ 2.5 V [c] 293 K [d]	156

Table 10. Continuation

Cell system	Discharge capacity (mAh g^{-1}) / discharge-charge efficiency (%) in first cycle	Discharge capacity (mAh g^{-1}) and discharge/charge efficiency (%)	Experimental conditions	Ref.
Li / $1.28BuMePyrr^+Tf_2N^-$-$P(EO)_{20}Li^+Tf_2N^-$ / V_2O_5	– / –	132 (40 cycles) / –	0.063 mA cm^{-2} [b] 2.0 ~ 3.6 V [c] 313 K [d]	492
Li / $1.92BuMePyrr^+Tf_2N^-$-$P(EO)_{20}Li^+Tf_2N^-$ / V_2O_5	– / –	126 (80 cycles) / –	0.054 mA cm^{-2} [b] 2.0 ~ 3.6 V [c] 313 K [d]	492
Li / $0.96BuMePyrr^+Tf_2N^-$-$P(EO)_{10}Li^+Tf_2N^-$ / $LiFePO_4$	134 / 98	30 (340 cycles) / –	0.02*C* [b] 2.0 ~ 4.0 V [c] 293 K[d]	497
Li / $1.73MePrPyrr^+Tf_2N^-$-$P(EO)_{20}Li^+Tf_2N^-$ / $LiFePO_4$	148 / –	127 (240 cycles) / 0.992 (240 cycles)	0.05*C* [b] 2.0 ~ 4.0 V [c] 313 K [d]	498
Li / $DEtDMeIm^+Tf_2N^-$-$LiTf_2N$ (0.8 M) / $LiCoO_2$	– / –	100 (unknown) / –	0.1 mA [b] 3.0 ~ 4.2 V [c]	493
Li / $DMePrIm^+Tf_2N^-$-$LiTf_2N$ (0.32 mol Kg^{-1}) / $LiCoO_2$	130 / –	121 (50 cycles) / 99.5 (~ 5 cycles)	0.05 mA cm^{-2} [b] 3.0 ~ 4.2 V [c] 303 K [d]	494
Li / $DMePrIm^+Tf_2N^-$-$LiTf_2N$ (0.32 mol Kg^{-1}) / $LiCoO_2$	133 / 0.93	120 (50 cycles) / 99 (~ 10 cycles)	0.05 mA cm^{-2} [b] 3.0 ~ 4.2 V [c] RT[d]	495

Table 10. Continuation

Cell system	Discharge capacity (mAh g^{-1}) / discharge-charge efficiency (%) in first cycle	Discharge capacity (mAh g^{-1}) and discharge/charge efficiency (%)	Experimental conditions	Ref.
Li / $EtMeIm^+Tf_2N^-$-$LiTf_2N$ (0.32 mol Kg^{-1}) / $LiCoO_2$	124 / –	107 (50 cycles) / 96 (~ 5 cycles)	0.05 mA cm^{-2} [b] 3.0 ~ 4.2 V [c] 303 K [d]	494
Li / $BuMePip^+Tf_2N^-$-$LiTf_2N$ (1 M) / S	1055 / –	770 (5 cycles) / –	50 mA g^{-1} [b] 1.5 ~ 3.0 V [c]	396
Li / Acetamide-$LiTf_2N$ (20 mol %) / MnO_2	243 / –	– / –	*C*/10 [b] 2.0 ~ 3.5 V [c] 298 K [d]	261

$MePrPip^+$: 1-methyl-1-*n*-propylpiperidinium$^+$; $EtMeIm^+$: 1-ethyl-3-methylimidazolium$^+$; G($EtMeIm^+BF_4^-$–$Li^+BF_4^-$): poly(ethylene glycol) diacrylate with $EtMeImBF_4$–$LiBF_4$; $HexMe_3N^+$: *n*-hexyltrimethylammonium$^+$; $DMFP^+$: 1,2-dimethyl-4-fluoropyrazolium$^+$; EMP^+: 1-ethyl-3-methylpyrazolium$^+$; $DEME^+$: *N*,*N*-diethyl-*N*-methyl-*N*-(2-methoxyethyl)ammonium$^+$; $DEMPyr123^+$: 1,2-diethyl-3-methylpyrazolium$^+$; $BuMePyrr^+$: 1-*n*-butyl-1-methylpyrrolidinium$^+$; $MePrPyrr^+$: 1-methyl-1-*n*-propylpyrrolidinium$^+$; $DEtDMeIm^+$: 1,2-diethyl-3,4(5)-dimethylimidazolium$^+$; $BuMePip^+$: 1-*n*-butyl-methylpiperidinium$^+$; $DMePrIm^+$: 1,2-dimethyl-3-propylimidazolium$^+$; Tf_2N^-: $(CF_3SO_2)_2N^-$.
[a] Li^+ / (Li^+ + $MePrPip^+$). [b] Operating current density; [c] Operating voltage; [d] Temperature.

film that is produced on the anode during the charge/discharge process is inferior in stability to that obtained in conventional organic solvents. This incompatibility problem limits the cycling efficiency of such cells. Very recently, MacFarlane et al.[506] have succeeded in elucidating the mechanism of film formation on Li negative electrode in RTILs based on *N*-alkyl-*N*-methylpyrrolidinium ions and Tf_2N^-. On a negative note, Dahn et al.[508] have concluded that RTILs are not necessarily safer solvents for Li-ion batteries.

Intercalation of the organic cation into a graphite electrode has also been observed.[509] Thus, research and development involving RTIL-based Li batteries is no less difficult than that involving conventional solvents. As shown in Table 10, it is likely that the low capacity of the RTIL-based cells is caused by these undesirable solvent reactions. However, research with Li batteries based on ionic liquid electrolytes is in its infancy, and future success may be just around the corner.

Although there are difficulties with rechargeable Li batteries, we expect that Li-air[510] and Li-seawater[511] primary batteries will achieve practical use without great difficulty. These batteries are not novel systems, but the use of hydrophobic RTILs greatly improves their performance. It is notable that the capacity of the Li-air cell using $EtMeIm^+Tf_2N^-$ is 5360 mAh g^{-1} and that this performance figure exceeds that of similar cells that are based on conventional organic solvents. Figure 18 (top), which is reproduced from this article, shows changes in the cell voltage as a function of time with the measured temperature and humidity. The cell was discharged continuously for 56 days under an air atmosphere. The discharge profiles of this cell are shown in Fig. 18 (bottom) for the first 40 hours of operation. The cell performance drops off rapidly at temperatures of 253 K or less, but the performance is quite good when the operating temperature is 293 ~ 373 K. This result implies that the cell can be used effectively at temperatures of more than 373 K. In this investigation, it was also revealed that $EtMeIm^+$ plays an important role in preventing hydrolysis of the Li negative electrode. It is likely that $EtMeIm^+$ contributes to the formation of a stable SEI film on the Li negative electrode.

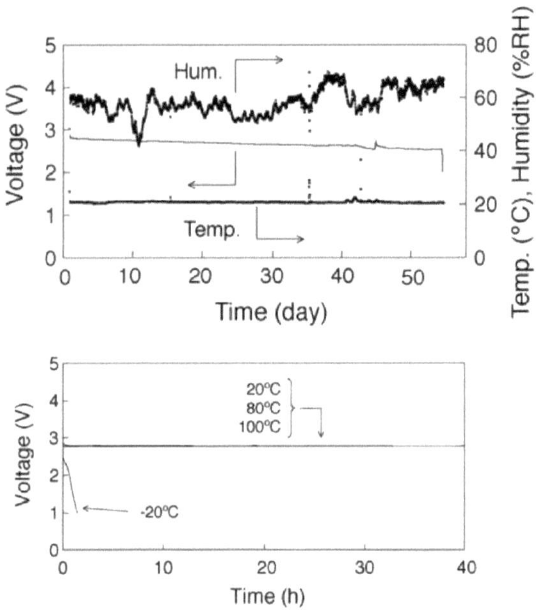

Figure 18. Top: Effect of humidity on the discharge profile in air for a Li-air cell with a $EtMeIm^+Tf_2N^-$ + $LiTf_2N$ (0.5 M) electrolyte. The operating current density was 0.01 mA cm^{-2}. Bottom: Discharge curves for this cell at different temperatures under the same experimental conditions. Reproduced from Ref. 510 with permission of Elsevier Ltd.

4. Nuclear Waste Treatment

As described at the beginning of this article, RTILs have a good chemical and electrochemical stability. Recently, the effect of radiation on the stability of several RTILs was investigated. These studies were conducted with different objectives. Some research groups were interested in utilizing RTILs as solvents for radiolytic studies, whereas others were interested in using them for the processing of nuclear materials. The radiation-induced reactions in these solvents were found to be complex,[512-518] but several salts based on the n-$BuMeIm^+$ cation were declared to be excellent solvents for the generation of radical cations and anions.[513] However, the tetraalkylammonium-based RTILs such as n-$Bu_3MeN^+Tf_2N^-$, tend to show the best stability toward solvated electrons,[515-517] and are superior to RTILs based on dialkylimidazolium salts.[518]

(i) *Removal of $^{137}Cs^+$ and $^{90}Sr^{2+}$ from Spent Reactor Fuel*

The reprocessing of spent nuclear reactor fuel results in large volumes of liquid waste that contains many insidious by-products, notably $^{137}Cs^+$ and $^{90}Sr^{2+}$. There are many on-going efforts to develop an effective process to capture and concentrate these waste products, including liquid-liquid extraction with hydrophobic organic solvents, e.g., 1,2-dichloroethane, dichloromethane, toluene, xylene, containing macrocyclic ligands.[519] Given their favorable solvent properties, it is not surprising that hydrophobic RTILs have also been investigated for this purpose.[89,289,295,296,368,520-533] For example, calix[4]arenes such as calix[4]arene-bis(t-octylbenzo-crown-6) (BOBCalixC6) dissolved in 1-C_n-3-methylimidazolium bis[(trifluoromethyl)sulfonyl]imide ionic liquids were found to be an extremely effective extraction system for Cs^+.[527] Likewise, Sr^{2+} could be extracted into these RTILs with dicycohexano-18-crown-6 (DCH18C6).[89,368,520-526,528-532] Figure 19 (top) shows the selectivity ratio of Sr^{2+}/alkali metal ions in different RTILs for two different macrocyclic ligands. This figure indicates that the Sr^{2+}/alkali metal selectivity ratio is not good enough for DCH18C6 in these dialkylimidazolium-based RTILs to form the basis of a practical system. However, the related crown ether, *N*-octyl-aza-18-crown-6, provides the needed selectivity as shown in Fig. 19 (bottom).[526] Presently in order to improve the extraction efficiency, several approaches are being studied by modification of hydrophobic RTILs for this treatment [89,531] and through molecular dynamics simulations.[533]

Unfortunately, any practical extraction process for Cs^+ and Sr^{2+} that is based on RTILs must include a provision for recycling the expensive extraction solvent. Recycling methods based on the electrochemical removal of the ionophore-coordinated Cs^+ and Sr^{2+} have been investigated and found to be feasible in tetraalkylammonium-based RTILs such as n-$Bu_3MeN^+Tf_2N^-$.[150,289,295,368,532] The related dialkylimidazolium-based RTILs are not compatible with electrochemical recycling strategies owing to their pronounced tendency to undergo reduction at the same potential at which the ionophore-coordinated alkali metal ions are reduced. A schematic model of the proposed processing cycle for Cs^+ is shown in Fig. 20.[289]

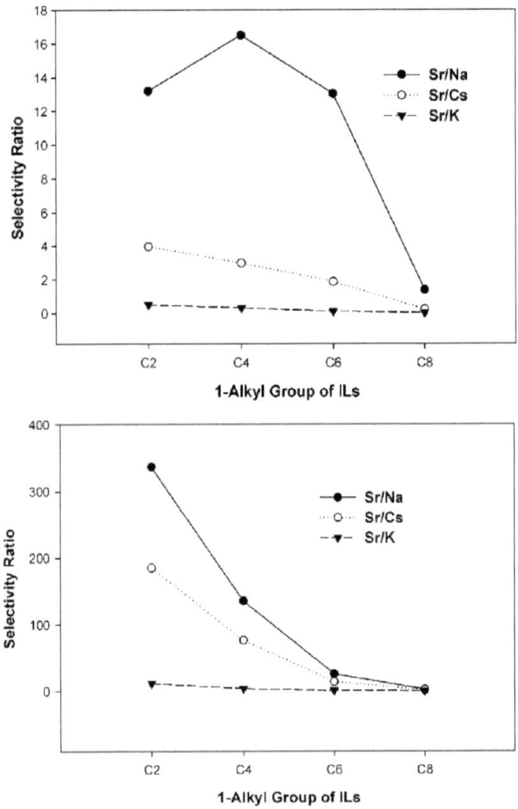

Figure 19. Effect of the alkyl chain length in $C_nMeIm^+Tf_2N^-$ on Sr^{2+}/Na^+, Sr^{2+}/Cs^+, and Sr^{2+}/K^+ selectivities during metal cation extraction from aqueous solutions with (top) DCH18C6 (0.1 M) or (bottom) *N*-octyl-aza-18-crown-6 (0.1M). Reprinted with permission from Ref. 526. Copyright (2004) American Chemical Society.

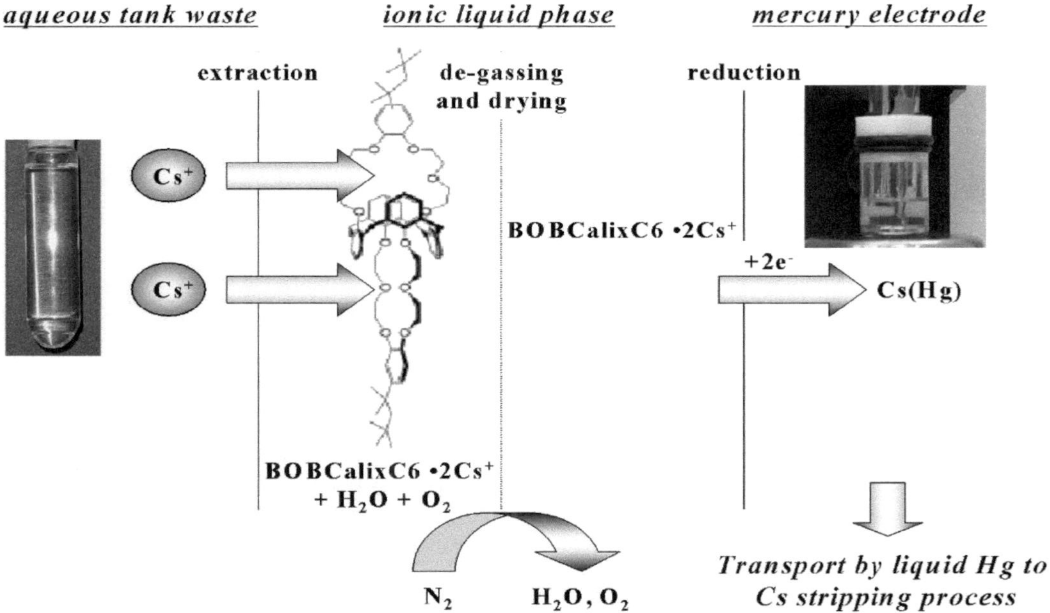

Figure 20. Overall cesium extraction model using n-$Bu_3MeN^+Tf_2N^-$ + BOBCalixC6. Reproduced from Ref. 289 by permission of ECS—The Electrochemical Society.

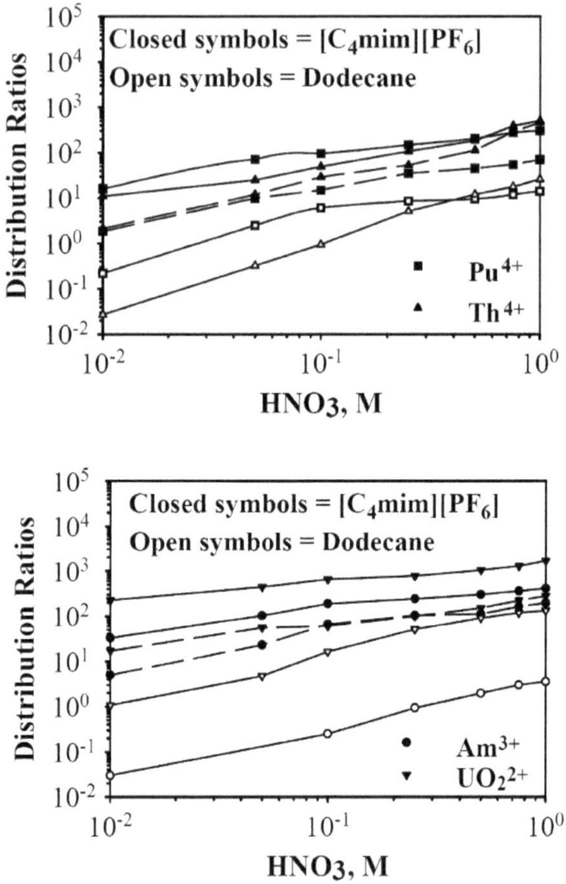

Figure 21. Distribution ratios for $^{241}Am^{3+}$, $^{233}UO_2^{2+}$, $^{238}Pu^{4+}$, and $^{230}Th^{4+}$ in the liquid/liquid extraction systems: *n*-BuMeIm$^+$PF$_6^-$ /aqueous solution (solid symbols) or dodecane/aqueous solution (open symbols). The extracting phase is either (- - -) 0.1 M octyl(phenyl)-*N,N*-diisobutylcarbamoylmethyl phosphine oxide (CMPO) in *n*-BuMeIm$^+$PF$_6^-$ or (——) 0.1 M CMPO + 1 M tri-*n*-butylphosphate (TBP) in *n*-BuMeIm$^+$PF$_6^-$. Reproduced from Ref. 534 with permission of Elsevier Ltd.

(ii) Recovery of Actinides with RTILs

The high level waste resulting from the treatment of spent nuclear reactor fuel also contains actinides that must be eliminated from the processing solution. Chemical extraction experiments have been conducted to remove actinide cations using imidazolium-based RTILs with octyl(phenyl)-*N*,*N*-diisobutylcarbamoylmethyl phosphine oxide (CMPO) and tri-*n*-butylphosphate (TBP).[534-536] The CMPO works as an extractant, and the TBP is a phase modifier. As shown in Fig. 21, the distribution ratio increases with the HNO_3 concentration independently of the solvent and obviously improves by the use of a RTIL in comparison to a conventional dodecane-based system.
There are a limited number of papers about the electrochemical behavior of actinides in the non-chloroaluminate RTILs,[363,364,385,386,537] probably because the solubility of the actinide salts is much lower in these solvents. To overcome this issue, the synthesis of actinide complexes consisting of organic cations and actinide chloride anions have been attempted.[363,364,538-540] Undoubtedly, more research on this subject will appear in the near future.

SYMBOLS (USUAL UNITS)

A	area (cm^2)
C^*	bulk concentration (mol cm^{-3})
C_j^*	bulk concentration of species j (mol cm^{-3})
D	diffusion coefficient (cm^2 s^{-1})
D_j	diffusion coefficient of species j (cm^2 s^{-1})
E	electrode potential versus some reference electrode (V)
$E_{1/2}$	half-wave potential (V)
$E^{0'}$	formal potential (V)
E_p	peak potential (V)
$E_{p/2}$	potential at $i = i_p/2$ (V)
E_{pa}	anodic peak potential (V)
E_{pc}	cathodic peak potential (V)
E_{eq}	equilibrium potential in a solution containing a redox couple (V)
F	Faraday constant (96485 C)

i	current (A)
$i(t)$	current as a function of time (A)
i_{p}	peak current (A)
$i_{\mathrm{p/2}}$	half-peak peak current (A)
i_{pa}	anodic peak current (A)
i_{pc}	cathodic peak current (A)
i_{sp}	current at the switching potential (A)
i_{la}	anodic limiting current (A)
i_{lc}	cathodic limiting current (A)
i_{k}	kinetically limited current (A)
j	current density (A cm^{-2})
j_{m}	maximum current density (A cm^{-2})
$k_{\mathrm{f}}(E)$	potential dependent heterogeneous rate constant for forward reaction (cm s^{-1})
k^0	standard heterogeneous rate constant (cm s^{-1})
n	number of electrons in an electrode reaction
R	molar gas constant (8.314 J mol^{-1} K^{-1})
T	absolute temperature (K)
t	time (s)
t_{m}	time at maximum current density (s)
t_0	nucleation delay time (s)
t'	corrected time = $t - t_0$ (s)
t_{m}'	corrected time = $t_{\mathrm{m}} - t_0$ (s)
υ	kinematic viscosity (cm^2 s^{-1})
x	dimensionless time of the maximum
α	(i) dimensionless parameter; (ii) transfer coefficient
η	overpotential (V)
v	potential scan rate (V s^{-1})
ω	angular frequency of rotation (s^{-1})

REFERENCES

[1]K. R. Seddon, *Nature Mater.* **2** (2003) 363.
[2]R. D. Rogers and K. R. Seddon, *Science* **302** (2003) 792.
[3]J. S. Wilkes, *Green Chem.* **4** (2002) 73.
[4]K. Ui, M. Ueda, R. Hagiwara, and M. Mizuhata, *Molten Salts* **47** (2004) 114 (in Japanese).
[5]V. C. Reinsborough, *Rev. Pure and Appl. Chem.* **18** (1968) 281.

[6]H. L. Chum and R. A. Osteryoung, "Chemical and Electrochemical Studies in Room Temperature Aluminum-Halide-Containing Melts," in: *Ionic Liquids*, p. 407, D. Inman and D. G. Lovering, eds., Plenum Press, New York, 1981.
[7]C. L. Hussey, "Room Temperature Molten Salt Systems," in: *Advances in Molten Salt Chemistry*, Vol. 5, p. 185, G. Mamantov and C. B. Mamantov, eds., Elsevier, Amsterdam, 1983.
[8]D. G. Lovering and R. J. Gale, "Introduction," in: *Molten Salt Techniques*, Vol. 1, p. 1, D. G. Lovering and R. J. Gale, eds., Plenum Press, New York, 1983.
[9]R. J. Gale and R. A. Osteryoung, "Haloaluminates," in: *Molten Salt Techniques*, Vol. 1, p. 55, D. G. Lovering and R. J. Gale, eds., Plenum Press, New York, 1983.
[10]P. Tissot, "Ionized Organic Salts," in: *Molten Salt Techniques*, Vol. 1, p. 137, D. G. Lovering and R. J. Gale, eds., Plenum Press, New York, 1983.
[11]R. T. Carlin and J. S. Wilkes, "Chemistry and Speciation in Room-Temperature Chloroaluminate Molten Salts," in: *Chemistry of Nonaqueous Solutions*, G. Mamantov and A. I. Popov, eds., p. 277, VCH Publishers, Inc., New York, 1994.
[12]K. R. Seddon, *J. Chem. Tech. Biotechnol.* **68** (1997) 351.
[13]S. Takahashi, N. Koura, S. Kohara, M.-L. Saboungi, and L. A. Curtiss, *Plasmas & Ions* **2** (1999) 91.
[14]R. Hagiwara and Y. Ito, *J. Fluorine Chem.* **105** (2000) 221; T. Tsuda and R. Hagiwara, *J. Fluorine Chem.* in press (10.1016/j.jfluchem.2007.10.004.).
[15]S. A. Forsyth, J. M. Pringle, and D. R. MacFarlane, *Aust. J. Chem.* **57** (2004) 113; P. J. Scammells, J. L. Scott, and R. D. Singer, *Aust. J. Chem.* **58** (2005) 155.
[16]F. Endres and S. Z. E. Abedin, *Phys. Chem. Chem. Phys.* **8** (2006) 2101.
[17]Y. Ito and T. Nohira, *Electrochim. Acta* **45** (2000) 2611; H. Xue, R. Verma, and J. M. Shreeve, *J. Fluorine Chem.* **127** (2006) 159; J. Dupont and P. A. Z. Suarez, *Phys. Chem. Chem. Phys.* **8** (2006) 2441; Z. C. Zhang, *Adv. Catal.* **49** (2006) 153; S. Zhang, N. Sun, X. He, X. Lu, and X. Zhang, *J. Phys. Chem. Ref. Data* **35** (2006) 1475; K. Matsumoto and R. Hagiwara, *J. Fluorine Chem.* **128** (2007) 317; R. Hagiwara and J. S. Lee, *Electrochemistry* **75** (2007) 23.
[18]T. Tsuda and C. L. Hussey, *Interface* **16(1)** (2007) 42.
[19]C. F. Poole, K. G. Furton, and B. R. Kersten, *J. Chromatogr. Sci.* **24** (1986) 400.
[20]C. F. Poole, *J. Chromatogr. A* **1037** (2004) 49.
[21]G. A. Baker, S. N. Baker, S. Pandey, and F. V. Bright, *Analyst (Cambridge, U.K.)* **130** (2005) 800.
[22]J.-F. Liu, J. Å. Jönsson, and G.-B. Jiang, *Trends Anal. Chem.* **24** (2005) 20.
[23]S. Pandey, *Anal. Chim. Acta* **556** (2006) 38; J. L. Anderson, D. W. Armstrong, and G.-T. Wei, *Anal. Chem.* **78** (2006) 2892; V. A. Cocalia, K. E. Gutowski, and R. D. Rogers, *Coord. Chem. Rev.* **250** (2006) 755.
[24]A. Webber and G. E. Blomgren, "Ionic Liquids for Lithium Ion And Related Batteries," in: *Advances in Lithium-Ion Batteries*, W. A. van Schalkwijk and B. Scrosati, eds., p. 185, Kluwer Academic / Plenum Publishers, New York, 2002.
[25]J. S. Wilkes, "The Past, Present and Future of Ionic Liquids as Battery Electrolytes," in: *Green Industrial Applications of Ionic Liquids*, R. D. Rogers, K. R. Seddon, and S. Volkov, eds., p. 295, NATO Science Series, Vol. 92, Kluwer Academic Publishers, Dordrecht, Netherlands, 2002.
[26]P. Wasserscheid and W. Keim, *Angew. Chem. Int. Ed.* **39** (2000) 3772.
[27]R. Sheldon, *Chem. Commun.* (2001) 2399.
[28]J. Dupont, R. F. de Souza, and P. A. Z. Suarez, *Chem. Rev.* **102** (2002) 3667.

[29]U. Kragl, M. Eckstein, and N. Kaftzik, *Curr. Opin. Biotechnol.* **13** (2002) 565.
[30]H. O.-Bourbigou and L. Magna, *J. Mol. Catal. A*, **182-183** (2002) 419.
[31]P. J. Dyson, *Transition Met. Chem.* **27** (2002) 353.
[32]F. van Rantwijk, R. M. Lau, and R. A. Sheldon, *Trends Biotechnol.* **21** (2003) 131.
[33]S. Park and R. J. Kazlauskas, *Curr. Opin. Biotechnol.* **14** (2004) 432.
[34]T. Welton, *Coord. Chem. Rev.* **248** (2004) 2459.
[35]J. S. Wilkes, *J. Mol. Catal. A* **214** (2004) 11.
[36]G. Mamantov, C. L. Hussey, and R. Marassi, "An Introduction to Electrochemistry in Molten Salts," in: *Techniques for Characterization of Electrodes and Electrochemical Processes*, p. 471, R. Varma and J. R. Selman, eds., John Wiley & Sons, Inc., New York, 1991; C. L. Hussey, "The Electrochemistry of Room-Temperature Haloaluminate Molten Salts," in: *Chemistry of Nonaqueous Solutions*, G. Mamantov and A. I. Popov, eds., p. 227, VCH Publishers, Inc., New York, 1994.
[37]M. Galiński, A. Lewandowski, and I. Stępniak, *Electrochim. Acta* **51** (2006) 5597; A. P. Abbott and K. J. McKenzie, *Phys. Chem. Chem. Phys.* **8** (2006) 4265.
[38]J. Zhang and A. M. Bond, *Analyst (Cambridge, U.K.)* **130** (2005) 1132.
[39]G. R. Stafford and C. L. Hussey, "Electrodeposition of Transition Metal-Aluminum Alloys from Chloroaluminate Molten Salts," in: *Advances in Electrochemical Science and Engineering*, R.C. Alkire and D. M. Kolb, eds., Vol. 7, p. 275, Wiley-VCH, Weinheim, 2002.
[40]F. Endres, *ChemPhysChem* **3** (2002) 144.
[41]M. Antonietti, D. Kuang, B. Smarsly, and Y. Zhou, *Angew. Chem. Int. Ed.* **43** (2004) 4988; I. J. B. Lin and C. S. Vasam, *J. Organomet. Chem.* **690** (2005) 3498.
[42]T. Welton, *Chem Rev.* **99** (1999) 2071.
[43]M. J. Earle and K. R. Seddon, *Pure Appl. Chem.* **72** (2000) 1391.
[44]C. C. Tzschucke, C. Markert, W. Bannwarth, S. Roller, A. Hebel, and R. Haag, *Angew. Chem. Int. Ed.* **41** (2002) 3964.
[45]J. P. Canal, T. Ramnial, D. A. Dickie, and J. A. C. Clyburne, *Chem. Commun.* (2006) 1809; S. Zhang, Y. Chen, F. Li, X. Lu, W. Dai, and R. Mori, *Catal. Today* **115** (2006) 61; S. Chowdhury, R. S. Mohan, and J. L. Scott, *Tetrahedron* **63** (2007) 2363.
[46]S. V. Dzyuba and R. A. Bartsch, *Angew. Chem. Int. Ed.* **42** (2003) 148.
[47]S. Zhu, Y. Wu, Q. Chen, Z. Yu, C. Wang, S. Jin, Y. Ding, and G. Wu, *Green Chem.* **8** (2006) 325.
[48]*Transport Properties in Ionic Liquids*, J. L. Copeland, Gordon and Breach Science Publishers, New York, 1974.
[49]*Ionic Liquids*, D. Inman and D. G. Lovering, eds., Plenum Press, New York, 1981.
[50]*Modern Electrochemistry Second Edition, Vol. 1: Ionics*, J. O'M. Bockris and A. K. N. Reddy, Plenum Press, New York, 1998.
[51]*Ionic Liquids: Industrial Applications for Green Chemistry*, R. D. Rogers and K. R. Seddon, eds., ACS Symposium Series 818, American Chemical Society, Washington, DC, 2002.
[52]*Green Industrial Applications of Ionic Liquids*, R. D. Rogers, K. R. Seddon, and S. Volkov, eds., NATO Science Series, Vol. 92, Kluwer, Dordrecht, Netherlands, 2002.

[53]*Ionic Liquids as Green Solvents: Progress and Prospects*, R. D. Rogers and K. R. Seddon, eds., ACS Symposium Series 856, American Chemical Society, Washington, DC, 2003.
[54]*Ionic Liquids in Synthesis*, P. Wasserscheid and T. Welton, eds., Wiley-VCH, Weinheim, 2003.
[55]*Ionic Liquids III A: Fundamentals, Progress, Challenges, and Opportunities*, R. D. Rogers and K. R. Seddon, eds., ACS Symposium Series 901, American Chemical Society, Washington, DC, 2005.
[56]*Ionic Liquids III B: Fundamentals, Progress, Challenges, and Opportunities*, R. D. Rogers and K. R. Seddon, eds., ACS Symposium Series 902, American Chemical Society, Washington, DC, 2005.
[57]*Ionic Liquids in Polymer Systems: Solvents, Additives, and Novel Applications*, C. S. Brazel and R. D. Rogers, eds., ACS Symposium Series 913, American Chemical Society, Washington, DC, 2005.
[58]*Electrochemical Aspects of Ionic Liquids*, H. Ohno, ed., Wiley-Interscience, New Jersey, 2005.
[59]*Metal Catalysed Reactions in Ionic Liquids*, P. J. Dyson and T. J. Geldbach, Springer, Netherlands, 2006.
[60]*Proceeding Series of the International Symposium on Molten Salts*, I ~ XIV, The Electrochemical Society, Inc., Pennington, NJ, 1976~2006.
[61]R. Hagiwara, T. Hirashige, T. Tsuda, and Y. Ito, *J. Fluorine Chem.* **99** (1999) 1.
[62]R. Hagiwara, T. Nohira, T. Shimada, T. Fujinaga, S. Konno, and T. Tsuda, *ECS Transactions* **3(35)** (2007) 187.
[63]R. Hagiwara, T. Hirashige, T. Tsuda, and Y. Ito, *J. Electrochem. Soc.* **149** (2002) D1.
[64]R. Hagiwara, K. Matsumoto, T. Tsuda, Y. Ito, S. Kohara, K. Suzuya, H. Matsumoto, and Y. Miyazaki, *J. Non-Cryst. Solids* **312-314** (2002) 414.
[65]K. Matsumoto, T. Tsuda, R. Hagiwara, Y. Ito, and O. Tamada, *Solid State Sci.* **4** (2002) 23.
[66]R. Hagiwara, K. Matsumoto, Y. Nakamori, T. Tsuda, Y. Ito, H. Matsumoto, and K. Momota, *J. Electrochem. Soc.* **150** (2003) D195.
[67]K. Matsumoto, R. Hagiwara, and Y. Ito, *Electrochem. Solid-State Lett.* **7** (2004) E41.
[68]Y. Shodai, S. Kohara, Y. Ohishi, M. Inaba, and A. Tasaka, *J. Phys. Chem. A* **108** (2004) 1127.
[69]K. Matsumoto and R. Hagiwara, *Electrochemistry* **73** (2005) 730.
[70]Y. Saito, K. Hirai, K. Matsumoto, R. Hagiwara, and Y. Miyazaki, *J. Phys. Chem. B.* **109** (2005) 2942.
[71]R. Hagiwara, Y. Nakamori, K. Matsumoto, and Y. Ito, *J. Phys. Chem. B.* **109** (2005) 5445.
[72]M. Salanne, C. Simon, and P. Turq, *J. Phys. Chem. B* **110** (2006) 3504.
[73]T. G. Coker, J. Ambrose, and G. J. Janz, *J. Am. Chem. Soc.* **92** (1970) 5293.
[74]B. K. M. Chan, N.-H. Chang, and M. R. Grimmett, *Aust. J. Chem.* **30** (1977) 2005.
[75]J. S. Wilkes, J. A. Levisky, R. A. Wilson, and C. L. Hussey, *Inorg. Chem.* **21** (1982) 1263.
[76]C. F. Poole, B. R. Kersten, S. J. Ho, M. E. Coddens, and K. G. Furton, *J. Chromatogr.* **352** (1986) 407.
[77]J. Pernak, K. Sobaszkiewicz, and I. Mirska, *Green Chem.* **5** (2003) 52.

[78] T. Mizumo, E. Marwanta, N. Matsumi, and H. Ohno, *Chem. Lett.* **33** (2004) 1360.
[79] R. E. D. Sesto, C. Corley, A. Robertson, and J. S. Wilkes, *J. Organomet. Chem.* **690** (2005) 2536.
[80] J. M. S. S. Esperança, H. J. R. Guedes, M. Blesic, and L. P. N. Rebelo, *J. Chem. Eng. Data* **51** (2006) 237.
[81] J. E. Gordon and G. N. SubbaRao, *J. Am. Chem. Soc.* **100** (1978) 7445.
[82] E. I. Cooper and C. A. Angell, *Solid State Ionics* **18-19** (1986) 570.
[83] W. Xu, E. I. Cooper, and C. A. Angell, *J. Phys. Chem. B* **107** (2003) 6170.
[84] T. A. Zawodzinski, Jr. and R. A. Osteryoung, *Inorg. Chem.* **27** (1988) 4383.
[85] J. L. E. Campbell and K. E. Johnson, *Inorg. Chem.* **32** (1993) 3809.
[86] J. L. E. Campbell, K. E. Johnson, and J. R. Torkelson, *Inorg. Chem.* **33** (1994) 3340.
[87] M. Deetlefs, K. R. Seddon, and M. Shara, *New J. Chem.* **30** (2006) 317.
[88] A. Bagno, C. Butts, C. Chiappe, F. D'Amico, J. C. D. Lord, D. Pieraccini, and F. Rastrelli, *Org. Biomol. Chem.* **3** (2005) 1624.
[89] H. Luo, S. Dai, P. V. Bonnesen, and A. C. Buchanan, III, *J. Alloy Compd.* **418** (2006) 195.
[90] M. Gorlov, H. Pettersson, A. Hagfeldt, and L. Kloo, *Inorg. Chem.* **46** (2007) 3566.
[91] E. I. Cooper and C. A. Angell, *Solid State Ionics* **9-10** (1983) 617.
[92] S. A. Forsyth and D. R. MacFarlane, *J. Mater. Chem.* **13** (2003) 2451.
[93] Y. L. Yagupolskii, T. M. Sokolenko, K. I. Petko, and L. M. Yagupolskii, *J. Fluorine Chem.* **126** (2005) 669.
[94] O. D. Gupta, P. D. Armstrong, and J. M. Shreeve, *Tetrahedron Lett.* **44** (2003) 9367.
[95] F. Mazille, Z. Fei, D. Kuang, D. Zhao, S. M. Zakeeruddin, M. Grätzel, and P. J. Dyson, *Inorg. Chem.* **45** (2006) 1585.
[96] H. Paulsson, M. Berggrund, E. Svantesson, A. Hagfeldt, and L. Kloo, *Sol. Energy Mater. Sol. Cells* **82** (2004) 345.
[97] M. Deetlefs, K. R. Seddon, and M. Shara, *Phys. Chem. Chem. Phys.* **8** (2006) 642.
[98] J.-P. Mikkola, P. Virtanen, and R. Sjöholm, *Green Chem.* **8** (2006) 250.
[99] J. S. Wilkes and M. J. Zaworotko, *J. Chem. Soc., Chem. Commun.* (1992) 965.
[100] A. B. McEwen, H. L. Ngo, K. LeCompte, and J. L. Goldman, *J. Electrochem. Soc.* **146** (1999) 1687.
[101] A. Noda, K. Hayamizu, and M. Watanabe, *J. Phys. Chem. B* **105** (2001) 4603.
[102] S. Forsyth, J. Golding, D. R. MacFarlane, and M. Forsyth, *Electrochim. Acta* **46** (2001) 1753.
[103] Z.-B. Zhou, H. Matsumoto, and K. Tatsumi, *Chem. Lett.* **33** (2004) 886.
[104] H. Ohno, "Neutralized Amines," in: *Electrochemical Aspects of Ionic Liquids*, p. 237, H. Ohno, ed., Wiley-Interscience, New Jersey, 2005.
[105] G.-H. Tao, L. He, N. Sun, and Y. Kou, *Chem. Commun.* (2005) 3562.
[106] N. K. Sharma, M. D. Tickell, J. L. Anderson, J. Kaar, V. Pino, B. F. Wicker, D. W. Armstrong, J. H. Davis, Jr., and A. J. Russell, *Chem. Commun.* (2006) 646.
[107] Z.-B. Zhou, H. Matsumoto, and K. Tatsumi, *Chem. Eur. J.* **12** (2006) 2196.
[108] S. Guo, Z. Du, S. Zhang, D. Li, Z. Li, and Y. Deng, *Green Chem.* **8** (2006) 296.
[109] D. Kuang, P. Wang, S. Ito, S. M. Zakeeruddin, and M. Grätzel, *J. Am. Chem. Soc.* **128** (2006) 7732.
[110] W. T. Ford, R. J. Hauri, and D. J. Hart, *J. Org. Chem.* **38** (1973) 3916.

[111]W. T. Ford and D. J. Hart, *J. Am. Chem. Soc.* **96** (1974) 3261.
[112]W. T. Ford, *Anal. Chem.* **47** (1975) 1125.
[113]W. T. Ford and D. J. Hart, *J. Phys. Chem.* **80** (1976) 1002.
[114]P. Wasserscheid, M. Sesing, and W. Korth, *Green Chem.* **4** (2002) 134.
[115]N. Nishi, S. Imakura, and T. Kakiuchi, *Anal. Chem.* **78** (2006) 2726.
[116]Z.-B. Zhou, H. Matsumoto, and K. Tatsumi, *ChemPhysChem* **6** (2005) 1324.
[117]Z.-B. Zhou, M. Takeda, and M. Ue, *J. Fluorine Chem.* **123** (2003) 127.
[118]Z.-B. Zhou, H. Matsumoto, and K. Tatsumi, *Chem. Eur. J.* **10** (2004) 6581.
[119]Z.-B. Zhou, H. Matsumoto, and K. Tatsumi, *Chem. Eur. J.* **11** (2005) 752.
[120]Z.-B. Zhou, H. Matsumoto, and K. Tatsumi, *Chem. Lett.* **33** (2004) 680.
[121]Z.-B. Zhou, H. Matsumoto, and K. Tatsumi, *Chem. Lett.* **33** (2004) 1636.
[122]Z.-B. Zhou, M. Takeda, and M. Ue, *J. Fluorine Chem.* **125** (2004) 471.
[123]D. Zhao, Z. Fei, C. A. Ohlin, G. Laurenczy, and P. J. Dyson, *Chem. Commun.* (2004) 2500.
[124]J. Fuller, R. T. Carlin, H. C. De Long, and D. Haworth, *J. Chem. Soc., Chem. Commun.* (1994) 299; Z. Mu, W. Liu, S. Zhang, and F. Zhou, *Chem. Lett.* **33** (2004) 524.
[125]V. R. Koch, L. A. Dominey, C. Nanjundiah, and M. J. Ondrechen, *J. Electrochem. Soc.* **143** (1996) 798.
[126]J. Golding, N. Hamid, D. R. MacFarlane, M. Forsyth, C. Forsyth, C. Collins, and J. Huang, *Chem. Mater.* **13** (2001) 558.
[127]P. Wasserscheid, A. Bösmann, and C. Bolm, *Chem. Commun.* (2002) 200.
[128]J. R. Harjani, T. Friščić, L. R. MacGillivray, and R. D. Singer, *Inorg. Chem.* **45** (2006) 10025.
[129]J. Zhang and A. M. Bond, *J. Phys. Chem. B* **108** (2004) 7363.
[130]N. V. Ignat'ev, U. W.-Biermann, A. Kucheryna, G. Bissky, and H. Willner, *J. Fluorine Chem.* **126** (2005) 1150.
[131]K. Matsumoto, R. Hagiwara, R. Yoshida, Y. Ito, Z. Mazej, P. Benkič, B. Žemva, O. Tamada, H. Yoshino, and S. Matsubara, *Dalton Trans.* (2004) 144.
[132]Y. Chauvin, L. Mussmann, and H. Olivier, *Angew. Chem. Int. Ed.* **34** (1995) 2698.
[133]C. E. Song, W. H. Shim, E. J. Roh, and J. H. Choi, *Chem. Commun.* (2000) 1695.
[134]J. L. Anderson, J. Ding, T. Welton, and D. W. Armstrong, *J. Am. Chem. Soc.* **124** (2002) 14247.
[135]M. Ue, M. Takeda, T. Takahashi, and M. Takehara, *Electrochem. Solid-State Lett.* **5** (2002) A119.
[136]K. Matsumoto, R. Hagiwara, and Y. Ito, *J. Fluorine Chem.* **115** (2002) 133.
[137]K. Matsumoto and R. Hagiwara, *J. Fluorine Chem.* **126** (2005) 1095.
[138]N. Bicak, *J. Mol. Liq.* **116** (2005) 15.
[139]P. Bonhôte, A.-P. Dias, N. Papageorgiou, K. Kalyanasundaram, and M. Grätzel, *Inorg. Chem.* **35** (1996) 1168.
[140]M. J. Earle, P. B. McCormac, and K. R. Seddon, *Green Chem.* **1** (1999) 23.
[141]E. I. Cooper and E. J. M. O'Sullivan, in: *Proceedings of the Eighth International Symposium on Molten Salts*, p. 386, R. J. Gale, G. Blomgren, and H. Kojima, eds., **PV92-16**, The Electrochemical Society, Inc., Pennington, NJ, 1992.
[142]A. R. Katritzky, H. Yang, D. Zhang, K. Kirichenko, M. Smiglak, J. D. Holbrey, W. M. Reichert, and R. D. Rogers, *New J. Chem.* **30** (2006) 349.
[143]S. A. Forsyth, K. J. Fraser, P. C. Howlett, D. R. MacFarlane, and M. Forsyth, *Green Chem.* **8** (2006) 256.

[144]H. Matsumoto, H. Sakaebe, K. Tatsumi, M. Kikuta, E. Ishiko, and M. Kono, *J. Power Sources* **160** (2006) 1308.
[145]M. Ishikawa, T. Sugimoto, M. Kikuta, E. Ishiko, and M. Kono, *J. Power Sources* **162** (2006) 658.
[146]J. Sun, M. Forsyth, and D. R. MacFarlane, *J. Phys. Chem. B* **102** (1998) 8858; D. R. MacFarlane, P. Meakin, J. Sun, N. Amini, and M. Forsyth, *J. Phys. Chem. B* **103** (1999) 4164.
[147]M. Yoshizawa, W. Ogihara, and H. Ohno, *Electrochem. Solid-State Lett.* **4** (2001) E25.
[148]H. Sakaebe and H. Matsumoto, *Electrochem. Commun.* **5** (2003) 594.
[149]K.-S. Kim, S. Choi, D. Demberelnyamba, H. Lee, J. Oh, B.-B. Lee, and S.-J. Mun, *Chem. Commun.* (2004) 828.
[150]P.-Y. Chen and C. L. Hussey, *Electrochim. Acta* **49** (2004) 5125.
[151]H. Tokuda, K. Hayamizu, K. Ishii, M. A. B. H. Susan, and M. Watanabe, *J. Phys. Chem. B* **109** (2005) 6103.
[152]N. Nishimura and H. Ohno, "Ionic Liquidized DNA," in: *Electrochemical Aspects of Ionic Liquids*, p. 337, H. Ohno, ed., Wiley-Interscience, New Jersey, 2005.
[153]M. Y.-Fujita, D. R. MacFarlane, P. C. Howlett, and M. Forsyth, *Electrochem. Commun.* **8** (2006) 445.
[154]J. S. Lee, N. D. Quan, J. M. Hwang, J. Y. Bae, H. Kim, B. W. Cho, H. S. Kim, and H. Lee, *Electrochem. Commun.* **8** (2006) 460.
[155]C.-M. Jin, C. Ye, B. S. Phillips, J. S. Zabinski, X. Liu, W. Liu, and J. M. Shreeve, *J. Mater. Chem.* **16** (2006) 1529.
[156]Y. A.-Lebdeh, A. Abouimrane, P.-J. Alarco, and M. Armand, *J. Power Sources* **154** (2006) 255.
[157]M. L. Patil, C. V. L. Rao, K. Yonezawa, S. Takizawa, K. Onitsuka, and H. Sasai, *Org. Lett.* **8** (2006) 227.
[158]J.-F. Huang, H. Luo, and S. Dai, *J. Electrochem. Soc.* **153** (2006) J9; M. Y.-Fujita, K. Johansson, P. Newman, D. R. MacFarlane, and M. Forsyth, *Tetrahedron Lett.* **47** (2006) 2755.
[159]J.-F. Huang, G. A. Baker, H. Luo, K. Hong, Q.-F. Li, N. J. Bjerrum, and S. Dai, *Green Chem.* **8** (2006) 599.
[160]J. M. Pringle, J. Golding, K. Baranyai, C. M. Forsyth, G. B. Deacon, J. L. Scott, and D. R. MacFarlane, *New J. Chem.* **27** (2003) 1504.
[161]H. Matsumoto, H. Kageyama, and Y. Miyazaki, *Chem. Commun.* (2002) 1726.
[162]P. Walden, *Bulletin de l'Académie Impériale des Sciences de St.-Pétersbourg* (1914) 405.
[163]S. Sugden and H. Wilkins, *J. Chem. Soc.* (1929) 1291.
[164]G.-H. Tao, L. He, W.-S. Liu, L. Xu, W. Xiong, T. Wang, and Y. Kou, *Green Chem.* **8** (2006) 639.
[165]S. I. Lall, D. Mancheno, S. Castro, V. Behaj, J. I. Cohen, and R. Engel, *Chem. Commun.* (2000) 2413.
[166]S. Lall, V. Behaj, D. Mancheno, R. Casiano, M. Thomas, A. Rikin, J. Gaillard, R. Raju, A. Scumpia, S. Castro, R. Engel, and J. I. Cohen, *Synthesis* **11** (2002) 1530.
[167]B. S. Lalia and S. S. Sekhon, *Chem. Phys. Lett.* **425** (2006) 294.
[168]W. Ogihara, M. Yoshizawa, and H. Ohno, *Chem. Lett.* (2002) 880.
[169]P. Kölle and R. Dronskowski, *Inorg. Chem.* **43** (2004) 2803.

[170]T. B. Scheffler and M. S. Thomson, in: *Proceedings of the Seventh International Symposium on Molten Salts*, p. 281, C. L. Hussey, S. N. Flengas, J. S. Wilkes, and Y. Ito, eds., **PV90-17**, The Electrochemical Society, Inc., Pennington, NJ, 1990.
[171]P. B. Hitchcock, R. J. Lewis, and T. Welton, *Polyhedron* **12** (1993) 2039.
[172]M. Morimitsu, T. Matsuo, and M. Matsunaga, in: *Proceedings of the Twelfth International Symposium on Molten Salts*, p. 117, P. C. Trulove, H. C. De Long, G. R. Stafford, and S. Deki, eds., **PV99-41**, The Electrochemical Society, Inc., Pennington, NJ, 1999.
[173]G. F. Reynolds and C. J. Dymek, Jr., *J. Power Sources* **15** (1985) 109.
[174]Y. Katayama, I. Konishiike, T. Miura, and T. Kishi, *J. Power Sources* **109** (2002) 327.
[175]Y. Yoshida, A. Otsuka, G. Saito, S. Natsume, E. Nishibori, M. Takata, M. Sakata, M. Takahashi, and T. Yoko, *Bull. Chem. Soc. Jpn.* **78** (2005) 1921.
[176]M. S. Sitze, E. R. Schreiter, E. V. Patterson, and R. G. Freeman, *Inorg. Chem.* **40** (2001) 2298.
[177]J.-Z. Yang, W.-G. Xu, Q.-G. Zhang, Y. Jin., and Z.-H. Zhang, *J. Chem. Thermodyn.* **35** (2003) 1855.
[178]A. P. Abbott, G. Capper, D. L. Davies, and R. Rasheed, *Inorg. Chem.* **43** (2004) 3447.
[179]Q.-G. Zhang, J.-Z. Yang, X.-M. Lu, J.-S. Gui, and M. Huang, *Fluid Phase Equilib.* **226** (2004) 207.
[180]S. Hayashi and H. Hamaguchi, *Chem. Lett.* **34** (2005) 740.
[181]Y. Yoshida, J. Fujii, K. Muroi, A. Otsuka, G. Saito, M. Takahashi, and T. Yoko, *Synth. Mat.* **153** (2005) 421.
[182]Y. Yoshida and G. Saito, *J. Mater. Chem.* **16** (2006) 1254.
[183]S. Hayashi, S. Saha, and H. Hamaguchi, *IEEE Trans. Magn.* **42** (2006) 12.
[184]P. B. Hitchcock, K. R. Seddon, and T. Welton, *J. Chem. Soc., Dalton Trans.* (1993) 2639.
[185]J. T. Yoke III, J. F. Weiss, and G. Tollin, *Inorg. Chem.* **2** (1963) 1210.
[186]D. D. Axtell, B. W. Good, W. W. Porterfield, and J. T. Yoke, *J. Am. Chem. Soc.* **95** (1973) 4555.
[187]D. D. Axtell and J. T. Yoke, *Inorg. Chem.* **12** (1973) 1265.
[188]S. A. Bolkan and J. T. Yoke, *J. Chem. Eng. Data* **31** (1986) 194.
[189]N. Koura, A, Suzuki, and S. Ito, in: *Proceedings of the Ninth International Symposium on Molten Salts*, p. 70, M.-L. Saboungi and H. Kojima, eds., **PV93-9**, The Electrochemical Society, Inc., Pennington, NJ, 1993.
[190]E. R. Schreiter, J. E. Stevens, M. F. Ortwerth, and R. G. Freeman, *Inorg. Chem.* **38** (1999) 3935.
[191]M. Hasan, I. V. Kozhevnikov, M. R. H. Siddiqui, A. Steiner, and N. Winterton, *Inorg. Chem.* **38** (1999) 5637.
[192]A. J. Easteal and C. A. Angell, *J. Phys. Chem.* **74** (1970) 3987.
[193]A. P. Abbott, G. Capper, D. L. Davies, H. L. Munro, R. K. Rasheed, and V. Tambyrajah, *Chem. Commun.* (2001) 2010.
[194]V. Lecocq, A. Graille, C. C. Santini, A. Baudouin, Y. Chauvin, J. M. Basset, L. Arzel, D. Bouchu, and B. Fenet, *New J. Chem.* **29** (2005) 700.
[195]S. P. Wicelinski, R. J. Gale, and J. S. Wilkes, *J. Electrochem. Soc.* **134** (1987) 262.
[196]M. W. Verbrugge and M. K. Carpenter, *AIChE J.* **36** (1990) 1097.
[197]M. K. Carpenter and M. W. Verbrugge, *J. Electrochem. Soc.* **137** (1990) 123.

[198]J.-Z. Yang, P. Tian, W.-G. Xu, B. Xu, and S.-Z. Liu, *Thermochim. Acta* **412** (2004) 1.
[199]M. K. Carpenter and M. W. Verbrugge, *US Patent* **US 5,264,111A** (1993).
[200]M. K. Carpenter and M. W. Verbrugge, *J. Mater. Res.* **9** (1994) 2584.
[201]S.-L. Zang, Q.-G. Zhang, M. Huang, B. Wang, and J.-Z. Yang, *Fluid Phase Equilib.* **230** (2005) 192.
[202]W. Guan, J.-Z. Yang, L. Li, H. Wang, and Q. G. Zhang, *Fluid Phase Equilib.* **239** (2006) 161.
[203]J.-Z. Yang, Q.-G. Zhang, and F. Xue, *J. Mol. Liq.* **128** (2006) 81.
[204]G. W. Parshall, *J. Am. Chem. Soc.* **94** (1972) 8716.
[205]F. N. Jones, *J. Org. Chem.* **32** (1967) 1667.
[206]G. Ling and N. Koura, *Denki Kagaku* **65** (1997) 149 (in Japanese).
[207]N. Koura, T. Umebayashi, Y. Idemoto, and G. Ling, *Electrochemistry* **67** (1999) 684.
[208]P. Wasserscheid and H. Waffenschmidt, *J. Mol. Catal. A* **164** (2000) 61.
[209]H. Matsuzawa, R. Nakai, K. Ui, N. Koura, and G. Ling, *Electrochemistry* **73** (2005) 715.
[210]Y. Yoshida, K. Muroi, A. Otsuka, G. Saito, M. Takahashi, and T. Yoko, *Inorg. Chem.* **43** (2004) 1458.
[211]Y. Yoshida, J. Fujii, G. Saito, T. Hiramatsu, and N. Sato, *J. Mater. Chem.* **16** (2006) 724.
[212]D. R. MacFarlane, J. Golding, S. Forsyth, M. Forsyth, and G. B. Deacon, *Chem. Commun.* (2001) 1430.
[213]D. R. MacFarlane, S. A. Forsyth, J. Golding, and G. B. Deacon, *Green Chem.* **4** (2002) 444.
[214]S. A. Forsyth, S. R. Batten, Q. Dai, and D. R. MacFarlane, *Aust. J. Chem.* **57** (2004) 121.
[215]D. Gerhard, S. C. Alpaslan, H. J. Gores, M. Uerdingen, and P. Wasserscheid, *Chem. Commun.* (2005) 5080.
[216]G. J. Janz, R. D. Reeves, and A. T. Ward, *Nature* **204** (1964) 1188.
[217]J. E. Gordon, *J .Am. Chem. Soc.* **87** (1965) 4347.
[218]T. G. Coker, B. Wunderlich, and G. J. Janz, *Trans. Faraday Soc.* **73** (1969) 3361.
[219]J. M. Pringle, J. Golding, C. M. Forsyth, G. B. Deacon, M. Forsyth, and D. R. MacFarlane, *J. Mater. Chem.* **12** (2002) 3475.
[220]P. Wang, S. M. Zakeeruddin, J.-E. Moser, R. H.-Baker, and M. Grätzel, *J. Am. Chem. Soc.* **126** (2004) 7164.
[221]H. S. Kim, Y. J. Kim, H. Lee, K. Y. Park, C. Lee, and C. S. Chin, *Angew. Chem. Int. Ed.* **41** (2002) 4300.
[222]W. Xu, L.-M. Wang, R. A. Nieman, and C. A. Angell, *J. Phys. Chem. B* **107** (2003) 11749.
[223]C. G. Swain, A. Ohno, D. K. Roe, R. Brown, and T. Maugh, II, *J. Am. Chem. Soc.* **89** (1967) 2648.
[224]Y. Fukaya, A. Sugimoto, and H. Ohno, *Biomacromolecules* **7** (2006) 3295.
[225]R. P. Seward, *J. Am. Chem. Soc.* **73** (1951) 515.
[226]P. Wasserscheid, R. van Hal, and A. Bösmann, *Green Chem.* **4** (2002) 400.
[227]J. D. Holbrey, W. M. Reichert, R. P. Swatloski, G. A. Broker, W. R. Pitner, K. R. Seddon, and R. D. Rogers, *Green Chem.* **4** (2002) 407.
[228]J.-Z. Yang, X.-M. Lu, J.-S. Gui, and W.-G. Xu, *Green Chem.* **6** (2004) 541.
[229]S. Baj, A. Chrobok, and S. Derfla, *Green Chem.* **8** (2006) 292.

[230]A. Oehlke, K. Hofmann, and S. Spange, *New J. Chem.* **30** (2006) 533; S. Himmler, S. Hörmann, R. van Hal, P. S. Schulz, and P. Wasserscheid, *Green Chem.* **8** (2006) 887.
[231]T. Mukai, M. Yoshio, T. Kato, and H. Ohno, *Chem. Lett.* **33** (2004) 1630.
[232]N. Nishi, T. Kawakami, F. Shigematsu, M. Yamamoto, and T. Kakiuchi, *Green Chem.* **8** (2006) 349.
[233]W. Ogihara, M. Yoshizawa, and H. Ohno, *Chem. Lett.* **33** (2004) 1022.
[234]K. Fukumoto, M. Yoshizawa, and H. Ohno, *J. Am. Chem. Soc.* **127** (2005) 2398; J. Kagimoto, K. Fukumoto, and H. Ohno, *Chem. Commun.* (2006) 2254; J.-Z. Yang, Q.-G. Zhang, B. Wang, and J. Tong, *J. Phys. Chem. B* **110** (2006) 22521; K. Fukumoto and H. Ohno, *Chem. Commun.* (2006) 3081.
[235]A. S. Larsen, J. D. Holbrey, F. S. Tham, and C. A. Reed, *J. Am. Chem. Soc.* **122** (2000) 7264.
[236]J. van den Broeke, F. Winter, B.-J. Deelman, and G. van Koten, *Org. Lett.* **4** (2002) 3851; J. van den Broeke, M. Stam, M. Lutz, H. Kooijman, A. L. Spek, B.-J. Deelman, and G. van Koten, *Eur. J. Inorg. Chem.* (2003) 2798.
[237]A. P. Abbott, D. Boothby, G. Capper, D. L. Davies, and R. K. Rasheed, *J. Am. Chem. Soc.* **126** (2004) 9142.
[238]O. D. Gupta, B. Twamley, and J. M. Shreeve, *Tetrahedron Lett.* **45** (2004) 1733; O. D. Gupta, B. Twamley, and J. M. Shreeve, *J. Fluorine Chem.* **126** (2005) 1222; O. D. Gupta, B. Twamley, and J. M. Shreeve, *J. Fluorine Chem.* **127** (2006) 263.
[239]H. Shobukawa, H. Tokuda, S.-I. Tabata, and M. Watanabe, *Electrochim. Acta* **50** (2004) 305.
[240]H. Shobukawa, H. Tokuda, M. A. B. H. Susan, and M. Watanabe, *Electrochim. Acta* **50** (2005) 3872.
[241]X. Jin, L. Yu, D. Garcia, R. X. Ren, and X. Zeng, *Anal. Chem.* **78** (2006) 6980.
[242]M. Yoshizawa and H. Ohno, *Chem. Commun.* (2004) 1828.
[243]M. Yoshizawa and H. Ohno, *Chem. Lett.* **33** (2004) 1594.
[244]M. Yoshizawa, A. Narita, and H. Ohno, "Zwitterionic Liquids," in: *Electrochemical Aspects of Ionic Liquids*, p. 245, H. Ohno, ed., Wiley-Interscience, New Jersey, 2005.
[245]A. Narita, W. Shibayama, K. Sakamoto, T. Mizumo, N. Matsumi, and H. Ohno, *Chem. Commun.* (2006) 1926.
[246]A. Narita, W. Shibayama, and H. Ohno, *J. Mater. Chem.* **16** (2006) 1475; A. Narita, W. Shibayama, K. Sakamoto, T. Mizumo, N. Matsumi, and H. Ohno, *Chem. Commun.* (2006) 1926.
[247]S. A. A. Zaidi and Z. A. Siddiqi, *J. Inorg. Nucl. Chem.* **37** (1975) 1806.
[248]M. Gambino and J. P. Bros, *Thermochim. Acta* **127** (1988) 223.
[249]H. Liang, H. Li, Z. Wang, F. Wu, L. Chen, and X. Huang, *J. Phys. Chem. B* **105** (2001) 9966.
[250]Y. Tong, P. Liu, L. Liu, and Q. Yang, *J. Rare Earths* **19** (2001) 275.
[251]A. P. Abbott, G. Capper, D. L. Davies, R. K. Rasheed, and V. Tambyrajah, *Chem. Commun.* (2003) 70.
[252]G. Imperato, E. Eibler, J. Niedermarier, and B. König, *Chem. Commun.* 1170 (2005).
[253]A. P. Abbott, G. Capper, D. L. Davies, R. K. Rasheed, and P. Shikotra, *Inorg. Chem.* **44** (2005) 6497.
[254]J.-H. Liao, P.-C. Wu, and Y.-H. Bai, *Inorg. Chem. Commun.* **8** (2005) 390.

[255]G. E. McManis, A. N. Fletcher, D. E. Bliss, and M. H. Miles, *J. Electroanal. Chem.* **190** (1985) 171.
[256]R. Chen, F. Wu, H. Liang, L. Li, and B. Xu, *J. Electrochem. Soc.* **152** (2005) A1979.
[257]P. Liu, Y. Du, Q. Yang, Y. Tong, and G. A. Hope, *J. Electrochem. Soc.* **153** (2006) C57.
[258]R. A. Wallace and P. F. Bruins, *J. Electrochem. Soc.* **114** (1967) 209.
[259]R. A. Wallace and P. F. Bruins, *J. Electrochem. Soc.* **114** (1967) 212.
[260]V. Bartocci, M. Gusteri, R. Marassi, F. Pucciarelli, and P. Cescon, *J. Electroanal. Chem.* **94** (1978) 153.
[261]Y. Hu, H. Li, X. Huang, and L. Chen, *Electrochem. Commun.* **6** (2004) 28.
[262]Y. Hu, Z. Wang, H. Li, X. Huang, and L. Chen, *J. Electrochem. Soc.* **151** (2004) A1424.
[263]Y. Hu, Z. Wang, X. Huang, and L. Chen, *Solid State Ionics* **175** (2004) 277.
[264]Y. Hu, Z. Wang, H. Li, X. Huang, and L. Chen, *Vib. Spectrosc.* **37** (2005) 1.
[265]Y. Hu, Z. Wang, H. Li, X. Huang, and L. Chen, *Spectrochim. Acta A* **61** (2005) 403.
[266]Y. Hu, Z. Wang, H. Li, X. Huang, and L. Chen, *Spectrochim. Acta A* **61** (2005) 2009.
[267]G. B. Appetecchi, S. Scaccia, C. Tizzani, F. Alessandrini, and S. Passerini, *J. Electrochem. Soc.*, **153** (2006) A1685.
[268]M. J. Earle, C. M. Gordon, N. V. Plechkova, K. R. Seddon, and T. Welton, *Anal. Chem.* **79** (2007) 758.
[269]M. E. van Valkenburg, R. L. Vaughn, M. Williams, and J. S. Wilkes, in: *Proceedings of the Thirteenth International Symposium on Molten Salts*, p. 112, P. C. Trulove, H. C. De Long, R. A. Mantz, G. R. Stafford, and M. Matsunaga, eds., **2002-19**, The Electrochemical Society, Inc., Pennington, NJ, 2002; M. E.Van Valkenburg, R. L. Vaughn, M. Williams, and J. S. Wilkes, *Thermochim. Acta* **425** (2005) 181.
[270]D. M. Fox, W. H. Awad, J. W. Gilman, P. H. Maupin, H. C. De Long, and P. C. Trulove, *Green Chem.* **5** (2003) 724.
[271]K. J. Baranyai, G. B. Deacon, D. R. MacFarlane, J. M. Pringle, and J. L. Scott, *Aust. J. Chem.* **57** (2004) 145.
[272]M. Kosmulski, J. Gustafsson, and J. B. Rosenholm, *Thermochim. Acta* **412** (2000) 47.
[273]D. M. Fox, J. W. Gilman, H. C. De Long, and P. C. Trulove, *J. Chem. Thermodyn.* **37** (2005) 900.
[274]T. J. Wooster, K. M. Johanson, K. J. Fraser, D. R. MacFarlane, and J. L. Scott, *Green Chem.* **8** (2006) 691.
[275]W. H. Awad, J. W. Gilman, M. Nyden, R. H. Harris, Jr., T. E. Sutto, J. Callahan, P. C. Trulove, H. C. De Long, and D. M. Fox, *Thermochim. Acta* **409** (2004) 3.
[276]H. L. Ngo, K. LeCompte, L. Hargens, and A. B. McEwen, *Thermochim. Acta* **357-358** (2000) 97.
[277]J. E. Gordon, *J. Org. Chem.* **30** (1965) 2760.
[278]S. J. Abraham and W. J. Criddle, *J. Anal. Appl. Pyrolysis* **7** (1985) 337.
[279]S. J. Abraham and W. J. Criddle, *J. Anal. Appl. Pyrolysis* **9** (1985) 65.
[280]C. P. Fredlake, J. M. Crosthwaite, D. G. Hert, S. N. V. K. Aki, and J. F. Brennecke, *J. Chem. Eng. Data* **49** (2004) 954.
[281]A. G. Glenn and P. B. Jones, *Tetrahedron Lett.* **45** (2004) 6967.
[282]K. R. Seddon, A. Stark, and M.-J. Torres, *Pure Appl. Chem.* **72** (2000) 2275.

[283]J. G. Huddleston, A. E. Visser, W. M. Reichert, H. D. Willauer, G. A. Broker, and R. D. Rogers, *Green Chem.* **3** (2001) 156.
[284]B. D. Fitchett, T. N. Knepp, and J. C. Conboy, *J. Electrochem. Soc.* **151** (2004) E219.
[285]J. Jacquemin, P. Husson, A. A. H. Padua, and V. Majer, *Green Chem.* **8** (2006) 172.
[286]J. A. Widegren, A. Laesecke, and J. W. Magee, *Chem. Commun.* (2005) 1610.
[287]T. Katase, T. Onishi, S. Imashuku, K. Murase, T. Hirato, and Y. Awakura, *Electrochemistry* **73** (2005) 686.
[288]H. Matsumoto, "Electrochemical Windows of Room-Temperature Ionic Liquids," in: *Electrochemical Aspects of Ionic Liquids*, p. 35, H. Ohno, ed., Wiley-Interscience, New Jersey, 2005.
[289]T. Tsuda, C. L. Hussey, H. Luo, and S. Dai, *J. Electrochem. Soc.* **153** (2006) D171.
[290]S. Sahami and R. A. Osteryoung, *Anal. Chem.* **55** (1983) 1970.
[291]L. Cammarata, S. G. Kazarian, P. A. Salter, and T. Welton, *Phys. Chem. Chem. Phys.* **3** (2001) 5192.
[292]C. D. Tran, S. H. D. P. Lacerda, and D. Oliveira, *Appl. Spectrosc.* **57** (2003) 152.
[293]R. P. Swatloski, J. D. Holbrey, and R. D. Rogers, *Green Chem.* **5** (2003) 361.
[294]S. Saha and H. Hamaguchi, *J. Phys. Chem. B* **110** (2006) 2777.
[295]T. Tsuda, C. L. Hussey, H. Luo, and S. Dai, *ECS Transactions* **1(13)** (2006) 25.
[296]A. E. Visser, R. P. Swatloski, W. M. Reichert, S. T. Griffin, and R. D. Rogers, *Ind. Eng. Chem. Res.* **39** (2000) 3596.
[297]P. Wasserscheid. R. van Hal, and A. Bösmann, in: *Proceedings of the Thirteenth International Symposium on Molten Salts*, p. 146, P. C. Trulove, H. C. De Long, R. A. Mantz, G. R. Stafford, and M. Matsunaga, eds., **2002-19**, The Electrochemical Society, Inc., Pennington, NJ, 2002.
[298]C. Villagrán, M. Deetlefs, W. R. Pitner, and C. Hardacre, *Anal. Chem.* **76** (2004) 2118.
[299]E. Amigues, C. Hardacre, G. Keane, M. Migaud, and M. O'Neill, *Chem. Commun.* (2006) 72.
[300]H. R. Clark and M. M. Jones, *J. Am. Chem. Soc.* **92** (1970) 816.
[301]S. Radosavljević, V. Šćepanović, S. Stević, and D. Milojković, *J. Fluorine Chem.* **13** (1979) 465.
[302]P. C. Trulove and R. A. Mantz, "Electrochemical Properties of Ionic Liquids," in: *Ionic Liquids in Synthesis*, p. 103, P. Wasserscheid and T. Welton, eds., Wiley-VCH, Weinheim, 2003.
[303]J. H. Davis, Jr., C. M. Gordon, C. Hilgers, and P. Wasserscheid, "Synthesis and Purification of Ionic Liquids," in: *Ionic Liquids in Synthesis*, p. 7, P. Wasserscheid and T. Welton, eds., Wiley-VCH, Weinheim, 2003.
[304]I. Billard, G. Moutiers, A. Labet, A. E. Azzi, C. Gaillard, C. Mariet, and K. Lützenkirchen, *Inorg. Chem.* **42** (2003) 1726.
[305]Y. Katayama, H. Onodera, M. Yamagata, and T. Miura, *J. Electrochem. Soc.* **151** (2004) A59.
[306]R. G. Evans, O. V. Klymenko, S. A. Saddoughi, C. Hardacre, and R. G. Compton, *J. Phys. Chem. B* **108** (2004) 7878.
[307]I. M. AlNashef, M. L. Leonard, M. C. Kittle, M. A. Matthews, and J. W. Weidner, *Electrochem. Solid-State Lett.* **4** (2001) D16.
[308]M. C. Buzzeo, O. V. Klymenko, J. D. Wadhawan, C. Hardacre, K. R. Seddon, and R. G. Compton, *J. Phys. Chem. A* **107** (2003) 8872.

[309]D. Zhang, T. Okajima, F. Matsumoto, and T. Ohsaka, *J. Elctrochem. Soc.* **151** (2004) D31.
[310]M. C. Buzzeo, C. Hardacre, and R. G. Compton, *Anal. Chem.* **76** (2004) 4583.
[311]Y. Katayama, K. Sekiguchi, M. Yamagata, and T. Miura, *J. Electrochem. Soc.* **152** (2005) E247.
[312]K. Ding, T. Okajima, and T. Ohsaka, *Electrochemistry* **73** (2007) 35.
[313]M. T. Carter, C. L. Hussey, S. K. D. Strubinger, and R. A. Osteryoung, *Inorg. Chem.* **30** (1991) 1149.
[314]K. N. Marsh and E. Juhasz, *Pure Appl. Chem.* **53** (1981) 1841.
[315]J. Braunstein and G. D. Robbins, *J. Chem. Ed.* **48** (1971) 52.
[316]H. Every, A. G. Bishop, M. Forsyth, and D. R. MacFarlane, *Electrochim. Acta* **45** (2000) 1279; H. Tokuda, K. Hayamizu, K. Ishii, M. A. B. H. Susan, and M. Watanabe, *J. Phys. Chem. B* **108** (2004) 16593.
[317]K. Hayamizu, Y. Aihara, H. Nakagawa, T. Nukuda, and W. S. Price, *J. Phys. Chem. B* **108** (2004) 19527; H. Tokuda, K. Ishii, M. A. B. H. Susan, S. Tsuzuki, K. Hayamizu, and M. Watanabe, *J. Phys. Chem. B* **110** (2006) 2833.
[318]P. C. Trulove and R. A. Mantz, "Electrochemical Properties of Ionic Liquids," in: *Ionic Liquids in Synthesis*, p. 103, P. Wasserscheid and T. Welton, eds., Wiley-VCH, Weinheim, 2003.
[319]S.-Y. Lee, H. H. Yong, Y. J. Lee, S. K. Kim, and S. Ahn, *J. Phys. Chem. B* **109** (2005) 13663.
[320]M. J. Earle, J. M. S. S. Esperança, M. A. Gilea, J. N. C. Lopes, L. P. N. Rebelo, J. W. Magee, K. R. Seddon, and J. A. Widegren, *Nature* **439** (2006) 831.
[321]*Electrochemical Methods: Fundamentals and Applications*, A. J. Bard and L. R. Faulkner, John Wiley & Sons, New York, 2001.
[322]M. Johnston, J.-J. Lee, G. S. Chottiner, B. Miller, T. Tsuda, C. L. Hussey, and D. A. Scherson, *J. Phys. Chem. B* **109** (2005) 11296.
[323]G. A. Snook, A. S. Best, A. G. Pandolfo, and A. F. Hollenkamp, *Electrochem. Commun.* **8** (2006) 1405.
[324]G. Gritzner and J. Kůta, *Pure Appl. Chem.* **56** (1984) 461.
[325]Z. J. Karpinski, C. Nanjundiah, and R. A. Osteryoung, *Inorg. Chem.* **23** (1984) 3358; A. I. Bhatt, A. M. Bond, D. R. MacFarlane, J. Zhang, J. L. Scott, C. R. Strauss, P. I. Iotov, and S. V. Kalcheva, *Green Chem.* **8** (2006) 161.
[326]T. Tsuda, C. L. Hussey, T. Nohira, and R. Hagiwara, unpublished data (2006).
[327]"Potentiometry in Non-Aqueous Solutions," in: *Electrochemistry in Nonaqueous Solutions*, p. 167, K. Izutsu, Wiley-VCH, Weinheim, 2002.
[328]K. Murase, K. Nitta, T. Hirato, and Y. Awakura, *J. Appl. Electrochem.* **31** (2001) 1089.
[329]H. Matsumoto, M. Yanagida, K. Tanimoto, M. Nomura, Y. Kitagawa, and Y. Miyazaki, *Chem. Lett.* (2000) 922.
[330]H. Matsumoto, H. Sakaebe, and K. Tatsumi, *J. Power Sources* **146** (2005) 45.
[331]A. A. Fannin, Jr., D. A. Floreani, L. A. King, J. S. Landers, B. J. Piersma, D. J. Stech, R. L. Vaughn, J. S. Wilkes, and J. L. Williams, *J. Phys. Chem.* **88** (1984) 2614.
[332]S. Carda-Broch, A. Berthod, and D. W. Armstrong, *Anal. Bioanal. Chem.* **375** (2003) 191.
[333]H. Matsumoto, T. Matsuda, and Y. Miyazaki, *Chem Lett.* (2000) 1430.
[334]T. Goto and Y. Ito, *J. Electrochem. Soc.* **144** (1997) 2271; T. Goto and Y. Ito, *Electrochim. Acta* **43** (1998) 3379.

[335]R. Bilewicz, K. Wikiel, R. Osteryoung, and J. Osteryoung, *Anal. Chem.* **61** (1989) 965.

[336]R. S. Nicholson and I. Shain, *Anal. Chem.* **36** (1964) 706.

[337]R. S. Nicholson, *Anal. Chem.* **38** (1966) 1406.

[338]Q. Zhu and C. L. Hussey, *J. Electrochem. Soc.* **148** (2001) C395; B. J. Tierney, W. R. Pitner, J. A. Mitchell, C. L. Hussey, and G. R. Stafford, *J. Electrochem. Soc.* **145** (1998) 3110; J. A. Mitchell, W. R. Pitner, C. L. Hussey, and G. R. Stafford, *J. Electrochem. Soc.* **143** (1996) 3448.

[339]P. C. Andricacos, J. Tabib, and L. T. Romankiw, *J. Electrochem. Soc.* **135** (1988) 1172; K. H. Wong and P. C. Andricacos, *J. Electrochem. Soc.* **137** (1990) 1087; J. Horkans, I-C. H. Chang, P. C. Andricacos, and E. J. Podlaha, *J. Electrochem. Soc.* **138** (1991) 411.

[340]F. G. Cottrell, *Z. Phys. Chem.* **42** (1903) 385.

[341]P.-Y. Chen and I-W. Sun, *Electrochim. Acta* **45** (2000) 3163.

[342]J.-F. Huang and I-W. Sun, *J. Electrochem. Soc.* **149** (2002) E348.

[343]*Techniques for Characterization of Electrodes and Electrochemical Processes*, R. Varma and J. R. Selman, eds., Wiley, New York, 1991.

[344]B. Scharifker and G. Hills, *Electrochim. Acta* **28** (1983) 879.

[345]J.-J. Lee, B. Miller, X. Shi, R. Kalish, and K. A. Wheeler, *J. Electrochem. Soc.* **147** (2000) 3370.

[346]C. L. Hussey and X. Xu, *J. Electrochem. Soc.* **138** (1991) 1886.

[347]X.-H. Xu and C. L. Hussey, *J. Electrochem. Soc.* **139** (1992) 1295.

[348]X.-H. Xu and C. L. Hussey, *J. Electrochem. Soc.* **139** (1992) 3103.

[349]W. R. Pitner and C. L. Hussey, *J. Electrochem. Soc.* **144** (1997) 3095.

[350]J.-J. Lee, B. Miller, X. Shi, R. Kalish, and K. A. Wheeler, *J. Electrochem. Soc.* **148** (2001) C183.

[351]X.-H. Xu and C. L. Hussey, *J. Electrochem. Soc.* **140** (1993) 618.

[352]X.-H. Xu and C. L. Hussey, *J. Electrochem. Soc.* **140** (1993) 1226.

[353]W. R. Pitner, C. L. Hussey, and G. R. Stafford, *J. Electrochem. Soc.* **143** (1996) 130.

[354]T. Tsuda, L. Boyd, and C. L. Hussey, unpublished data (2006).

[355]P.-Y. Chen and C. L. Hussey, *Electrochim. Acta* **52** (2007) 1857.

[356]B. R. Scharifker and J. Mostany, *J. Electroanal. Chem.* **177** (1984) 13.

[357]V. Tsakova and A. Milchev, *J. Electroanal. Chem.* **235** (1987) 237.

[358]*Electrocrystallization: Fundamentals of Nucleation and Growth*, A. Milchev, Kluwer Academic Publishers, Boston, Massachusetts, 2002.

[359]T. Tsuda, C. L. Hussey, and G. R. Stafford, *ECS Transactions* **3(35)** (2007) 217.

[360]J. Fuller, R. T. Carlin, and R. A. Osteryoung, *J. Electrochem. Soc.* **144** (1997) 3881.

[361]K. Kubo, N. Hirai, T. Tanaka, and S. Hara, *Surf. Sci.* **565** (2004) L271.

[362]D. L. Boxall and R. A. Osteryoung, *J. Electrochem. Soc.* **149** (2002) E185.

[363]S. I. Nikitenko, C. Cannes, C. Le Naour, P. Moisy, and D. Trubert, *Inorg. Chem.* **44** (2005) 9497.

[364]S. I. Nikitenko and P. Moisy, *Inorg. Chem.* **45** (2006) 1235.

[365]S. Z. E. Abedin, N. Borissenko, and F. Endres, *Electrochem. Commun.* **6** (2004) 422.

[366]M. Matsumiya, M. Terazono, and K. Tokuraku, *Electrochim. Acta* **51** (2006) 1178.

[367]M. Yamagata, N. Tachikawa, Y. Katayama, and T. Miura, *Electrochim. Acta* **52** (2007) 3317.

[368]P.-Y. Chen and C. L. Hussey, *Electrochim. Acta* **50** (2005) 2533.
[369]M. Matsunaga, T. Matsuo, and M. Morimitsu, in: *Proceedings of the Twelfth International Symposium on Molten Salts*, p. 931, P. C. Trulove, H. C. De Long, R. A. Mantz, G. R. Stafford, and M. Matsunaga, eds., **PV2002-19**, The Electrochemical Society, Inc., Pennington, NJ, 2002.
[370]J. Zhang, A. M. Bond, D. R. MacFarlane, S. A. Forsyth, J. M. Pringle, A. W. A. Mariotti, A. F. Glowinski, and A. G. Wedd, *Inorg. Chem.* **44** (2005) 5123.
[371]M. Yamagata, N. Tachikawa, Y. Katayama, and T. Miura, *Electrochemistry* **73** (2005) 564.
[372]B. M. Quinn, Z. Ding, R. Moulton, and A. J. Bard, *Langmuir* **18** (2002) 1734.
[373]R. Fukui, Y. Katayama, and T. Miura, *Electrochemistry* **73** (2005) 567.
[374]B. K. Sweeny and D. G. Peters, *Electrochem. Commun.* **3** (2001) 712.
[375]S.-I Hsiu, C.-C. Tai, and I-W. Sun, *Electrochim. Acta* **51** (2006) 2607.
[376]C.-C. Tai, F.-Y. Su, and I-W. Sun, *Electrochim. Acta* **50** (2005) 5504.
[377]P.-Y. Chen and I-W. Sun, *Electrochim. Acta* **45** (1999) 441.
[378]Y. Katayama, S. Dan, T. Miura, and T. Kishi, *J. Electrochem. Soc.* **148** (2001) C102.
[379]Y. Katayama, T. Morita, M. Yamagata, and T. Miura, *Electrochemistry* **71** (2003) 1033.
[380]F.-Y. Su, J.-F. Huang, and I-W. Sun, *J. Electrochem. Soc.* **151** (2004) C811.
[381]M-H. Yang and I-W. Sun, *J. Appl. Electrochem.* **33** (2003) 1077.
[382]Y. Zhang and J. B. Zheng, *Electrochim. Acta* **52** (2007) 4082.
[383]M. Yamagata, Y. Katayama, and T. Miura, *J. Electrochem. Soc.* **153** (2006) E5.
[384]R. Nagaishi, M. Arisaka, T. Kimura, and Y. Kitatsuji, *J. Alloys Compd.* **431** (2007) 221.
[385]A. I. Bhatt, N. W. Duffy, D. Collison, I. May, and R. G. Lewin, *Inorg. Chem.* **45** (2006) 1677.
[386]P. Giridhar, K. A. Venkatesan, T. G. Srinivasan, and P. R. V. Rao, *Electrochim. Acta* **52** (2007) 3006.
[387]C. A. Brooks and A. P. Doherty, *J. Phys. Chem. B* **109** (2005) 6276.
[388]P. Liu, Y.-P. Du, Q.-Q. Yang, G.-R. Li, and Y.-X. Tong, *Electrochim. Acta* **52** (2006) 710.
[389]F. H. Hurley and T. P. Wier, Jr., *J. Electrochem. Soc.* **98** (1951) 203.
[390]F. H. Hurley, *US Patent* **US 2,446,331** (1948); F. H. Hurley and T. P. Wier, Jr., *US Patent* **US 2,446,349** (1948).
[391]H. Zheng, K. Jiang, T. Abe, and Z. Ogumi, *Carbon* **44** (2006) 203.
[392]M. Egashira, S. Okada, J. Yamaki, D. A. Dri, F. Bonadies, and B. Scrosati, *J. Power Sources* **138** (2004) 240.
[393]M. Egashira, M. Nakagawa, I. Watanabe, S. Okada, and J. Yamaki, *J. Power Sources* **146** (2005) 685.
[394]P. C. Howlett, D. R. MacFarlane, and A. F. Hollenkamp, *Electrochem. Solid-State Lett.* **7** (2004) A97.
[395]N. Byrne, P. C. Howlett, D. R. MacFarlane, and M. Forsyth, *Adv. Mater.* **17** (2005) 2497.
[396]L. X. Yuan, J. K. Feng, X. P. Ai, Y. L. Cao, S. L. Chen, and H. X. Yang, *Electrochem. Commun.* **8** (2006) 610.
[397]J. Xu, J. Yang, Y. NuLi, J. Wang, and Z. Zhang, *J. Power Sources* **160** (2006) 621.
[398]V. Baranchugov, E. Markevich, E. Pollak, G. Salitra, and D. Aurbach, *Electrochem. Commun.* **9** (2007) 796.

[399]Y. NuLi, J. Yang, and R. Wu, *Electrochem. Commun.* **7** (2005) 1105.
[400]Y. NuLi, J. Yang, J. Wang, J. Xu, and P. Wang, *Electrochem. Solid-State Lett.* **8** (2005) C166.
[401]I. Mukhopadhyay, C. L. Aravinda, D. Borissov, and W. Freyland, *Electrochim. Acta* **50** (2005) 1275.
[402]S. Z. E. Abedin, H. K. Farag, E. M. Moustafa, U. W.-Biermann, and F. Endres, *Phys. Chem. Chem. Phys.* **7** (2005) 2333.
[403]J.-F. Huang and I-W. Sun, *J. Electrochem. Soc.* **151** (2004) C8.
[404]N. Koura, N. Mitsuta, T. Endoh, and S. Itoh, *J. Surf. Fin. Soc. Jpn.* **46** (1995) 752 (in Japanese).
[405]N. Koura, T. Endo, and Y. Idemoto, *J. Non-Cryst. Solids* **205-207** (1996) 650.
[406]N. Koura, S. Matsumoto, and Y. Idemoto, *J. Surf. Fin. Soc. Jpn.* **49** (1998) 1215 (in Japanese).
[407]P.-Y. Chen and I-W. Sun, *Electrochim. Acta* **46** (2001) 1169.
[408]J.-K. Chang, W.-T. Tsai, P.-Y. Chen, C.-H. Huang, F.-H. Yeh, and I-W. Sun, *Electrochem. Solid-State Lett.* **10** (2007) A9.
[409]N. Koura, M. Iwai, K. Ueda, and A. Suzuki, *J. Surf. Fin. Soc. Jpn.* **44** (1993) 439 (in Japanese).
[410]P.-Y. Chen, M.-C. Lin, and I-W. Sun, *J. Electrochem. Soc.* **147** (2000) 3350.
[411]S. Z. E. Abedin, A. Y. Saad, H. K. Farag, N. Borisenko, Q. X. Liu, and F. Endres, *Electrochim. Acta* **52** (2007) 2746.
[412]L. Aldous, D. S. Silvester, C. Villagrán, W. R. Pitner, R. G. Compton, M. C. Lagunas, and C. Hardacre, *New J. Chem.* **30** (2006) 1576.
[413]A. P. Abbott, G. Capper, D. L. Davies, R. K. Rasheed, and V. Tambyrajah, *Trans. IMF* **79** (2001) 204.
[414]Y.-F. Lin and I-W. Sun, *Electrochim. Acta* **44** (1999) 2771.
[415]S.-I Hsiu, J.-F. Huang, I-W. Sun, C.-H. Yuan, and J. Shiea, *Electrochim. Acta* **47** (2002) 4367.
[416]J.-F. Huang and I-W. Sun, *Chem. Mater.* **16** (2004) 1829.
[417]F.-H. Yeh, C.-C. Tai, J.-F. Huang, and I-W. Sun, *J. Phys. Chem. B* **110** (2006) 5215.
[418]Y.-W. Lin, C.-C. Tai, and I-W. Sun, *J. Electrochem. Soc.* **154** (2007) D316.
[419]T. Iwagishi, H. Yamamoto, K. Koyama, H. Shirai, and H. Kobayashi, *Electrochemistry* **70** (2002) 671 (in Japanese).
[420]K. Koyama, T. Iwagishi, H. Yamamoto, H. Shirai, and H. Kobayashi, *Electrochemistry* **70** (2002) 178.
[421]J.-G. Wang, J. Tang, Y.-C. Fu, Y.-M. Wei, Z.-B. Chen, and B.-W. Mao, *Electrochem. Commun.* **9** (2007) 633.
[422]H. Yamamoto, H. Kinoshita, M. Kimura, H. Shirai, and K. Koyama, *Electrochemistry* **74** (2006) 370 (in Japanese).
[423]S. Z. E. Abedin, E. M. Moustafa, R. Hempelmann, H. Natter, and F. Endres, *Electrochem. Commun.* **7** (2005) 1111.
[424]M.-H. Yang, M.-C. Yang, and I-W. Sun, *J. Electrochem. Soc.* **150** (2003) C544.
[425]M. Morimitsu, Y. Nakahara, and M. Matsunaga, *Electrochemistry* **73** (2005) 754.
[426]Y. Katayama, M. Yokomizo, T. Miura, and T. Kishi, *Electrochemistry* **69** (2001) 834.
[427]S. Z. E. Abedin, N. Borissenko, and F. Endres, *Electrochem. Commun.* **6** (2004) 510.
[428]N. Borissenko, S. Z. E. Abedin, and F. Endres, *J. Phys. Chem. B* **110** (2006) 6250.

[429]F. Endres and C. Schrodt, *Phys. Chem. Chem. Phys.* **2** (2000) 5517.
[430]F. Endres, *Phys. Chem. Chem. Phys.* **3** (2001) 3165.
[431]F. Endres and S. Z. E. Abedin, *Chem. Commun.* (2002) 892.
[432]F. Endres, *Electrochem. Solid-State Lett.* **5** (2002) C38.
[433]F. Endres and S. Z. E. Abedin, *Phys. Chem. Chem. Phys.* **4** (2002) 1640.
[434]F. Endres and S. Z. E. Abedin, *Phys. Chem. Chem. Phys.* **4** (2002) 1649.
[435]W. Freyland, C. A. Zell, S. Z. E. Abedin, and F. Endres, *Electrochim. Acta* **48** (2003) 3053.
[436]M. Morimitsu, Y. Nakahara, Y. Iwaki, and M. Matsunaga, *J. Min. Met.* **39B** (2003) 59.
[437]M.-C. Lin, P.-Y. Chen, and I-W. Sun, *J. Electrochem. Soc.* **148** (2001) C653.
[438]D. D. Shivagan, P. J. Dale, A. P. Samantilleke, and L. M. Peter, *Thin Solid Films* **515** (2007) 5899.
[439]P. J. Dale, A. P. Samantilleke, D. D. Shivagan, and L. M. Peter, *Thin Solid Films* **515** (2007) 5751.
[440]T. Iwagishi, K. Sawada, H. Yamamoto, K. Koyama, and H. Shirai, *Electrochemistry* **71** (2003) 318 (in Japanese).
[441]T. Iwagishi, Y. Nakatsuka, H. Yamamoto, K. Koyama, and H. Shirai, *Electrochemistry* **72** (2004) 618 (in Japanese).
[442]N. Koura and T. Endo, *J. Surf. Fin. Soc. Jpn.* **46** (1995) 1191 (in Japanese).
[443]N. Koura, T. Endo, and Y. Idemoto, *J. Surf. Fin. Soc. Jpn.* **49** (1998) 913 (in Japanese).
[444]N. Koura, Y. Suzuki, Y. Idemoto, and F. Matsumoto, *J. Surf. Fin. Soc. Jpn.* **52** (2001) 116 (in Japanese).
[445]N. Koura, Y. Suzuki, Y. Idemoto, T. Kato, and F. Matsumoto, *Surf. Coat. Technol.* **169-170** (2003) 120.
[446]H.-Y. Hsu and C.-C. Yang, *Z. Naturforsch.* **58b** (2003) 1055.
[447]J.-F. Huang and I-W. Sun, *Electrochim. Acta* **49** (2004) 3251.
[448]H.-Y. Hsu and C.-C. Yang, *Z. Naturforsch.* **58b** (2003) 139.
[449]T. Katase, R. Kurosaki, K. Murase, T. Hirato, and Y. Awakura, *Electrochem. Solid-State Lett.* **9** (2006) C69.
[450]N. Koura, G. Ling, and H. Ito, *J. Surf. Fin. Soc. Jpn.* **46** (1995) 1162 (in Japanese).
[451]G. Ling and N. Koura, *J. Surf. Fin. Soc. Jpn.* **48** (1997) 454 (in Japanese).
[452]N. Koura, K. Shibano, F. Matsumoto, H. Matsuzawa, T. Katou, Y. Idemoto, and G. Ling, *J. Surf. Fin. Soc. Jpn.* **52** (2001) 645 (in Japanese).
[453]N. Koura, N. Tanabe, S. Seiki, S. Takahashi, M.-L. Saboungi, L. A. Curtiss, and K. Suzuya, in: *Proceedings of the Tenth International Symposium on Molten Salts*, p. 492, R. T. Carlin, S. Deki, M. Matsunaga, D. S. Newman, J. R. Selman, and G. R. Stafford, eds., **PV96-7**, The Electrochemical Society, Inc., Pennington, NJ, 1996.
[454]S. Takahashi, L. A. Curtiss, D. Gosztola, N. Koura, and M.-L. Saboungi, *Inorg. Chem.* **34** (1995) 2990.
[455]Q. Zhu and C. L. Hussey, *J. Electrochem. Soc.* **149** (2002) C268.
[456]F. H. Hurley and T. P. Wier, Jr., *J. Electrochem. Soc.* **98** (1951) 207.
[457]Q. Liao, W. R. Pitner, G. Stewart, C. L. Hussey, and G. R. Stafford, *J. Electrochem. Soc.* **144** (1997) 936.
[458]T. Tsuda, T. Nohira, and Y. Ito, *Electrochim. Acta* **47** (2002) 2817.
[459]K. Ui, T. Yatsushiro, M. Futamura, Y. Idemoto, and N. Koura, *J. Surf. Fin. Soc. Jpn.* **55** (2004) 409.

[460]F. A. Ludwig, R. A. Osteryoung, C. W. Townsend, and A. Kindler, *US Patent* **US 5,208,112A** (1993).
[461]M. Yoshizawa, A. Narita, and H. Ohno, "Fuel Cell," in: *Electrochemical Aspects of Ionic Liquids*, p. 199, H. Ohno, ed., Wiley-Interscience, New Jersey, 2005.
[462]T. Tsuda, C. L. Hussey, T. Nohira, Y. Ikoma, K. Yamauchi, R. Hagiwara, and Y. Ito, *Electrochemistry* **73** (2005) 644.
[463]T. Oi, N. Yamauchi, R. Hagiwara, T. Nohira, K. Matsumoto, Y. Tamba, and Y. Ito, *World Patent*, **WO 2005/086,266 A1** (2005).
[464]R. Hagiwara, T. Nohira, K. Matsumoto, and Y. Tamba, *Electrochem. Solid-State Lett.* **8** (2005) A231.
[465]P. G. Zambonin, E. Desimoni, F. Palmisano, and L. Sabbatini, "Hydrogen in Ionic Liquids: A Review," in: *Ionic Liquids*, p. 249, D. Inman and D. G. Lovering, eds., Plenum Press, New York, 1981.
[466]A. Noda, M. A. B. H. Susan, K. Kudo, S. Mitsushima, K. Hayamizu, and M. Watanabe, *J. Phys. Chem. B* **107** (2003) 4024.
[467]M. A. B. H. Susan, A. Noda, S. Mitsushima, and M. Watanabe, *Chem. Commun.* (2003) 938.
[468]K. Kudo, S. Mitsushima, N. Kamiya, and K.-I. Ota, *Electrochemistry* **73** (2005) 272.
[469]K. Kudo, S. Mitsushima, N. Kamiya, and K.-I. Ota, *Electrochemistry* **73** (2005) 668.
[470]M. A. B. H. Susan, M. Yoo, H. Nakamoto, and M. Watanabe, *Chem. Lett.* (2003) 836.
[471]M. Watanabe, S. Mitsushima, T. Takeoka, A. Noda, K. Kudo, and R. Sakamoto, *Jpn Patent*, **JP 2003,123,791** (2003); M. Watanabe, A. Noda, T. Osawa, T. Kishi, and T. Matsuda, *World Patent*, **WO 03/083,981 A1** (2003).
[472]M. A. B. H. Susan, A. Noda, N. Ishibashi, and M. Watanabe, in *Proceedings of the Ninth International Conference on Solid State Ionics*, p. 899, B. V. R. Chowdari, H.-L. Too, G. M. Choi, and J.-H. Lee, eds., World Scientific Publishing Co., Singapore, 2004.
[473]C. A. Angell, W. Xu, J.-P. Belieres, and M. Yoshizawa, *World Patent*, **WO 2004/114,445 A1** (2004).
[474]R. F. de Souza, J. C. Padilha, R. S. Gonçalves, and J. Dupont, *Electrochem. Commun.* **5** (2003) 728.
[475]S. S. Sekhon, B. S. Lalia, J.-S. Park, C.-S. Kim, and K. Yamada, *J. Mater. Chem.* **16** (2006) 2256.
[476]S. S. Sekhon, P. Krishnan, B. Singh, K. Yamada, and C. S. Kim, *Electrochim. Acta* **52** (2006) 1639; A. Goto, Y. Kawagoe, Y. Katayama, and T. Miura, *Electrochemistry* **75** (2007) 231; H. Nakamoto, A. Noda, K. Hayamizu, S. Hayashi, H. Hamaguchi, and M. Watanabe, *J. Phys. Chem. C* **111** (2007) 1541.
[477]T. Tsuda, T. Nohira, Y. Nakamori, K. Matsumoto, R. Hagiwara, and Y. Ito, *Solid State Ionics* **149** (2002) 295.
[478]C. M. Lang, K. Kim, and P. A. Kohl, *Electrochem. Solid-State Lett.* **9** (2006) A545.
[479]*Lithium Batteries: Science and Technology*, G.-A. Nazri and G. Pistoia, eds., Kluwer Academic Publishers, Norwell, Massachusetts, USA, 2004.
[480]*Lithium Battery Technology*, H. V. Venkatasetty, ed., John Wiley and Sons, New York, 1984.

[481]H. Sakaebe and H. Matsumoto, "Application of Ionic Liquids to Li Batteries," in: *Electrochemical Aspects of Ionic Liquids*, p. 173, H. Ohno, ed., Wiley-Interscience, New Jersey, 2005.
[482]K. Ui, K. Ishikawa, T. Furuta, Y. Idemoto, and N. Koura, *Electrochemistry* **73** (2005) 120.
[483]K. Ui, T. Minami, K. Ishikawa, Y. Idemoto, and N. Koura, *Electrochemistry* **73** (2005) 279 (in Japanese).
[484]K. Ui, T. Minami, K. Ishikawa, Y. Idemoto, and N. Koura, *J. Power Sources* **146** (2005) 698.
[485]H. Nakagawa, S. Izuchi, K. Kuwana, T. Nukuda, and Y. Aihara, *J. Electrochem. Soc.* **150** (2003) A695.
[486]H. Zheng, H. Zhang, Y. Fu, T. Abe, and Z. Ogumi, *J. Phys. Chem. B* **109** (2005) 13676.
[487]J. Caja, T. D. J. Dunstan, D. M. Ryan, and V. Katovic, in: *Proceedings of the Twelfth International Symposium on Molten Salts*, p. 150, P. C. Trulove, H. C. De Long, G. R. Stafford, and S. Deki, eds., **PV99-41**, The Electrochemical Society, Inc., Pennington, NJ, 1999.
[488]J. Caja, T. D. J. Dunstan, and V. Katovic, in: *Proceedings of the Thirteenth International Symposium on Molten Salts*, p. 1014, P. C. Trulove, H. C. De Long, R. A. Mantz, G. R. Stafford, and M. Matsunaga, eds., **2002-19**, The Electrochemical Society, Inc., Pennington, NJ, 2002.
[489]S. Seki, Y. Kobayashi, H. Miyashiro, Y. Ohno, A. Usami, Y. Mita, M. Watanabe, and N. Terada, *Chem. Commun.* (2006) 544.
[490]S. Seki, Y. Kobayashi, H. Miyashiro, Y. Ohno, Y. Mita, A. Usami, N. Terada, and M. Watanabe, *Electrochem. Solid-State* **8** (2005) A577.
[491]B. Garcia, S. Lavallée, G. Perron, C. Michot, and M. Armand, *Electrochim. Acta* **49** (2004) 4583.
[492]J.-H. Shin, W. A. Henderson, G. B. Appetecchi, F. Alessandrini, and S. Passerini, *Electrochim. Acta* **50** (2005) 3859.
[493]K. Hayashi, Y. Nemoto, K. Akuto, and Y. Sakurai, *J. Power Sources* **146** (2005) 689.
[494]S. Seki, Y. Kobayashi, H. Miyashiro, Y. Ohno, A. Usami, Y. Mita, N. Kihira, M. Watanabe, and N. Terada, *J. Phys. Chem. B* **110** (2006) 10228.
[495]S. Seki, Y. Ohno, Y. Kobayashi, H. Miyashiro, A. Usami, Y. Mita, H. Tokuda, M. Watanabe, K. Hayamizu, S. Tsuzuki, M. Hattori, and N. Terada, *J. Electrochem. Soc.* **154** (2007) A173.
[496]M. Egashira, M. T.-Nakagawa, I. Watanabe, S. Okada, and H. Yamaki, *J. Power Sources* **160** (2006) 1387.
[497]J.-H. Shin, W. A. Henderson, C. Tizzani, S. Passerini, S.-S. Jeong, and K.-W. Kim, *J. Electrochem. Soc.* **153** (2006) A1649.
[498]J.-H. Shin, W. A. Henderson, S. Scaccia, P. P. Prosini, and S. Passerini, *J. Power Sources* **156** (2006) 560.
[499]M. Holzapfel, C. Jost, and P. Novák, *Chem. Commun.* (2004) 2098.
[500]T. Sato, T. Maruo, S. Marukane, and K. Takagi, *J. Power Sources* **138** (2004) 253.
[501]A. Chagnes, M. Diaw, B. Carré, P. Willmann, and D. Lemordant, *J. Power Sources* **145** (2005) 82.
[502]M. Diaw, A. Chagnes, B. Carré, P. Willmann, and D. Lemordant, *J. Power Sources* **146** (2005) 682.

[503]Y. Katayama, M. Yukumoto, and T. Miura, *Electrochem. Solid-State Lett.* **6** (2003) A96.
[504]H. Zheng, B. Li, Y. Fu, T. Abe, and Z. Ogumi, *Electrochim. Acta* **52** (2006) 1556.
[505]H. Sakaebe, H. Matsumoto, and K. Tatsumi, *J. Power Sources* **146** (2005) 693.
[506]P. C. Howlett, N. Brack, A. F. Hollenkamp, M. Forsyth, and D. R. MacFarlane, *J. Electrochem. Soc.* **153** (2006) A595.
[507]F. F. C. Bazito, Y. Kawano, and R. M. Torresi, *Electrochim. Acta* **52** (2007) 6427.
[508]Y. Wang, K. Zaghib, A. Guerfi, F. F. C. Bazito, R. M. Torresi, and J. R. Dahn, *Electrochim. Acta* **52** (2007) 6346.
[509]T. E. Sutto, P. C. Trulove, and H. C. De Long, *Electrochem. Solid-State Lett.* **6** (2003) A50.
[510]T. Kuboki, T. Okuyama, T. Ohsaki, and N. Takami, *J. Power Sources* **146** (2005) 766.
[511]Y. Zhang and M. U.-Macdonald, *J. Power Sources* **144** (2005) 191.
[512]D. Behar, C. Gonzalez, and P. Neta, *J. Phys. Chem. A* **105** (2001) 7607.
[513]A. Marcinek, J. Zielonka, J. Gębicki, C. M. Gordon, and I. R. Dunkin, *J. Phys. Chem. A* **105** (2001) 9305.
[514]D. Allen, G. Baston, A. E. Bradley, T. Gorman, A. Haile, I. Hamblett, J. E. Hatter, M. J. F. Healey, B. Hodgson, R. Lewin, K. V. Lovell, B. Newton, W. R. Pitner, D. W. Rooney, D. Sanders, K. R. Seddon, H. E. Sims, and R. C. Thied, *Green Chem.* **4** (2002) 152.
[515]D. Behar, P. Neta, and C. Schultheisz, *J. Phys. Chem. A* **106** (2002) 3139.
[516]J. Grodkowski and P. Neta, *J. Phys. Chem. A* **106** (2002) 5468.
[517]J. F. Wishart and P. Neta, *J. Phys. Chem. B* **107** (2003) 7261.
[518]L. Berthon, S. I. Nikitenko, I. Bisel, C. Berthon, M. Faucon, B. Saucerotte, N. Zorz, and P. Moisy, *Dalton Trans.* (2006) 2526.
[519]*Nuclear Wastes: Technologies for Separations and Transmutation*, N. C. Rasmussen, National Academy Press, Washington, 1996.
[520]S. Dai, Y. H. Ju, and H. Luo, in: *Proceedings of International George Papatheodorou Symposium*, p. 254, Patras Science Park, Greece (1999).
[521]S. Dai, Y. H. Ju, and C. E. Barnes, *J. Chem. Soc., Dalton Trans.* (1999) 1201.
[522]S. Chun, S. V. Dzyuba, and R. A. Bartsch, *Anal. Chem.* **73** (2001) 3737.
[523]M. L. Dietz and J. A. Dzielawa, *Chem. Commun.* (2001) 2124.
[524]M. P. Jensen, J. A. Dzielawa, P. Rickert, and M. L. Dietz, *J. Am. Chem. Soc.* **124** (2002) 10664.
[525]M. L. Dietz, J. A. Dzielawa, I. Laszak, B. A. Young, and M. P. Jensen, *Green Chem.* **5** (2003) 682.
[526]H. Luo, S. Dai, and P. V. Bonnesen, *Anal. Chem.* **76** (2004) 2773.
[527]H. Luo, S. Dai, P. V. Bonnesen, A. C. Buchanan, III, J. D. Holbrey, N. J. Bridges, and R. D. Rogers, *Anal. Chem.* **76** (2004) 3078.
[528]D. C. Stepinski, M. P. Jensen, J. A. Dzielawa, and M. L. Dietz, *Green Chem.* **7** (2005) 151.
[529]P. Vayssière, A. Chaumont, and G. Wipff, *Phys. Chem. Chem. Phys.* **7** (2005) 124.
[530]H. Heitzman, B. A. Young, D. J. Rausch, P. Rickert, D. C. Stepinski, and M. L. Dietz, *Talanta* **69** (2006) 527.
[531]H. Luo, M. Yu, and S. Dai, *Z. Naturforsch.* **62a** (2007) 281.
[532]P.-Y. Chen, *Electrochim. Acta* **52** (2007) 5484.

[533]N. Sieffert and G. Wipff, *J. Phys. Chem. B* **110** (2006) 19497.
[534]A. E. Visser and R. D. Rogers, *J. Solid State Chem.* **171** (2003) 109.
[535]A. E. Visser, M. P. Jensen, I. Laszak, K. L. Nash, G. R. Choppin, and R. D. Rogers, *Inorg. Chem.* **42** (2003) 2197.
[536]A. Chaumont and G. Wipff, *Phys. Chem. Chem. Phys.* **8** (2006) 494.
[537]M.-O. Sornein, C. Cannes, C. Le Naour, G. Lagarde, E. Simoni, and J.-C. Berthet, *Inorg. Chem.* **45** (2006) 10419.
[538]P. B. Hitchcock, T. J. Mohammed, K. R. Seddon, J. A. Zora, C. L. Hussey, and E. H. Ward, *Inorg. Chim. Acta* **113** (1986) L25.
[539]A. E. Bradley, J. E. Hatter, M. Nieuwenhuyzen, W. R. Pitner, K. R. Seddon, and R. C. Thied, *Inorg. Chem.* **41** (2002) 1692.
[540]M. Deetlefs, C. L. Hussey, T. J. Mohammed, K. R. Seddon, J.-A. van den Berg, and J. A. Zora, *Dalton Trans.* (2006) 2334.

3

Electrochemical Step Edge Decoration (ESED): A Versatile Tool for the Nanofabrication of Wires

Reginald M. Penner

Institute For Surface and Interface Science and Department of Chemistry, University of California, Irvine, CA 92697-2025

I. INTRODUCTION

The properties of nanowires composed of metals and semiconductors became a source of tremendous interest to materials chemists and physicists as soon as it became possible to make them – in the late 1980's. Among the first methods used for this purpose was electrodeposition into porous templates, such as ultrafiltration membranes (e.g., Nuclepore®) and anodic porous alumina (e.g., Anopore™)—a technique that would become known at template synthesis. This family of methods, pioneered by Martin,[1-8] Moskovits,[9-14] and Searson[15-21] can be used to prepare nanowires that were up to 50 μm in length and as small as 20 nm in diameter. Nanowires are formed in the pores of the template material in a *vertical* orientation, *perpendicular* to the plane of the electrode (Fig. 1a). The wire shape and diameter are predetermined by the corresponding properties of the pores present in the template.

R.E. White (ed.), *Modern Aspects of Electrochemistry, No. 45*, Modern Aspects of Electrochemistry 45, DOI 10.1007/978-1-4419-0655-7_3,

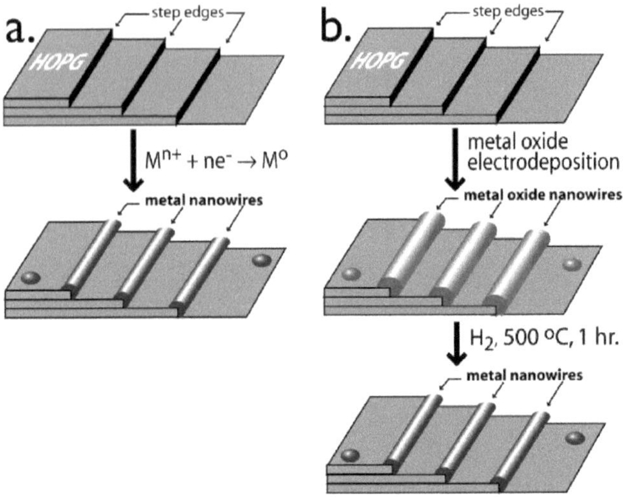

Figure 1. Schematic diagram of the preparation by ESED of metal nanowires by two methods. a) Direct metal electrodeposition at step edges, b) electrodeposition of metal nanowires in two steps involving, first, the electrodeposition of nanowires composed of a precursor material (e.g., MoO_2), and second, the reduction of this material to form the metal of interest (e.g., Mo^o).

Template synthesis is, by far, the most widely used method for fabricating nanowires but it has several serious limitations. First, the length of the nanowires that can be produced is limited by the thickness of the templates and for templates of all types, this limiting thickness is less than 50 μm. Second, with a few exceptions, template synthesis produces polycrystalline not single crystalline nanowires. Third, electrical contacts are difficult or impossible to establish to nanowires while they are confined within these templates because very little wire area is presented at the surface of the *filled* template. As luck would have it, when the nanowires are released from the templates by dissolution of the template material, electrical contacts are difficult to apply for entirely different reasons: the freed nanowires must be collected onto a surface, oriented, and anchored prior to the application of electrical contacts. This problem has stimulated some innovative ideas for *wir-*

ing the nanowires prepared by template synthesis but these methods have not become widely adopted because of their complexity. The difficulty associated with electrically contacting template-synthesized nanowires has impeded their use in devices that require electrical contacts for their function, including transistors, chemical sensors, and photonic devices.

Step edge decoration, pioneered by Himpsel,[22-25] Kern,[26] and others, is a horizontal, rather than a vertical templating method that involves the exploitation of step edges on single crystalline substrates as templates for the formation of nanowires. In the prior work cited above, the deposition method is physical vapor deposition (PVD) in vacuum and the deposition flux and the substrate temperature were adjusted to achieve nucleation of the deposited material at step edges to produce continuous *nanowires* of varied width. The nanowires produced by PVD were very thin, with a thickness in the direction perpendicular to the surface of just one or at most two atomic layers. Such nanowires could not be removed from the surfaces on which they were deposited and the electrical characterization of these nanowires was not reported. Behm[27,28] demonstrated the principle that individual atomic scale defects on a pristine, reconstructed Au(111) surface could be selectively decorated by the electrodeposition of metals at low overpotentials, but continuous metal nanowires were not obtained in these experiments.

The decoration of step edges on HOPG surfaces by electrodeposited silver or platinum nanoparticles was considered a nuisance in our experiments,[29-36] carried out from 1992 to 2000, aimed at preparing highly size monodisperse dispersions of metal nanoparticles because the diameter of the particles that nucleated at step edges were invariably smaller than those that nucleated on the graphite terraces between them. But in 1999, Mike Zach, a graduate student in my laboratory, showed that the propensity of step edges to nucleate material could be exploited to synthesize highly uniform nanowires composed of MoO_2, an electronically conductive oxide of molybdenum.[37] This constituted the first example of nanowire preparation using electrochemical step edge decoration or ESED and it is the jumping off point for the discussion of ESED in Section II below.

Highly oriented pyrolytic graphite (HOPG) holds the key to ESED: it is the only electrode material that has been successfully used for nanowire growth using this method. The HOPG (0001) plane has three properties that aid the ESED process. First, this surface has an extremely low surface free energy (35 dynes cm^{-1})[38] that insures that the electrodeposition of material will occur via a Volmer-Weber growth mode[39] involving prompt 3D growth. Second, it is characterized by a highly regular and predictable defect structure consisting of 0.1-1 mm diameter (for high quality ZYA or ZYB grade crystals) domains of approximately linear step edges that are oriented approximately parallel with one another. These domains are the exposed faces of columnar grains with axes oriented along (0001) that are fused to form this material. In other words, each grain present on the HOPG basal plane encompasses a set of thousands of linear step edges with total length 0.1-1.0 mm, on average. These step edges do not cross the grain boundaries: between grains are disordered regions where ordered step edges are rarely observed. Thus, the diameter of these grains fixes the maximum length of the nanowires that can be formed using the ESED method at ≈ 1 mm. Finally, these step edges are capable of catalyzing electron transfer to electron donor and acceptor species present in a contacting liquid. This was the conclusion of work carried out by McCreery and coworkers[40-43] who investigated the effect of defects on the observed rates of electron transfer at graphite electrodes. They also demonstrated that defect-free regions of the basal plane, in contrast, were highly unreactive electrochemically. The implication is that the linear step edges found on the HOPG basal plane act like linear microelectrodes. The pronounced electrochemical anisotropy of the HOPG surface further promotes the localization of an electrodeposited material at step edge defects.

II. METALS

Although MoO_2 nanowires were the first to be prepared using ESED,[37] their synthesis (described in Section III.1 below) is exceptional in many ways. From a purely electrochemical perspective, a much simpler electrodeposition reaction involves the 1 or 2 electron reduction of a hydrated metal ion to form a metal:

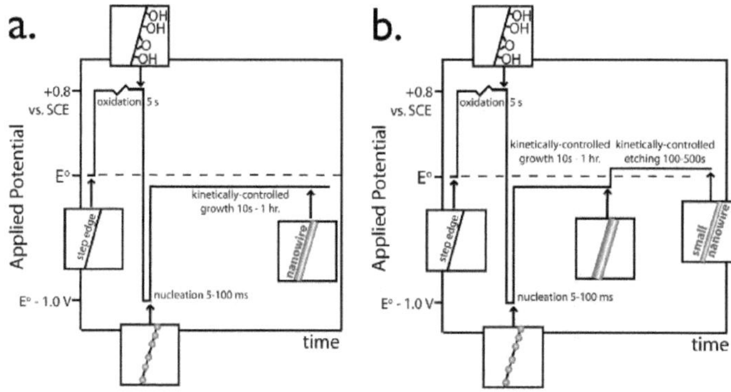

Figure 2. Potential programs used for the direct electrodeposition of metal nanowires, as shown in Fig. 1b. a) Three-step "canonical" program involving oxidation of the HOPG surface (+0.80 V × 5 s), nucleation (-1.0 V vs. $E^{o}_{Mn+/Mo}$ × 10-100 ms), and growth [(–100)-(–10 mV) vs. $E^{o}_{Mn+/Mo}$ × 10-1000 s]. b) Four step deposition program for metal nanowires incorporating, in addition to steps 1-3 in (a) nanowire thinning by kinetically-controlled electrooxidation (+10-100 mV vs. $E^{o}_{Mn+/Mo}$ × 10-1000 s).

$M^{n+}_{aq} + n\,e^{-} \rightleftharpoons M^{o}$. The ESED method used to prepare nanowires of M^{o} is the same for many different metals and can be considered to be canonical.

1. Canonical ESED of Metal Nanowires

The distribution of a metal electrodeposit produced on an HOPG cathode is strongly influenced by the potential program used to produce it and by the pretreatment received by the HOPG surface prior to electrodeposition. Three voltage pulses are applied in rapid succession to effect the growth of metal nanowires (Fig. 2a):

1. First, a +0.8 V vs. SCE × 5 s pulse is applied to mildly oxidize step edges.
2. Then a large, negative voltage pulse (–1.0 V vs. $E^{o}_{Mn+/Mo}$ × 10-100 ms) is applied to create metal nuclei on the HOPG surface—both at step edges and on terraces.
3. Finally, the growth of these nuclei is continued at a much slower pace [(–10)-(–100) mV vs. $E^{o}_{Mn+/Mo}$ × 1-1000 s].

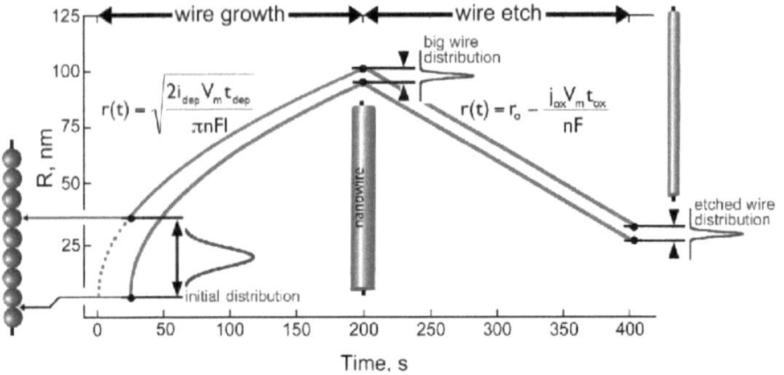

Figure 3. Schematic depiction of the dispersion of nanowire diameters as a function of growth time. The three step program depicted in Fig. 2a (and terminating at 200 s in this figure) causes a smoothing of the nanowire as its mean diameter increases because constrictions in the nanowire grow at a faster rate than bulges. The addition of a fourth voltage pulse, kinetically-controlled electrooxidation (Fig. 2b) causes a reduction in the mean diameter of the nanowire without inducing the roughening of its surface. Reprinted with permission from Ref. 80. Copyright (2006) American Chemical Society.

The original objective was to achieve kinetically-controlled growth in this time regime and this would be characterized by a current that increases in proportion to the area of the growing nanostructures. In practice, however, we have found that the current stabilizes at a pseudo-constant value after approximately 10 s, suggesting the presence of a transport limitation to the growth rate. This observation is general for the deposition of many different materials and this unexpected circumstance is fortuitous because the resulting growth law for nanowires is given by:

$$r(t) = \sqrt{\frac{2i_{dep}t_{dep}V_m}{\pi nFl}} \tag{1}$$

Equation (2) is an example of a convergent growth law – it predicts that the nanowire radius grows in proportion to $t^{1/2}$ resulting in a gradual smoothing of the nanowire as dr/dt is larger for constrictions than for bulges in the nanowire (Fig. 3). This effect is analogous to the convergent evolution of the diameter distribution for colloidal nanoparticles growing under conditions of diffusion control.[44]

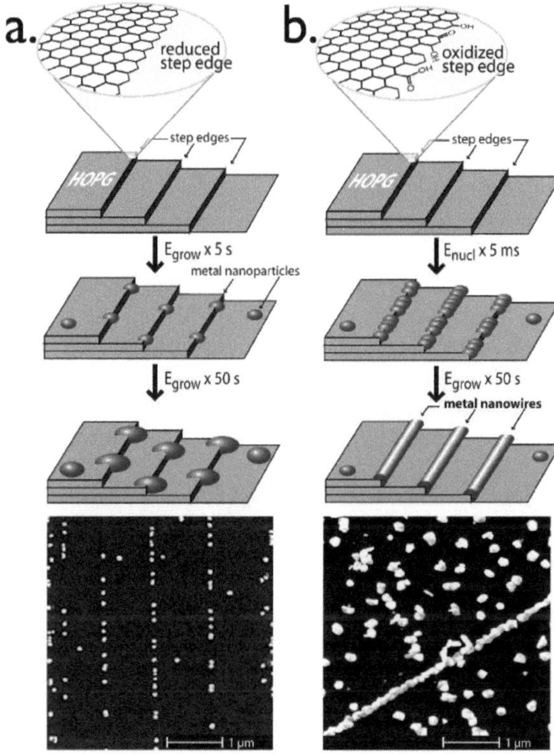

Figure 4. The effect of mild electrooxidation of the graphite surface is to increase the nucleation density at step edges for virtually all electrodeposited materials (b). In the absence of this oxidation step, the nucleation density at steps is insufficient to permit the formation of nanowires for many materials (a).

The effect of the oxidation pulse (Fig. 2a) is to increase the nucleation density along step edges relative to the nucleation density on terraces (Fig. 4b). If this oxidation step is omitted, a lower nucleation density at step edges is obtained which can be too low for the eventual formation of nanowires in the growth phase of this process (Fig. 4a). We discuss a technique for reducing the diameter of nanowires (as shown in Figs. 2b and 3) in Section IV of this chapter.

2. Coinage and Noble Metals

Nanowires composed of noble and coinage metals, including nickel, copper, silver, gold, platinum and palladium are readily prepared using the procedure depicted in Fig. 2a. Typical data for copper is shown in Fig. 5. Cyclic voltammograms for copper show that the onset for copper electrodeposition on the second voltammetric scan is shifted positively from that seen at the first scan. This is the general case seen for many different metals and it results from the incomplete removal of copper nuclei on the first positive-going stripping scan. Thus, the shift seen between the first and subsequent scans provides a good approximation of the nucleation overpotential for the metal in question on HOPG. In the ESED experiment, a growth potential within this interval is always selected to avoid further nucleation of copper on the HOPG surface (after the nucleation pulse) while providing for the growth of the preexisting nuclei (Fig. 5a). Suitable values for the nucleation potential and the oxidation potential are also indicated in Fig. 5a. The former is always selected based upon trial and error, and the later is usually $\approx +0.8$ V vs. SCE.

Equation (1) predicts that the nanowire diameter will be proportional to the (deposition time)$^{1/2}$ and this proportionality is usually recovered experimentally (Fig. 5b).[45-50] Scanning electron micrographs (SEMs) of copper nanowires clearly show the polycrystalline morphology of these wires. In the case of copper, the grain diameter is close to the wire diameter, and this is the generally observed case when complexing agents have not been added to the plating solution. The addition of additives such as ethylenediaminetetraacetic acid (EDTA) or saccharine produces a smaller grain diameter, in the 2-10 nm range, and a smoother nanowire morphology, in general. These additives are therefore useful particularly for achieving the smallest possible nanowire diameters using the ESED method.

3. Base Metals

Like noble and coinage metals, the base metals (Cd, Mo, Pb, etc.) may be produced by direct electrodeposition by direct electrodeposition (Fig. 1a), but after deposition, these metals spontaneously oxidize while in contact with water. Depending on the pH and the

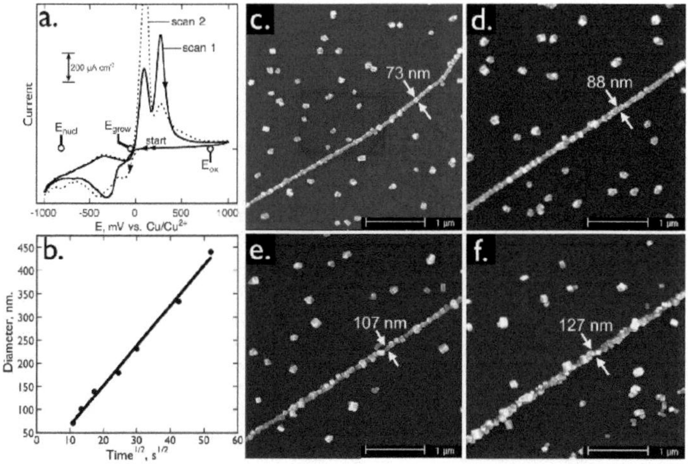

Figure 5. Typical results for the preparation by ESED of copper nanowires: (a) Cyclic voltammograms of a copper plating solution at an HOPG electrode showing the values for E_{ox}, E_{nucl}, and E_{grow} typically used for the preparation of copper nanowires. The solution was aqueous 2.0 mM $CuSO_4$*5 H_2O, 0.1 M Na_2SO_4. Two oxidation waves seen at +100 mV and +350 mV are assigned to copper stripping (+100 mV) and oxidation of Cu^o to CuO (+350 mV). (b) The effect of mild electrooxidation of the graphite surface is to increase the nucleation density at step edges for virtually all electrodeposited materials. Nanowire diameter versus (deposition time)$^{1/2}$ for the growth of copper nanowires. Each series of experiments for a particular metal were performed using a single graphite crystal in order to limit the variation in the step edge density from experiment to experiment (see Eq. 1). (c-f) Scanning electron microscope (SEM) images of copper nanowires prepared using different growth times ranging from 120 s (c) to 600 s (f).

particular metal involved, the product of this oxidation is either metal ions or an insoluble metal oxide. This oxidation does not occur while the incipient nanowire is cathodically protected during its electrodeposition, it commences upon release of the potential to open circuit. The final state of the nanowire is unpredictable under these circumstances. This problem is circumvented using the strategy shown in Fig. 1b. Nanowires of a metal oxide are first electrodeposited either cathodically or anodically, and these nanowires are then reduced in hydrogen at elevated temperature to produce nanowires of the parent metal.

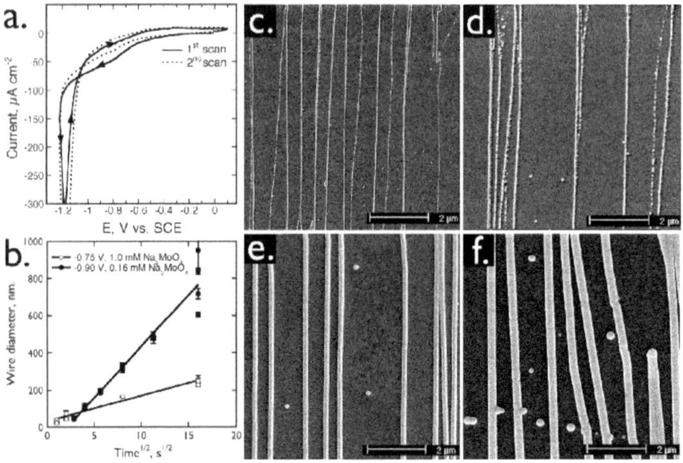

Figure 6. (a) Cyclic voltammogram at 20 mV s^{-1} for an HOPG working electrode in an aqueous plating solution containing 1 mM Na_2MoO_4, 1.0 M NaCl, 1.0 M NH_4Cl and adjusted to pH 8.5 with the addition of aqueous NH_3. (b) Plot of the MoO_2 nanowire diameter as a function of the square root of the deposition time. Linear plats, in accordance with Eq. (1), are obtained for two different sets of growth conditions (MoO_4^{2-} concentration and deposition potential, as indicated). (c-f) MoO_2 nanowires of various diameters prepared by varying the growth time as follows: 8 s (c), 16 s (d), 32 s (e), 64 s (f). All wires shown here were prepared at –0.95 V vs. SCE from a solution containing 7.0 mM MoO_4^{2-}, 1.0 M NaCl, 1.0 M NH_4Cl.

Our most successful implementation of this scheme has involved the production of molybdenum nanowires. Nanowires of MoO_2 are first electrodeposited from a basic aqueous electrolyte containing MoO_4^{2-}, according to:

$$MoO_4^{2-} + 2\,H_2O + 2\,e^- \rightleftharpoons MoO_2 + 4\,OH^- \qquad (2)$$

Subsequent hydrogen gas reduction produces nanowires of Mo:

$$MoO_2 + 2\,H_2 \rightleftharpoons Mo^0 + H_2O \qquad (3)$$

A second motivation for pursuing this approach in the case of molybdenum is the simple fact that there are no known methods for directly electroplating molybdenum metal, to our knowledge.

The formation of the MoO_2 precursor material, in this case, occurs via an irreversible reduction (Fig. 6a). Virtually any potential at the foot of the deposition wave produces nanowires on the HOPG surface, even when the nucleation pulse (Fig. 2a) is omitted. These wires increase in size in proportion to $t_{dep}^{1/2}$, as expected based upon Eq. (1) (Fig. 6b), and the resulting nanowires have a nearly perfect, hemicylindrical shape (Fig. 6c-f).[37,51]

The reduction of MoO_2 to Mo in flowing H_2 at 500 °C for one hour is accompanied by a significant change involve because the molar volume of Mo is much smaller than that of MoO_2 (9.4 moles cm^{-3} vs. 19.8 moles cm^{-3}) leading to shrinkage of the wires in length and diameter during the reduction process. This contraction is large enough to be visible in the SEM images shown in Fig. 7. The completeness of the reduction process is readily verified by x-ray photoelectron spectroscopy (XPS, Fig 7c-e). A key attribute of this approach, as with all ESED nanowire depositions, is the tremendous length of the nanowire arrays that are obtained (Fig. 8). Nanowires that are continuous for lengths of more than a millimeter are frequently observed.

A requirement for this approach is that the metal of interest has an electronically conductive oxide that can be electrodeposited. Other metals for which this *electrochemical/chemical* approach is practical include copper (using Cu_2O as an intermediate), silver (Ag_2O), and manganese (MnO_2), but there are surely many other candidates.

4. Nanowire Nucleation with *Seeds* Prepared Using Physical Vapor Deposition (PVD)

The minimum diameter for metal nanowires prepared using either of the ESED methods depicted in Fig. 1 is in the 70-90 nm range depending on the metal. In the case of some metal oxides including MoO_2, nanowires as small as 20 nm have been obtained.[37, 51] What determines the minimum diameter? The answer is apparent when one considers the mechanism of growth for nanowires prepared by ESED. After the nucleation of metal (Fig. 2a), one has a linear ensemble of metal nanoparticles arrayed along each step edge on the HOPG surface (Fig. 9). These 1-D particle ensembles evolve into a nanowire by coalescing into a continuous metal

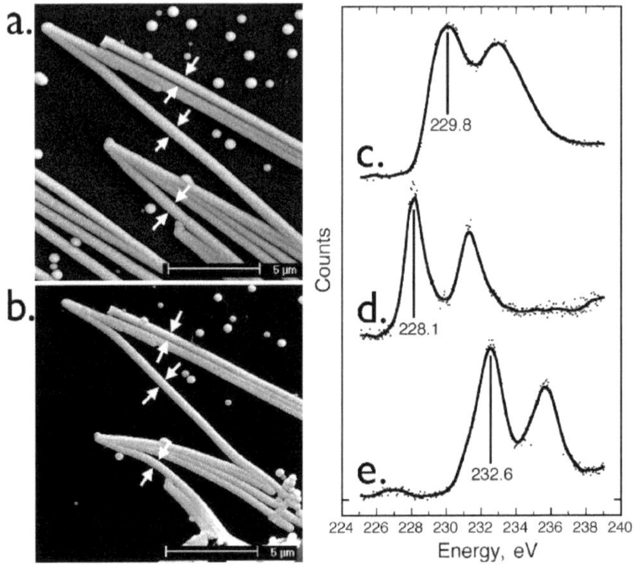

Figure 7. SEM micrographs of electrodeposited nanowires before (a) and after (b) reduction at 500°C for one hour. Arrows indicate the diameter of three wires in each image. Before reduction, this diameter for all three wires was 570 nm; after reduction the diameter is 460 nm. The graphs on the right show X-ray photoelectron spectra (Mo 3d region only) of graphite surfaces decorated with electrodeposited nanowires. All three samples were deposited for 256 s from a solution containing 2.6 mM MoO_4^{2-} (1M NaCl, 1M NH_4Cl, pH = 8.0-8.5pH) for 256 s and consisted of nanowires approximately 400 nm in diameter. The binding energy of the Mo $3d_{5/2}$ peak is indicated for each spectrum. (c) Immediately following electrodeposition a Mo $3d_{5/2}$ binding energy consistent with MoO_2 is observed. (d) After reduction of freshly deposited nanowires in hydrogen at 650°C for 3 hours a Mo $3d_{5/2}$ binding energy characteristic of molybdenum metal is seen. (e) After exposure to a laboratory air ambient for approximately 5 days.

structure as each individual particles increase uniformly in size. This mechanism leads to the conclusion that the minimum nanowire diameter, dia_{min}, is inversely proportional to the density of nuclei present on the step edges, δ:[52]

$$dia_{\min} = \frac{1}{\delta} \tag{4}$$

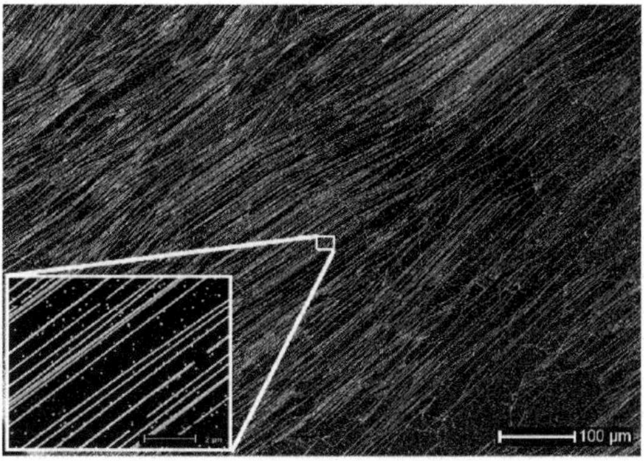

Figure 8. Low magnification SEM image of MoO_2 nanowires showing the parallel organization of step edges on the HOPG surface, and the tremendous length of these nanowire arrays.

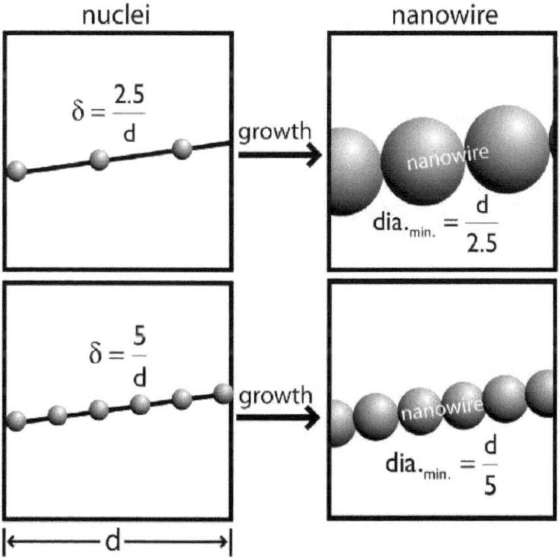

Figure 9. Schematic diagram illustrating the relationship between nucleation density, δ, and the minimum nanowires diameter, dia_{min}. As shown here, increasing the nucleation density by a factor of two reduces dia_{min} by one-half.

In the specific case of gold, the highest values for δ seen in purely electrochemical gold nanowire growth experiments is in the 10-13 μm^{-1} range which coincides with the minimum dimensions of 80-100 nm seen in these experiments.[49] These nucleation densities are obtained using the procedure shown in Fig. 2a and we have not discovered other electrochemical or chemical means to enhance δ beyond those summarized by this scheme.

A desire to prepare still smaller gold nanowires using the ESED method motivated us to look at the possibility of creating nuclei at step edges using physical vapor deposition (PVD).[52] After a systematic investigation of the conditions of gold flux and sample temperature that were optimal for this purpose, surface like those shown in Fig. 10a were obtained by thermally evaporating 4 nm of gold at a rate of ~0.3 nm/s onto HOPG surfaces heated to 673 K. These conditions lead to the highly selective decoration of step edges and other defects on the HOPG surface, with virtually no extraneous gold deposition on terraces—a far higher degree of defect selectivity that has been seen for gold[49] using the purely electrochemical ESED method of Fig. 2a. The δ values obtained using this approach were 30-40 μm^{-1} –much higher than is achievable for the electrochemical nucleation of gold using potentiostatic pulses.[47,50] However, we were not able to identify conditions of gold flux and sample temperature that permitted the preparation of nanowires by PVD alone.[52]

Instead, we substituted PVD gold deposition for the first two steps of the canonical ESED procedure (Fig. 2a) and we then electrodeposited additional gold with the goal of obtaining continuous gold nanowires using this hybrid approach. Cyclic voltammetry for clean (Fig. 10b, top) and PVD-decorated HOPG surfaces (Fig. 10b, bottom) showed that the PVD gold nuclei catalyzed the electrodeposition of gold (Fig. 10b), shifting the onset for gold electrodeposition positively by 300 mV. Using a deposition potential of 640 mV vs. SCE, we found that electrodeposited gold selectively decorated the preexisting PVD gold nuclei thereby producing nanowires with minimum diameters in the 70-90 nm range and lengths of 1.0 μm or more. In fact, these minimum dimensions are similar to those obtained in the purely electrochemical ESED grow of gold nanowires—a somewhat disappointing outcome. Higher δ values obtained using PVD do not lead immediately to smaller nanowires because of fluctuations in the nucleation density that

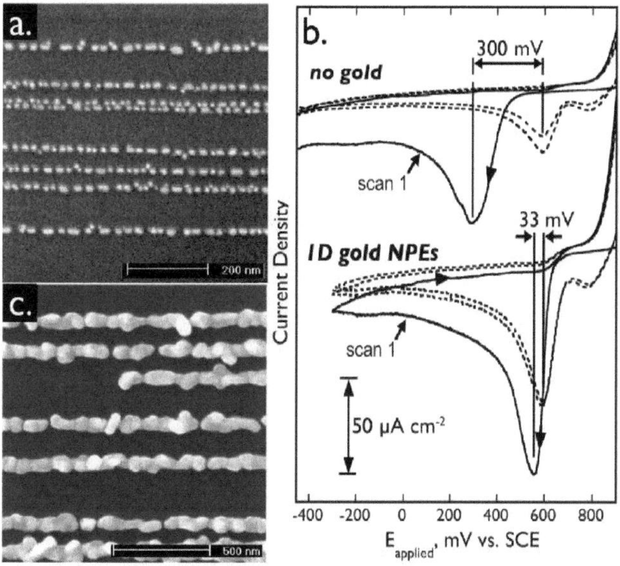

Figure 10. (a) SEM image of 1D Au NPEs prepared by depositing 4 nm of gold at 673 K. The particle diameter is in the 2-15 nm range. (b) Cyclic voltammograms (3 scans) in aqueous 0.2 mM Au(III)Cl_3, 0.1 M NaCl at 20 mV s^{-1} for two electrodes: freshly cleaved HOPG (top) and HOPG on which 3.3 nm of gold was deposited at 673K (bottom). In each case, the first voltammetric scan is shown with solid lines and the dashed lines indicate the second and third. (b) Current versus time transients at five deposition potentials, as indicated, for HOPG on which 3.3 nm of gold was deposited at 673K. (c) SEM image of 1D Au NPEs after the potentiostatic electrodeposition of gold for 150 s from aqueous 0.2 mM Au(III)Cl_3, 0.1 M NaCl at 640 mV (vs. SCE), $Q_{Au} = 3.1$ mC.

lead to gaps—locally lower values of δ—along the nanowires and these can only be filled by adding more gold thereby increasing the overall diameter of these wires over their entire lengths.

III. COMPOUND SEMICONDUCTORS

Research on semiconducting nanowires has concentrated on single crystalline wires and comparatively little work has been carried out on the synthesis and characterization of polycrystalline semiconductor nanowires. The exception to this generalization is the large body of work pertaining to template-synthesized nanorods already referenced earlier. Many unresolved questions pertaining to polycrystalline nanowires persist: Is the carrier mobility high enough to permit applications such as photoconductive switching and transistors? Is it possible that lower carrier lifetimes actually improve the performance of polycrystalline nanowires relative to single crystalline nanowires of the same material? Can such nanowires produce clean photoluminescence and electro-luminescence emission spectra, characterized by emission at the bandgap?

Two different approaches—analogous to those shown in Fig. 1—can be used to obtain nanowires composed of compound semiconductors. The first approach is to electrochemically synthesize the material of interest directly at step edges (analogous to Fig. 1a). The second approach is to deposit nanowires of a precursor material, and then react these nanowires *in-place* at elevated temperatures with a gas phase reagent to effect the chemical conversion to the desired material, hopefully retaining the nanowire morphology (analogous to Fig. 1b). We have employed both of these approaches to obtain semiconducting nanowires using the ESED method.

1. Stoichiometric Electrodeposition of CdSe and Bi_2Te_3 Nanowires using Cyclic Electrodeposition-Stripping in Concert with ESED

The electrodeposition of CdSe and most other binary semiconductors is complicated by the fact that elemental excesses of either component—Cd and Se—can be electrodeposited from the same plating solution. In fact, in a negative-going voltammetric scan, the onset for Cd and Se deposition coincides with the deposition of CdSe. Under these circumstances, how can one insure that stoichiometric CdSe is obtained? Cyclic electrodeposition/stripping is a strategy for circumventing this problem, and it is effective for a variety of different materials, deposited as nanowires, nanoparticles, or films.

In the specific case of CdSe, cyclic electrodeposition/stripping works as follows.[53] First, CdSe and a stoichiometric excess of cadmium and some elemental selenium are electrodeposited in a negative-going voltammetric scan. These first two reactions occur together with an onset near –0.80 V vs. SCE, while selenium deposition occurs somewhat positive of these with an onset at –0.60 V. The three reactions occurring concurrently at potentials negative of –0.80 V are:[53]

$$H_2SeO_3 + Cd^{2+} + 6\,e^- + 4\,H^+ \rightarrow CdSe + 3\,H_2O \qquad (5)$$

$$Cd^{2+} + 2\,e^- \rightleftharpoons Cd^o \qquad (6)$$

$$H_2SeO_3 + 6\,H^+ + 4\,e^- \rightleftharpoons Se^o + 2\,H^+ + 3\,H_2O \qquad (7)$$

At a negative limit of –1.0 V, the voltammetric scan is reversed and on the positive-going scan, reactions 6 and 7 occur in reverse, removing excess cadmium and selenium from the nascent CdSe electrodeposit. Over this range of potentials, the enthalpy of formation for CdSe (ΔH_F = –163 kJ mol^{-1}) imparts stability to the CdSe formed on the first negative-going scan relative to Cd^o and Se^o allowing these two *contaminants* to be selectively and quantitatively removed. The precise potentials for the processes indicated in Eqs. (5)-(7) are shown in Fig. 11. Reversal of the positive going scan at +1.0 V vs. SCE (Fig. 11) provides for the selective removal of both Cd^o and Se^o but it avoids CdSe stripping (the reverse of reaction 6). This deposition-stripping cycle is repeated as necessary to build up a stoichiometric CdSe deposit of the required thickness (Fig. 12).[53] In the case of the ESED experiment, CdSe nanowires with diameters from 50 to 300 nm can be prepared using this approach.[53] These nanowires are nanocrystalline with grain diameters in the 7 to 25-nm range, based on analysis by transmission electron microscopy.

The nanowires shown in Fig. 12 were obtained without applying a nucleation pulse, as shown in Fig. 2. Why then does nucleation occur with such high density along step edges in this experiment, which involves purely cyclic voltammetry? The answer can be seen in Fig. 11a: at the foot of the CdSe/Se/Cd deposition wave, near –0.6 V, a *pre-wave* is observed that corresponds to the electrodeposition of elemental selenium according to reaction 7. For reasons which are not entirely clear, this selenium deposition occurs with a very high degree of selectivity at step edges (data not shown),[53] and as the scan proceeds to more negative potentials, these selenium *seeds* then serve to efficiently nucleate CdSe, Se, and Cd.

The same cyclic electrodeposition/stripping strategy can be used to obtain nanowires of bismuth telluride (Bi_2Te_3), a material of technological importance for thermoelectric power generation and cooling/heating. Once again, Bi_2Te_3 electrodeposition (reaction 8) occurs concurrently with the deposition of excess elemental bismuth (reaction 9) and tellurium (reaction 10) (see Fig. 13b):

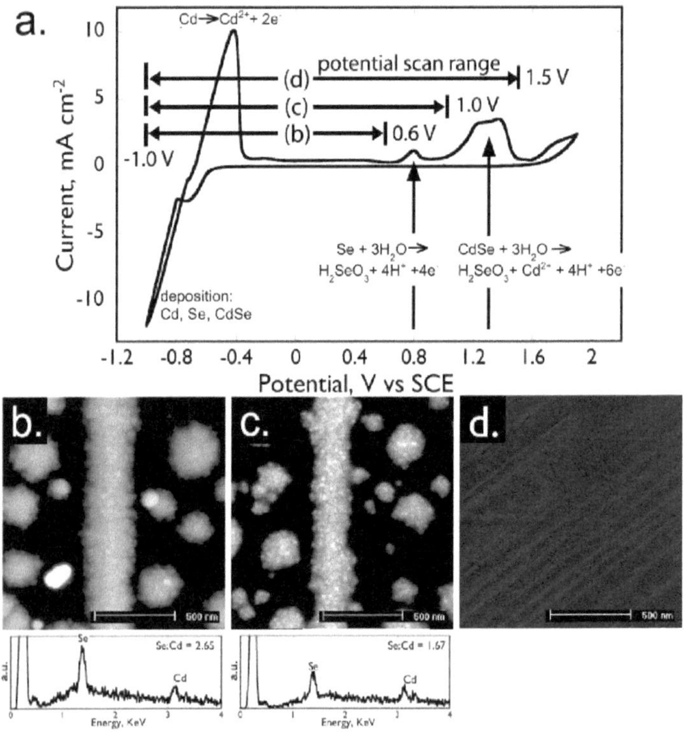

Figure 11. (a) Survey CV of a 120 mM $CdSO_4$/ 1m M SeO_2 solution at pH = 2.70. (b) SEM image of HOPG electrode after first scanning to –1V, then scanning back to 0.6 V for 3 cycles (inset: EDX spectrum); (c) SEM image of HOPG electrode after first scanning to –1V, then scanning back to 1 V for 3 cycles (inset: EDX spectrum); (d) SEM image of HOPG electrode after first scanning to –1 V, then scanning back to 1.5 V for 3 cycles.

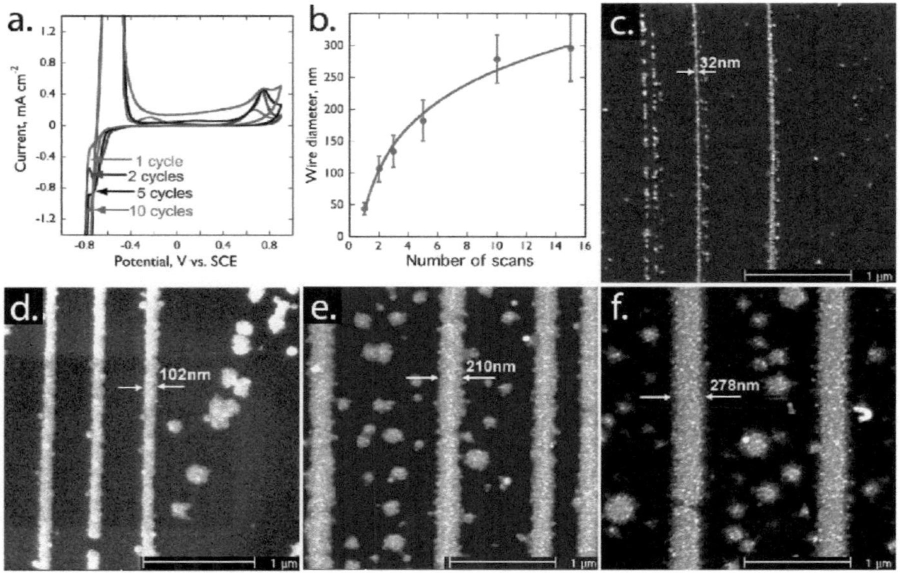

Figure 12. (a) CVs at 20 mV s^{-1} between –1.0 V and 0.8 V in aqueous 100 mM $CdSO_4$ and 1mM SeO_2 aqueous solution, pH = 2.7. (b) Plot of the CdSe nanowire diameter measured by SEM as a function of the number of deposition/stripping cycles. (c-f) SEM images of CdSe nanowires prepared using 1 (c), 2 (d), 5 (e), and 10 (f) electrodeposition/stripping cycles.

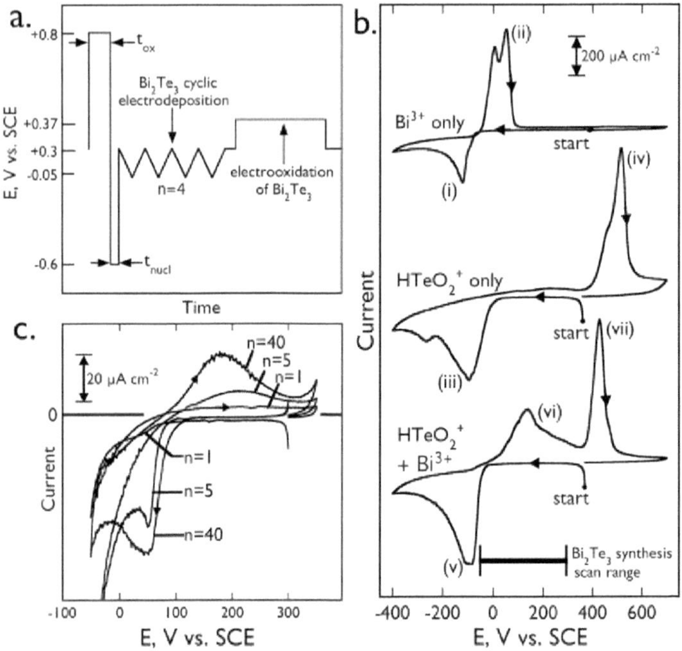

Figure 13. (a) Program used for the synthesis and electrooxidation of Bi_2Te_3 nanowires on HOPG electrodes. The electrodeposition of Bi_2Te_3 nanowires involves three steps: 1) Potentiostatic oxidation of step edges on the HOPG surface at E_{ox} = 0.80 V vs. SCE for a time, t_{ox} = 5 s; 2) Potentiostatic nucleation of Bi_2Te_3 at E_{nucl} = –0.75 V for t_{nucl} = 5 ms; and, 3) Cyclic electrodeposition of stoichiometric Bi_2Te_3 involving n potential cycles at 20 mV s^{-1} between a positive limit of $E_{(+)}$ (+0.35 V) and a negative limit of $E_{(-)}$ (–0.05 V). Electrooxidation adds a fourth step: potentiostatic oxidation at +0.37 V. b) Cyclic voltammograms at 20 mV s^{-1} for an HOPG electrode in contact with three different aqueous plating solutions. Top: 1.5 mM $Bi(NO_3)_3$ in 1 M HNO_3; middle: 1.0 mM TeO, in 1 M HNO_3; bottom: 1.5 mM $Bi(NO_3)_3$ and 1.0 mM TeO, in 1 M HNO_3. (c) CVs acquired during the growth of Bi_2Te_3 nanowires using cyclic electrodeposition/stripping. The CV is shown at scans 1, 5 and 40 as indicated.

$$2\,Bi^{3+} + 3\,HTeO_2^{+} + 9\,H^{+} + 18\,e^{-} \rightarrow Bi_2Te_3 + 6\,H_2O \qquad (8)$$

$$Bi^{3+} + 3\,e^{-} \rightarrow Bi^{o} \qquad (9)$$

$$HTeO_2^{+} + 3\,H^{+} + 4\,e^{-} \rightarrow Te + 2\,H_2O \qquad (10)$$

In contrast to the case of CdSe, however, it is possible to strip excess bismuth from the nascent deposit but not excess tellurium because its stripping potential is positive of that for Bi_2Te_3 itself (Fig. 13b). This means that the composition of the deposition solution must be skewed to insure that very little elemental Te is produced on the negative scan. In this case a nucleation pulse was applied before the cyclic electrodeposition/stripping is initiated (Fig. 13a and 13c) in order to as usual to produce a dense linear array of nuclei on step edges. Bi_2Te_3 nanowires such as those shown in Figs. 14a-14c are obtained. As always, the crystal structure must be confirmed by diffraction analyses and both x-ray diffraction and electron diffraction can be used for this purpose (Figs. 14d and 14e). Compositional analysis including purity and stoichiometry are confirmed using X-ray photoelectron spectroscopy (data not shown).

2. Electrochemical/Chemical Synthesis of CdS and MoS_2 Nanowires

Often electrodeposition-stripping cannot be employed for various reasons. In the case of metal sulfides such as CdS[54] and MoS_2,[55,56] an alternative is to use *electrochemical/chemical* or *E/C* synthesis that involves the electrodeposition of nanowires composed of a precursor material (Cd^{o} or MoO_2) and then exposing the precursor nanowires to H_2S at elevated temperature.

In the case of MoO_2, the electrodeposition and conversion reactions are:[55,56]

$$MoO_4^{2-} + 3\,H_2O + 2\,e^- \rightarrow MoO_2 + 4\,OH^- \tag{11}$$

$$MoO_2 + 2\,H_2S \rightarrow MoS_2 + 2\,H_2O \tag{12}$$

Reaction 12 is carried out at 800-1000 °C in order to achieve grain growth of the MoS_2 as it is formed. Durations of up to 72 hrs. were required to completely convert nanowires of MoO_2 (e.g., Fig. 6) to MoS_2 within this temperature range. As is apparent in Fig. 15, reaction 12 is accompanied by a radical restructuring of the nanowires as the isotropic MoO_2 crystal structure is transformed into

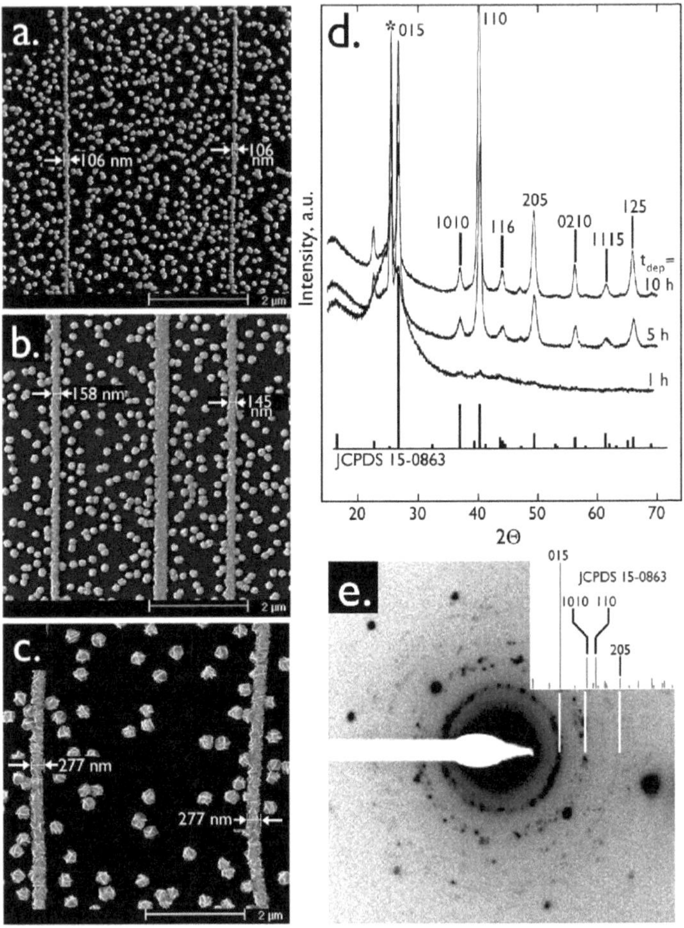

Figure 14. (a-c) Scanning electron micrographs (SEMs) of three Bi_2Te_3 nanowire samples prepared using 10, 20, and 50 electrodeposition/stripping scans. (d) Powder X-ray diffraction for three Bi_2Te_3 samples prepared by cyclic electrodeposition/stripping for 1 hr., 5 hrs. and 10 hrs. The 5 and 10-hr. samples were continuous films of Bi_2Te_3 whereas the 1-hr. sample consisted of wires with a mean diameter of approximately 500 nm. (e) Selected area electron diffraction pattern for a Bi_2Te_3 nanowire. White lines mark the expected position of diffractions for rhombohedral Bi_2Te_3, JCPDS card 15-0863, as shown. Diffractions from the crystalline graphite substrate are also indicated.

relatively large, platelet-like MoS_2 crystallites (Fig. 15, right). This transformation can be followed by SEM (Fig. 15).

The quality of the resulting nanowires is remarkable. As just one example, Fig. 16 shows absorption spectra for MoS_2 nanowires. The indirect and direct bandgaps of this material are clearly seen and two exciton transitions just above the near the direct bandgap (labeled A1 and B1) are well resolved. The energy of these two transitions shifts to higher values as the thickness of the MoS_2 nanowires is reduced (Figs. 16b and 16c), quantitatively as expected based upon carrier confinement.

3. Oxides

Two reasons exist for electrodepositing nanowires of oxides:

a) The first is to prepare nanowires of a precursor material that is to be chemically converted to a target material of interest (e.g., reaction 12).
b) The second reason is because the oxide material itself is of interest.

We have found that metal oxides are among the most forgiving materials in terms of adapting their electrodeposition to the ESED method. This means that the step edge-selective electrodeposition of metal oxides can usually be carried out potentiostatically, without using the preoxidation and nucleation pulses shown in Fig. 2.

The oxides we have deposited as nanowires and the corresponding half-reactions for each are the following:

AgO[57] $$Ag^+ + H_2O \rightarrow AgO + 2\,H^+ + 1\,e^- \quad (13)$$

Cu_2O $$2\,Cu^{2+} + 2\,e^- + H_2O \rightarrow Cu_2O + 2\,H^+ \quad (14)$$

MnO_2[58] $$MnO_4^- + 2\,H_2O + 3\,e^- \rightarrow MnO_2 + 4\,OH^- \quad (15)$$

MoO_2[37, 51] $$MoO_4^{2-} + 3\,H_2O + 2\,e^- \rightarrow MoO_2 + 4\,OH^- \quad (11)$$

SiO_2[59] $$SiCl_4 + 2\,H_2O \rightarrow SiO_2 + 4\,HCl \quad (16)$$

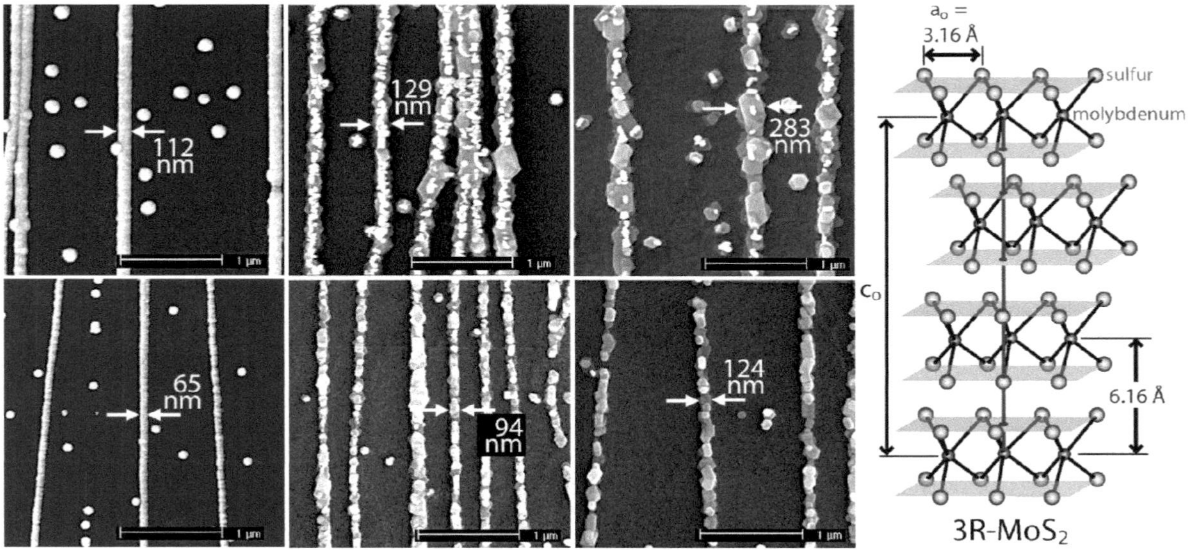

Figure 15. SEM images showing the evolution in morphology for MoO_2 nanowires (left images) following exposure to H_2S at 800°C in H_2S for 24 hrs. (center images) and 84 hrs. (right images) The MoO_2 nanowires were prepared using t_{dep} of 100 s (top) and 50 s (bottom). Also shown (right) is a schematic diagram of the 3R polymorph of MoS_2 showing the layered structure of this material.

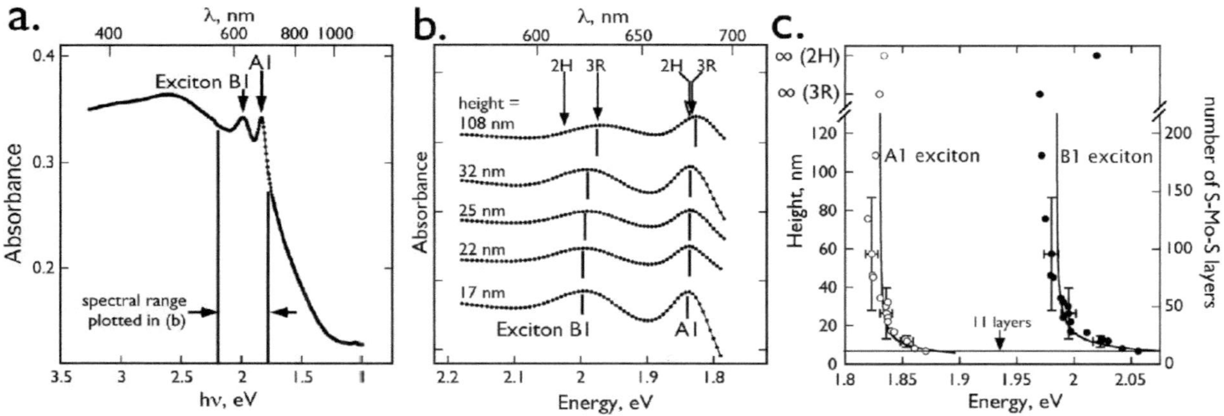

Figure 16. (a) Survey transmission-absorption spectrum for MoS_2 nanowires showing direct absorption edge and A1 and B1 exciton peaks. (b) Transmission-absorption spectra for the spectral interval shown in (a) for HT-MoS_2 nanowires having five different thicknesses, $L_{\|}$, from 17 nm to 180 nm. (c) Plot of $L_{\|}$ as a function of peak energies for the A1 and B1 excitons. $L_{\|}$ error bars are +1σ for AFM measurements on multiple wires in each sample; energy error bars correspond to +1.0 nm in wavelength. The solid lines are plots of equation 4 with $\mu_{\|}^{A1} = 0.16\ m_o$ and $\mu_{\|}^{B1} = 0.09\ m_o$.

Of course, the electrodeposition of metal oxides is only possible when these oxides possess some electronic conductivity. This is the case for all the oxides listed above except for SiO_2. In that case, we employed chemical vapor deposition, not electrodeposition, to obtain these nanowires. Briefly, $SiCl_4$ vapor was reacted with water that condensed preferentially at the hydrophilic step edges of a HOPG basal plane surface. The quantity of condensed water was controlled via the partial pressure of water in the reaction chamber.

IV. REDUCING THE DIAMETER OF ESED NANOWIRES BY KINETICALLY-CONTROLLED NANOWIRE ELECTROOXIDATION

The ESED method has two weaknesses:

1. The nanowires produced using ESED are always polycrystalline and often nanocrystalline; single crystalline nanowires can not be obtained, and
2. nanowires with diameters below 70 nm are obtained with difficulty.

In fact, metal oxides constitute the only case where sub-70 nm diameter nanowires can be routinely obtained by ESED. How can this minimum diameter be reduced? As already discussed (c.f., Fig. 9), the minimum nanowire diameter is directly related to the maximum density of nuclei that can be formed along step edges. This correlation immediately suggests a strategy for obtaining smaller nanowires is to increase the nucleation density, and we are working towards methods for accomplishing this objective (e.g., Section II.4) but our work in this direction has not yet yielded smaller nanowires. We discuss here a completely different strategy—kinetically controlled electrooxidation—[60] in this Section. Kinetically controlled electrooxidation involves electrochemically removing material from larger nanowires to make them smaller. For many materials, this material removal process can be carried out by electrochemically oxidizing the material to soluble products. We have successfully applied this approach to nanowires composed of antimony, gold, and Bi_2Te_3.[60] Electrooxidation is implemented using the potential program shown in Fig. 2b.

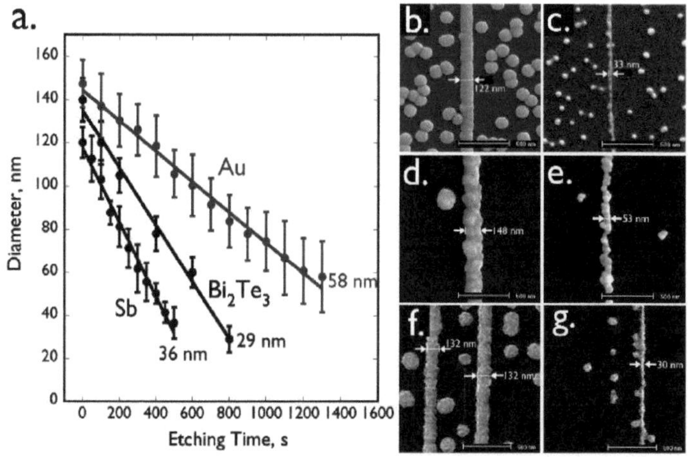

Figure 17. (a) Plots of mean nanowire diameter, <*dia.*>, versus t_{ox} for gold, antimony, and Bi_2Te_3. Each data point plotted in here represents a separate nanowire growth experiment involving, for a particular material, nucleation and growth under identical conditions, followed by electrooxidation for various *etching times*. (b)-(c) Antimony nanowires before (b) and after (c) electrooxidation for 500 s. (d)-(e) Gold nanowires before (d) and after (e) electrooxidation for 1300 s. (f)-(g) Bismuth telluride nanowires before (f) and after (g) electrooxidation for 800 s. Reprinted with permission from Ref. 60. Copyright (2006) American Chemical Society.

A key point is that the oxidation potential applied in the final step produces an extremely slow, kinetically-controlled oxidation of the nanowire. In this limit, wire diameter versus time plots (Fig. 17) were linear, in accordance with the equation:

$$r(t) = r_{o2} - \frac{j_{ox} t_{ox} V_m}{nF} \qquad (17)$$

where r_{o2} is the initial radius of a nanowire subjected to oxidation and j_{ox} and t_{ox}, are the oxidation current density and duration, respectively (Fig. 3). The derivation of Eq. (17) assumes that etching occurs by a kinetically controlled process. Electrooxidation permitted nanowires with an initial diameter of 100-150 nm to be reduced to a minimum diameter of 30 nm in the case of Bi_2Te_3 and antimony, and 50 nm in the case of gold. When a higher rate of electrooxidation was employed, nanowires decomposed rapidly as constrictions in these nanowires evolved rapidly into breaks because the rate of the nanowire narrowing, dr/dt is actually higher at constrictions when oxidation is not kinetically controlled.

We have applied the concept of kinetically-controlled electrooxidation to nanowires prepared using ESED but in principle, it is a completely general strategy for shrinking nanowires composed of a variety of materials and prepared by virtually any method.

V. SUMMARY

Electrochemical Step Edge Decoration (ESED) is a family of techniques for preparing nanowires composed of a variety of materials using electrodeposition. The common denominator in all these methods is the HOPG surface, with its characteristic and highly useful defect structure. By applying the electrooxidation strategy described in Section IV, ESED can produce nanowires with minimum diameters in the 20-30 nm range and the length of these nanowires can be more than 100 μm in many cases. Other advantages of ESED are the following:

1. nanowires are organized into parallel arrays containing hundreds or thousands of wires,
2. these nanowires interact weakly with the HOPG surface and can be removed by embedding them in cyanoacrylate, and
3. a highly diverse range of materials (Table 1) can be rendered as nanowires using ESED.

ESED also has several important limitations. It can produce only polycrystalline nanowires. Furthermore, ESED cannot be used to pattern nanowires on surfaces: nanowires are obtained at the location of step edges on the HOPG surface and the position of these steps cannot be controlled. We have sought to develop a new method that has many of the attributes of ESED, but which enables the patterning of the step edges on a dielectric. Our efforts have culminated in the development of a new method, called Lithographically Patterned Nanowire Electrodeposition or LPNE, that shows considerable promise.[61]

Table 1
Chronology of Nanowire Syntheses by Electrochemical Step Edge Decoration.

Material(s)	Method	Application	Refs.
MoO_2, Mo	Electrodeposition of MoO_2: $MoO_4^{2-} + 2\,H_2O + 2e^- \rightleftharpoons MoO_2 + 4\,OH^-$ Reduction in H_2 to Mo	None	37, 51
Pd	Direct electrodeposition of Pd	Hydrogen gas sensing	62, 63
Au, Cu, Ni	Direct electrodeposition of M^o	None	47, 49, 50
Ag	Direct electrodeposition of Ag	Amine vapor sensing	45, 46
AgO_x	$Ag^+ + H_2O \rightarrow AgO + 2\,H^+ + 1\,e^-$	Amine vapor sensing	57
CdS	Electrodeposition of Cd, followed by reaction with H_2S: $Cd + H_2S \rightleftharpoons CdS + H_2$	Photoconductive nanowires	54
Bi_2Te_3	Cyclic electrodeposition/ stripping with net reaction: $2\,Bi^{3+} + 3\,HTeO_2^+ + 3\,H_2O + 18\,e^- \rightleftharpoons Bi_2Te_3 + 9\,OH^-$	Thermoelectric energy generation	64, 65
α-MnO_2	$MnO_4^- + 2\,H_2O + 3\,e^- \rightleftharpoons MnO_2 + 4\,OH$	Li^+ battery cathode material	58
Pt, Pd, Au, Ag, Cu	Galvannic displacement using insoluble crystals of two ferrocene derivatives (Fc^o). Net reaction: $M^{n+} + n\,Fc^o + n\,X^- \rightleftharpoons Mo + n\,FcX$	None	66
CdSe	Cyclic electrodeposition/stripping with net reaction: $H_2SeO_3 + Cd^{2+} + 6\,e^- + 4\,H^+ \rightleftharpoons CdSe + 3\,H_2O$	None	53
MoS_2	Electrodeposition of MoO_2, followed by reaction with H_2S: $MoO_2 + 2\,H_2S \rightleftharpoons MoS_2 + 2\,H_2O$	None	55, 56
Au	Direct electrodeposition of Au on PVD-deposited Au seeds	None	52

ACKNOWLEDGEMENTS

The development of ESED was brought about by the efforts of many excellent coworkers to whom I am indebted: Hongtao Liu, Kwok Ng, Mike Zach, Dr. Fred Favier (CNRS Montpellier), Dr. Robert Dryfe (Univ. Manchester), Ben Murray, Qiguang Li, Erich Walter, Erik Menke, Cobey Cross, Michael Thompson, and Megan Bourg. Also crucial to the success of our work over a period of many years have been XPS analyses of nanowires carried out in a collaboration with the research group of Prof. John Hemminger and his students Koji Inuzu, John Newberg, and Matt Brown. The work described here was supported by a series of grants from the National Science Foundation and the Petroleum Research Fund of the American Chemical Society. Most of the graphite used for early studies was donated by Dr. Arthur Moore of GE Advanced Ceramics and I am very grateful to him for his generosity over an extended period of more than ten years. More recently, graphite was purchased with funds provided by EU Commission FP6 NMP-3 project 505457-1 ULTRA-1D.

REFERENCES

[1]C. A. Foss, M. J. Tierney, and C. R. Martin. *J. Phys. Chem.* **96** (1992) 9001.
[2]C. A. Foss, G. L. Hornyak, J. A. Stockert, and C. R. Martin. *J. Phys. Chem.* **98** (1994) 2963.
[3]L. Genzel, T. P. Martin, and U. Kreibig. *Z. Physik. B* **21** (1975) 339.
[4]S. A. Sapp, D. T. Mitchell, and C. R. Martin. *Chem Mater* **11** (1999) 1183.
[5]V. M. Cepak and C. R. Martin. *J Phys Chem B* **102** (1998) 9985.
[6]J. C. Hulteen, and C. R. Martin. *J Mater Chem* **7** (1997) 1075.
[7]C. R. Martin. *Chem Mater* **8** (1996) 1739.
[8]C. R. Martin. *Science* **266** (1994) 1961.
[9]D. N. Davydov, P. A. Sattari, D. AlMawlawi, A. Osika, T. L. Haslett, and M. Moskovits. *J Appl Phys* **86** (1999) 3983.
[10]D. N. Davydov, J. Haruyama, D. Routkevitch, B. W. Statt, D. Ellis, M. Moskovits, and J. M. Xu. *Phys Rev B-Condensed Matter* **57** (1998) 13550.
[11]A. A. Tager, J. M. Xu, and M. Moskovits. *Phys Rev B-Condensed Matter* **55** (1997) 4530.
[12]D. Routkevitch, A. A. Tager, J. Haruyama, D. Almawlawi, M. Moskovits, and J. M. Xu. *Ieee Trans Electron Devices* **43** (1996) 1646.
[13]D. Almawlawi, C. Z. Liu, and M. Moskovits. *J Mater Res* **9** (1994) 1014.
[14]C. K. Preston, and M. Moskovits. *J Phys Chem* **97** (1993) 8495.
[15]K. Liu, K. Nagodawithana, P. C. Searson, and C. L. Chien. *Phys Rev B-Condensed Matter* **51** (1995) 7381.

[16]G. Oskam, J. G. Long, A. Natarajan, and P. C. Searson. *J. Phys. E. Appl. Phys.* **31** (1998) 1927.

[17]K. Liu, C. L. Chien, and P. C. Searson. *Phys Rev B-Condensed Matter* **58** (1998) R14681.

[18]K. I. Liu, C. L. Chien, P. C. Searson, and K. YuZhang. *Appl Phys Lett* **73** (1998) 2222.

[19]K. Liu, C. L. Chien, P. C. Searson, and Y. Z. Kui. *Ieee Trans Magn* **34** (1998) 1093.

[20]L. Sun, P. C. Searson, and C. L. Chien. *Appl Phys Lett* **74** (1999) 2803.

[21]K. M. Hong, F. Y. Yang, K. Liu, D. H. Reich, P. C. Searson, C. L. Chien, F. F. Balakirev, and G. S. Boebinger. *J Appl Phys* **85** (1999) 6184.

[22]T. Jung, R. Schlittler, J. K. Gimzewski, and F. J. Himpsel. *Appl Phys a-Mat Sci Process* **61** (1995) 467.

[23]F. J. Himpsel, T. Jung, and J. E. Ortega. *Surf Rev Letters* **4** (1997) 371.

[24]D. Y. Petrovykh, F. J. Himpsel, and T. Jung. *Surface Sci* **407** (1998) 189.

[25]F. J. Himpsel, T. Jung, A. Kirakosian, J. L. Lin, D. Y. Petrovykh, H. Rauscher, and J. Viernow. *Mrs Bull* **24** (1999) 20.

[26]H. Röder, E. Hahn, H. Brune, J.-P. Bucher, and K. Kern. *Nature* **366** (1993) 141.

[27]S. Strbac, O. M. Magnussen, and R. J. Behm. *Phys Rev Lett* **83** (1999) 3246.

[28]F. A. Moller, O. M. Magnussen, and R. J. Behm. *Phys Rev Lett* **77** (1996) 5249.

[29]J. V. Zoval, P. Biernacki, and R. M. Penner. *Anal. Chem.* **68** (1996) 1585.

[30]J. V. Zoval, R. M. Stiger, P. R. Biernacki, and R. M. Penner. *J. Phys. Chem.* **100** (1996) 837.

[31]J. V. Zoval, J. Lee, S. Gorer, and R. M. Penner. *J. Phys. Chem.* **102** (1998) 1166.

[32]R. Stiger, B. Craft, and R. M. Penner. *Langmuir* **15** (1999) 790.

[33]H. Liu, and R. M. Penner. *J. Phys. Chem. B* **104** (2000) 9131.

[34]H. Liu, F. Favier, K. Ng, M. P. Zach, and R. M. Penner. *Electrochimica Acta* **47** (2001) 671.

[35]S. Gorer, H. Liu, R. M. Stiger, M. P. Zach, J. V. Zoval, and R. M. Penner, Electrodeposition of metal nanoparticles on graphite and silicon, in *Handbook of Metal Nanoparticles*, Eds. D. Feldheim, and C. Foss, Marcel-Dekker, New York, 2001.

[36]R. M. Penner. *J. Phys. Chem. B* **106** (2002) 3339.

[37]M. P. Zach, K. H. Ng, and R. M. Penner. *Science* **290** (2000) 2120.

[38]I. Morcos. *J. Chem. Phys.* **57** (1972) 1801.

[39]A. Zangwill, *Physics at Surfaces*, Cambridge University Press, Cambridge, 1988, p. 421.

[40]R. J. Bowling, R. T. Packard, and R. L. McCreery. *J. Am. Chem. Soc.* **111** (1991) 1217.

[41]K. K. Cline, M. T. McDermott, and R. L. McCreery. *J. Phys. Chem.* **98** (1994) 5314.

[42]R. S. Robinson, K. Sternitzke, M. T. McDermott, and R. L. McCreery. *J Electrochem Soc* **138** (1991) 2412.

[43]R. J. Bowling, R. L. McCreery, C. M. Pharr, and R. C. Engstrom. *Anal. Chem.* **61** (1989) 2763.

[44]H. Reiss. *J. Chem. Phys.* **19** (1954) 482.

[45]B. J. Murray, J. T. Newberg, E. C. Walter, Q. Li, J. C. Hemminger, and R. M. Penner. *Analytical Chemistry* **77** (2005) 5205.

[46]B. J. Murray, E. C. Walter, and R. M. Penner. *Nano Letters* **4** (2004) 665.

[47]R. M. Penner. *J Phys Chem B* **106** (2002) 3339.

[48]E. C. Walter, F. Favier, and R. M. Penner. *Analytical Chemistry* **74** (2002) 1546.
[49]E. C. Walter, B. J. Murray, F. Favier, G. Kaltenpoth, M. Grunze, and R. M. Penner. *Journal of Physical Chemistry B* **106** (2002) 11407.
[50]E. C. Walter, M. P. Zach, F. Favier, B. J. Murray, K. Inazu, J. C. Hemminger, and R. M. Penner. *Chemphyschem.* **4** (2003) 131.
[51]M. P. Zach, K. Inazu, K. H. Ng, J. C. Hemminger, and R. M. Penner. *Chem. Mater.* **14** (2002) 3206.
[52]C. E. Cross, J. C. Hemminger, and R. M. Penner. *Langmuir* (2007) in press.
[53]Q. Li,M. A. Brown,J. C. Hemminger, and R. M. Penner. *Chem. Mater.* **18** (2006) 3432.
[54]Q. G. Li, and R. M. Penner. *Nano Letters* **5** (2005) 1720.
[55]Q. Li, J. T. Newberg, E. C. Walter, J. C. Hemminger, and R. M. Penner. *Nano Letters* **4** (2004) 277.
[56]Q. Li,E. C. Walter, W. E. van der Veer, B. J. Murray, J. T. Newberg, E. W. Bohannan,J. A. Switzer,J. C. Hemminger, and R. M. Penner. *Journal of Physical Chemistry B* **109** (2005) 3169.
[57]B. J. Murray, J. C. Newberg, J. C. Hemminger, and R. M. Penner. *Chem. Mater.* **17** (2005) 6611.
[58]Q. G. Li,J. B. Olson, and R. M. Penner. *Chem. Mater.* **16** (2004) 3402.
[59]M. P. Zach,J. T. Newberg,L. Sierra,J. C. Hemminger, and R. M. Penner. *Journal of Physical Chemistry B* **107** (2003) 5393.
[60]M. A. Thompson,E. J. Menke,C. C. Martens, and R. M. Penner. *J. Phys. Chem. B* **110** (2006) 36.
[61]E. J. Menke, M. A. Thompson,C. Xiang, L. C. Yang, and R. M. Penner. *Nature-Materials* **5** (2006) 914.
[62]F. Favier, E. Walter, M. P. Zach, T. Benter, and R. M. Penner. *Science* **293** (2001) 2227.
[63]E. C. Walter, F. Favier, and R. M. Penner. *Anal Chem* **74** (2002) 1546.
[64]E. J. Menke, M. A. Brown, Q. Li, J. C. Hemminger, and R. M. Penner. *Langmuir* **22** (2006) 10564.
[65]E. J. Menke, Q. Li, and R. M. Penner. *Nano Letters* **4** (2004) 2009.
[66]R. A. W. Dryfe, E. C. Walter, and R. M. Penner. *Chemphyschem* **5** (2004) 1879.

4

Trends in Ammonia Electrolysis

Madhivanan Muthuvel and Gerardine G. Botte*

Department of Chemical and Biomolecular Engineering, Ohio University, Athens, Ohio 45701

I. INTRODUCTION

The world requires energy for transportation, machinery, buildings and various other operations. Energy has been an important issue for a long time. The energy requirement for our world's population has been provided by power plants and other small auxiliary power units. But there has been constant and increasing demand for more power and energy over the years. Different kinds of power plants are in operation in the present world; they are thermal, hydroelectric, nuclear, solar, wind and geo-thermal power plants. All of the power plants require a primary fuel to produce electric power. Most of the fuels used in the power plants are fossil fuels—coal, oil, and natural gas, whereas stored water and uranium are used in hydroelectric and nuclear power plants, respectively. Thermal, hydroelectric, and nuclear power plants are based on non-renewable fuels, while solar, wind, and geo-thermal power plants derive energy from renewable sources such as the sun, the wind and hot geysers, respectively.

Renewable sources for fuels are very important as more attention is given to the pollution and preservation of the environment.

R.E. White (ed.), *Modern Aspects of Electrochemistry, No. 45*, Modern Aspects of Electrochemistry 45, DOI 10.1007/978-1-4419-0655-7_4,

Among the fuels used in power plants, fossil fuels are the most destructive in nature since they generate green house gases on burning and their availability is constantly dwindling each day. Therefore, renewable energy sources (e.g., solar and wind) and fuels (e.g., biomass and biofuels) are in high demand. However, the power efficiency from renewable energy is lower than conventional power plants. Furthermore, these sources of energy are intermittent due to the availability of sunlight and wind. On the other hand, fuel cells offer a solution to the use of renewable energy sources by providing energy when an intermittent source is not available.

Fuel cells are electrochemical devices where hydrogen reacts with oxygen to produce electric power and water. The fuel cell was first demonstrated by Sir William Groove in 1839, but it has taken more than a century for fuel cells to be accepted as a genuine power unit. Various types of fuel cells with their operating temperature and mobile ions are listed in Table 1. Both hydrogen and oxygen gas are supplied to a fuel cell and the fuel cell can generate direct current as long as hydrogen is available as fuel. Since air contains 21% as oxygen, it is used instead of oxygen in most of the fuel cell applications.

Research in fuel cell systems has been devoted to areas such as electro-catalyst materials for anode and cathode, cell design, flow field plates, electrodes, hydrogen production and storage, and direct fuel supply systems. Cell design and electro-catalyst materials are being investigated because fuel cells are more expensive than internal combustion engines and existing power generators. However, critical areas of concern are the *source of fuel* (*hydrogen*) and *transportation and storage of the fuel*. Transportation and storage of hydrogen is not currently efficient and it is expected to take about 20 to 30 years of research to construct efficient and safe distribution networks and storage systems for hydrogen.

On the other hand, the major sources for hydrogen production are fossil fuels and water, which are discussed in the next Section. A novel approach *Ammonia Electrolysis* was introduced by researchers at Ohio University-Electrochemical Engineering Research Lab (EERL), where ammonia is electrochemically broken into hydrogen and nitrogen gas.[2,3] There are multiple advantages on using ammonia as a hydrogen carrier. Furthermore, the process of ammonia electrolysis allows compatibility with renewable energy sources (wind and solar energy). The ultimate goal of the process known as *Ammonia Electrolysis* is shown in Fig. 1 where ammonia-waste is sent into an ammonia electrolyzer operating with renewable energy sources to produce hydrogen and clean water. Because ammonia can be obtained from waste, it can be considered a renewable fuel. In this chapter, the process of ammonia electrolysis is explained, and the use of the technology as a solution to hydrogen storage and transportation, and even its production are presented.

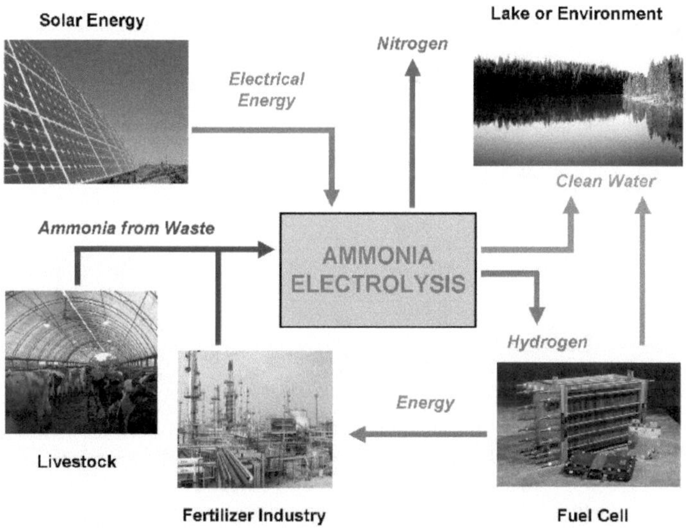

Figure 1. Diagram on the use of renewable energy by ammonia electrolyzer to produce hydrogen and clean water from livestock and industrial ammonia waste. As indicated in the picture, ammonia electrolysis can serve as a remediation process with hydrogen generation. More details about ammonia emissions and sources of emissions will be discussed in Section 3.1. Image Courtesy: www.nrel.gov (Fuel Cell), commons.wikimedia.org (Solar Energy) and Vaarok from en.wikipedia.org (Livestock).

Table 1
Different Types of Fuel Cells.[1]

Fuel cell type	Mobile ion	Operating temperature (°C)	Limitations	Applications[a]
Alkaline fuel cell (AFC)	OH^-	50-200	Air and fuel should be free from CO_2 to avoid carbonate formation in the electrolyte. Expensive catalysts are required	Used in space vehicles (Apollo, Shuttle).
Proton exchange membrane fuel cell (PEMFC)	H^+	30-100	Pure hydrogen and expensive catalysts are required	Vehicles and mobile applications, and for low power CHP systems.
Direct methanol fuel cell (DMFC)	H^+	20-90	Slow methanol oxidation rate, low power output and fuel crossover problems	Suitable for portable electronic systems of low power, running for long times.
Phosphoric acid fuel cell (PAFC)	H^+	~ 220	Fuel reformer needed: complex system and high cost	Large numbers of 200 kW CHP systems in use.
Molten carbonate fuel cell (MCFC)	CO_3^{2-}	~ 650	Requires molten electrolyte: hot and corrosive mixtures of lithium, potassium and sodium carbonate	Suitable for medium-to-large scale CHP systems, up to MW capacity.
Solid Oxide Fuel Cell (SOFC)	O^{2-}	500-1000	Ceramic electrolyte, complex and expensive accessory equipments	Suitable for all sizes of CHP systems, 2 kW to multi-MW.

[a]CHP: combined heat and power

II. HYDROGEN PRODUCTION METHODS

The operation of an efficient fuel cell requires pure hydrogen gas as fuel. In the world, hydrogen is abundantly available but in the form of different compounds and not as pure hydrogen gas. Compounds which have more hydrogen content are water, ammonia, and hydrocarbons such as natural gas, oil, coal, and methane. Fuel processing is the only option to get hydrogen gas from various compounds. The most widely used and well established techniques to generate hydrogen from hydrocarbons are described below.

1. Steam Reforming

One of the established methods for producing hydrogen gas from hydrocarbons is steam reforming.[1,4-6] The chemical reactions that a hydrocarbon (C_nH_m) undergoes in a reformer can be summarized as

$$C_nH_m + n\,H_2O \rightarrow n\,CO + (m/2 + n)\,H_2 \quad (1)$$

$$CO + H_2O \rightarrow CO_2 + H_2 \quad (2)$$

Reaction (2) is generally known as the water gas shift reaction where steam reacts with carbon monoxide to form carbon dioxide and hydrogen. In this method, steam is mixed with the hydrocarbon over a supported nickel catalyst at high temperatures, normally above 500 °C. The outlet gas from the reformer will contain carbon dioxide, carbon monoxide, hydrogen, and un-reacted steam and hydrocarbon. In order to obtain a clean supply of hydrogen, the reformed gas has to be purified to remove all gas components other than hydrogen before being sent to the fuel cells.[4] Hydrocarbons used in the steam reforming process are natural gas, methane and even methanol:[4-6]

$$CH_4 + H_2O \rightarrow CO + 3\,H_2 \quad (3)$$

$$CH_3OH + H_2O \rightarrow 3\,H_2 + CO_2 \quad (4)$$

The drawback of this method is that all the reforming reactions with most of the hydrocarbons are endothermic, which means energy has to be supplied to produce hydrogen gas. Even after the reforming process, the hydrogen gas has to be purified before its use in a fuel cell. In recent years, there is wide interest about using methanol directly as a fuel to the fuel cells, thereby both reforming and then electrochemically converting hydrogen to water for producing electricity; and this process has been studied. Such a setup is called a direct methanol fuel cell (DMFC).[1,7,8] The idea of using fuel reforming in a fuel cell is not only restricted to methanol but other fossil fuels. This process is generally known as internal reforming.[9-11]

2. Partial Oxidation Reforming

Another way of producing hydrogen gas from methane and other hydrocarbons is partial oxidation reforming. Typical operating temperature for partial oxidation lies in the range of 1200 to 1500 °C, and no catalyst is used during the fuel reforming process. The following partial oxidation reaction for methane conversion to hydrogen and carbon monoxide is an exothermic process:

$$CH_4 + \tfrac{1}{2} O_2 \rightarrow CO + 2H_2 \tag{5}$$

In comparison with reaction (3), less molecules of hydrogen are produced by the partial oxidation method than steam reforming suggesting that partial oxidation is less efficient, but the high operating temperature for partial oxidation makes it suitable to handle heavier petroleum fractions such as diesels and residual fractions.[1,6]

3. Coal Gasification

Gasification is another technology used for the large scale conversion of complex hydrocarbons into various products, which are further processed to enrich the hydrogen fraction.[12] This technology can use all carbon-based feedstock, coal, petroleum residues, biomass, and municipal wastes, and produce a variety of commodity products.[13] The process of coal gasification involves high temperature (1040-1540 °C) reactions with moderate pressure (5-10 bar) using steam and oxygen to produce gas products containing CO, CO_2, CH_4, H_2, N_2, and H_2S in varying proportions.[14]

Coal gasification is a complex process involving pyrolysis, steam reforming, partial oxidation, water gas shift (reaction 2), and methanation reactions. The basic reaction network in a steam and oxygen fed gasifier is given below:[13]

- fast pyrolysis:
$$C_nH_mO_y \rightarrow tar + H_2 + CO_2 + CH_4 + C_2H_4 + \ldots \quad (6)$$
- steam reforming:
$$tar + x\,H_2O \rightarrow x\,CO + y\,H_2 \quad (7)$$
- CO_2 reforming:
$$tar + CO_2 \rightarrow x\,CO + y\,H_2 \quad (8)$$
- partial oxidation:
$$C_nH_m + n/2\,O_2 \rightarrow n\,CO + m/2\,H_2 \quad (9)$$
- water-gas shift:
$$CO + H_2O \rightarrow CO_2 + H_2 \quad (2)$$
- methanation:
$$CO + 3\,H_2 \rightarrow CH_4 + 2\,H_2O \quad (10)$$

Reactions (7) and (9) are steam reforming and partial oxidation, respectively, for a generalized form of coal, whereas methane specific reactions are represented in reactions (3) and (5).

4. Water Electrolysis

The hydrogen produced in the world today is predominantly from reforming fossil fuels accounting for about ~95% and the rest are from other methods.[15] One of the alternate methods is electrolysis of water. Water electrolysis involves the supply of current through electrodes immersed in water thereby splitting water into hydrogen and oxygen gases. Hydrogen obtained from electrolysis is 100% pure and does not require any further purification treatment.

The reactions involved in electrolysis of water are the exact opposite to the reactions taking place inside a fuel cell. In a fuel cell, pure hydrogen and oxygen is allowed to react to form water and electricity is produced; whereas in water electrolysis current is used to split water into hydrogen and oxygen gases. Combining water electrolysis and fuel cell together might mean that the electricity and water produced from the fuel cell would be used in the electrolyzer to produce hydrogen and oxygen, which is not considered a power efficient method. Sometimes water electrolysis is looked at as a better way of producing hydrogen for fuel cell applications as there is no purification step involved. Different kinds of electrolyzers are available – alkaline electrolyzer, PEM electrolyzer, and solid oxide electrolyzer.

The first electrolyzer developed was an alkaline water electrolyzer. It used 30% potassium hydroxide (KOH) solution as the electrolyte with nickel or nickel compounds electrodes. Asbestos was used as a diaphragm or a membrane to provide OH^- ions exchange between the cathode and the anode sides. The operating conditions included temperatures ranging from 70° to 100°C with an applied high pressure of 1 to 30 bars.[16, 17]

Over the years another kind of electrolyzer for water was developed based on the proton exchange membrane (PEM). PEM electrolyzers use expensive electro-catalysts—platinum, on its electrodes. Water is filled in the anode side of the electrolyzer. Oxidation of water produces oxygen gas and protons. Protons are transferred to the cathode side by the selectively permeable PEM membrane for reduction of protons to hydrogen gas. A commonly used PEM material is a perfluoroalkyl sulphonic acid polymer, commercially known as a Nafion™ membrane.[15,18-20] The reactions for water electrolysis are given below

$$2\,H_2O \rightarrow O_2 + 4\,H^+ + 4\,e^- \tag{11}$$

$$4\,H^+ + 4\,e^- \rightarrow 2\,H_2 \tag{12}$$

where reactions (11) and (12) take place at the anode and cathode of the water electrolyzer, respectively. The overall reaction is given by

$$2\,H_2O \rightarrow O_2 + 2\,H_2 \tag{13}$$

Dissociation of water to hydrogen and oxygen requires a minimum voltage of 1.23 V at normal room temperature and atmospheric pressure, but at this voltage the rate of hydrogen production is less along with a low current density, which results in lesser current efficiency. In real applications, the voltage maintained over the electrolyzers is in the range of 1.6 to 2.0 V, which is higher than the thermoneutral voltage of 1.48 V for water electrolysis to generate hydrogen gas, and this voltage will facilitate current densities around 1.0 A cm^{-2}.[1,15] In addition to applying high cell voltage, the water electrolyzer also requires deionized water for its normal operation.[18]

5. Efficiency for Hydrogen Production Methods

The hydrogen gas produced from these methods will be used in a fuel cell to generate power. Traditionally, efficiency for any hydrogen production method is described as the ratio of energy derived from hydrogen over the energy contained in the raw material. But the calculation should include other energies involved in the process such as combustion, electrical heating, cooling, and purification step, so as to obtain the real efficiency for converting the raw material to final product – pure hydrogen gas. Using the values provided in a report by the National Research Council,[21] efficiency for hydrogen production methods are listed in Table 2.

The following assumptions were used for the calculation of the efficiencies shown in Table 2:

- Thermodynamically, one gram of H_2 can generate 33 Wh from a fuel cell at 100% efficiency.
- The standard density of hydrogen was 2.55 g ft^{-3} (1 atmosphere, 15°C) was used.

Table 2
Efficiency for Different Hydrogen Producing Methods

Methods	Efficiency
Steam methane reforming	63%
Coal gasification	64%
Water electrolysis	67%

- None of the efficiencies include compression of hydrogen after the purification step.
- In the case of steam methane reforming (SMR), the efficiency of the process is based on the lower heating value (LHV) of natural gas. On supplying natural gas of 3.794 MM (million) Btu (LHV) energy per hour to the methane reformer will produce 8,290 scf (standard cubic feet) of H_2 per hour, when 5-kW electrical power is consumed for heating the reformer. Overall efficiency for this process was estimated to be 63%.
- In the case of coal gasification, the efficiency was estimated for a large production facility, where 50,000 kg of H_2 per hour was produced from a gasifier by supplying it with coal at 8,399 MM Btu (LHV) per hour. The efficiency for the gasifier was assumed to be 75% and it required hot oxygen gas to combust coal to produce synthetic gas. With the help of 120,656 kW electrical power, air was heated into the hot oxygen gas. Synthetic gas has hydrogen and other components such as CO, and CO_2, which are converted to H_2 by water gas shift reaction to finally produce 50,000 kg of H_2 per hour. The efficiency for converting coal to hydrogen involving the electrical power and purification of the gas is 64%.
- Water electrolysis utilizes electricity to produce hydrogen gas and its raw material is water, which does not have any heating value. So to produce 8,290 scf of H_2 per hour, 180 kg of water is supplied per hour to a 75% efficient water electrolyzer. Electrical power consumed by the electrolyzer is 1050 kW and the overall efficiency for converting water to hydrogen gas in the water electrolyzer is 67%. It is important to mention that this efficiency does not include any penalties associated with the purification of water (deionized water is required for the process).

On comparing these three methods, electrolysis of water seems to have a slight edge over steam reforming and coal gasification.

III. AMMONIA

Ammonia is a pungent, colorless gas with a boiling point at 28°F and a freezing point of –107.86°F at atmospheric pressure. Ammonia is highly soluble in water and at 32°F, one volume of water can absorb 1.148 volumes of ammonia.[22] Ammonia has many attractive qualities; first, it is nonflammable and non-explosive. Second, it can easily be liquefied as its vapor pressure is 9.2 bar (~121 psig) at room temperature and it can be stored in simple pressure vessels. Third, hydrogen constitutes 17.65% mass of ammonia, which means liquid ammonia has about 45% more volumetric hydrogen density than liquid hydrogen.[23] In the United States, a safe and efficiently operating ammonia distribution system with more than 3000 miles of pipeline is already in existence.[24] The energy density for ammonia is 3.3 kWh kg^{-1}, which is better than water and methanol.[25] Using these properties, an investigation was performed to assess ammonia as a fuel to ammonia-oxygen fuel cell.[26] In recent years, ammonia has been considered as a fuel along with hydrogen for stationary application related fuel cells mainly in intermediate temperature solid oxide fuel cells.[27]

1. Sources

Ammonia is produced from hydrogen and nitrogen gas using the well established Haber-Bosch process. In this method, the gases are passed over an iron catalyst at high temperature (400-600°C) and pressure (200-400 atmospheres). Synthesis of ammonia is generally coupled with hydrogen production. When hydrogen is produced from natural gas during reforming process, water and nitrogen are introduced to form ammonia.

Ammonia can easily escape into the atmosphere because it is a gaseous compound. So entrapment of an ammonia emission is the next available source for ammonia. Ammonia emissions have their impact on vegetations and other organisms, which exhibits significant environmental concern.[28,29] Various industries and other operations are considered as emitters of ammonia. They are the fertilizer manufacture industry, livestock management, coke manufacture industry, fossil fuel combustion, and refrigeration methods.[22] Fossil fuel combustion is included in the list of ammonia emitters because of their control measure methods. Exhaust gas from a fossil fuel processing plant has nitrogen oxides, which is allowed to react with urea or ammonia to limit the emission of nitrogen oxides into the atmosphere. A portion of ammonia may exit along with the exhaust gas, which is known as an ammonia slip, and this is how fossil fuel combustion is one of the emitters.[30]

The U.S. is one of the leading producers of ammonia in the world. According to a report, in 2002 the U.S. saw over 5-million tons of ammonia emitted into the atmosphere from various sources.[31] The distribution of the sources for this ammonia air emission is given in Fig. 2.

On combining fertilizer, livestock and domestic sources, about 3.25 million tons of ammonia (65%) are emitted into the atmosphere. The production of ammonia in the U.S. in 2002 (using the Haber-Bosch process) was 10.1 million tons; this means that the emission of ammonia to air from these three sources was equivalent to 32.5% of the total production of ammonia in the U.S. Based on the EERL calculation, recovery of ammonia from the three sources—fertilizer, livestock and residential (domestic), and performing ammonia electrolysis using solar energy—should be able to generate enough hydrogen to power over 900,000 households per year (assuming an average electric power consumption of 12,000 kWh per year and a proton exchange membrane fuel cell operating with a 50% efficiency.)

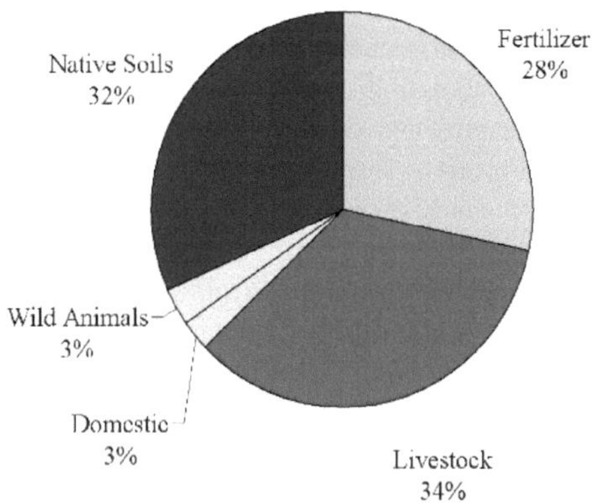

Figure 2. Distribution of ammonia air emission sources in 2002.[31]

2. Ammonia Cracking

Thermal decomposition of ammonia is commonly known as ammonia cracking. Decomposition of ammonia is an endothermic process, where ammonia is broken down into hydrogen and nitrogen at high temperatures (> 500°C). Ammonia cracking is a CO_x free process of producing hydrogen compared to steam reforming, partial oxidation, and coal gasification.[32] This has prompted researchers to use ammonia indirectly as fuel in an alkaline fuel cell.[33,34] Research on ammonia cracking began when synthesis of ammonia was widely investigated, and then using the information developed on ammonia synthesis, various catalysts were also developed for decomposition of ammonia. The catalysts tested for ammonia cracking are metals, alloys and compounds of noble metal. In recent decades, Ni and Ru based catalyst have been widely studied and these catalysts are supported over activated carbon or porous alumina.[25,32,35-37]

Ammonia cracking was introduced to fuel cell car technology by Kordesch and his co-workers in the 1960s.[38-40] Kordesch ran a hybrid car fitted with equipment to reform ammonia by thermal decomposition and the purified hydrogen gas was passed over an alkaline fuel cell to power the car. This could have been the first on board generation of hydrogen.

3. Ammonia Electrolysis

Production of hydrogen gas has been mainly concentrated with reforming fossil fuels and at times dissociation of water was considered. The key factors related to hydrogen gas as fuel are production, storage, and transportation. Storage of hydrogen has led to extensive research on finding new materials for hydrogen entrapment under high pressure conditions as well as forming new metal hydride compounds to quickly release hydrogen gas.[12] In order to overcome these concerns of producing hydrogen with less energy, no storage and transportation issues, ammonia electrolysis should be considered.

Electrolysis of ammonia, a new approach for hydrogen production, involves using an ammonia solution containing ammonium hydroxide (NH_4OH) and potassium hydroxide. A typical fuel and electrolyte composition for ammonia electrolysis is 1 M

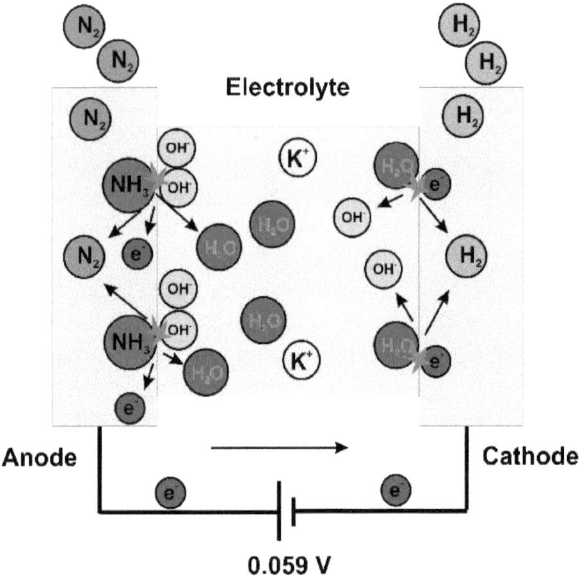

Figure 3. Schematic diagram of ammonia electrolysis with reactions at anode and cathode electrodes forming nitrogen and hydrogen gas, respectively.

NH_4OH in 5 M KOH. Ammonia is oxidized at the anode electrode to form nitrogen gas and water is reduced to hydrogen in the cathode side of the electrolytic cell. Hence, the overall reaction for the cell is dissociation of ammonia into hydrogen and nitrogen.[3,41] Reactions at each electrode and overall reaction are given below

- anode reaction:

$$2\,NH_3 + 6\,OH^- \rightarrow N_2 + 6\,H_2O + 6\,e^- \qquad E^0 = -0.77 \text{ V vs. SHE} \tag{14}$$

- cathode reaction

$$2\,H_2O + 2\,e^- \rightarrow H_2 + 2\,OH^- \qquad E^0 = -0.829 \text{ V vs. SHE} \tag{15}$$

- overall reaction

$$2\,NH_3 \rightarrow N_2 + 3\,H_2 \qquad E^0 = -0.059 \text{ V} \tag{16}$$

This ammonia electrolysis process has been illustrated in Fig. 3. There is a significant difference between 0.059 V and 1.23 V of cell voltage for ammonia and water electrolysis, respectively. Using these thermodynamic cell voltages, the energy required to produce one gram of H_2 is 1.55 Wh for NH_3 and 33 Wh for H_2O electrolysis. Cost for producing hydrogen is predicted, with \$0.21 per kWh for solar energy as source for electrolysis, to be \$0.90 for a kg of H_2 produced from NH_3, where the average cost of NH_3 was assumed to be \$175 per ton. The price for a kg of H_2 from ammonia electrolysis is well below the US Department of Energy (DOE) estimation – \$3. Assuming solar energy being used for water electrolysis, the cost for a kg of H_2 produced will be \$7.10. Energy expenditure and cost of H_2 production for ammonia electrolysis is almost 95% less energy consuming and 87% cheaper than water electrolysis. This demonstrates the importance of ammonia electrolysis and its possible impact in the present hydrogen economy.

4. Comparison of Ammonia Electrolysis and Ammonia Cracking

The two different approaches used to produce hydrogen from ammonia have been explained in the above Section. The efficiency for the two technologies under various conditions is listed in Table 3. Both methods are breaking down ammonia into hydrogen and nitrogen so their thermodynamic efficiencies are the same, and were determined by dividing the lower heating value of H_2 (242.7 kJ mol^{-1}) over the sum of ammonia lower heating value (320.1 kJ mol^{-1}) and either the electrical energy (NH_3 electrolysis) or thermal energy (NH_3 cracking).[3] In ammonia cracking, thermal energy is required to increase the temperature of the reactor and it is understood that at least 30% of H_2 heating value after H_2 production is used to heat the reactor.[42] The table lists efficiency more than 100% in theoretical terms for both processes because the energy to produce ammonia has not been included in the calculation. If NH_3 was commercially produced, then the energy required to make NH_3 should also be taken into consideration. This is one of the advantages of ammonia electrolysis vs. thermal cracking. Ammonia electrolysis allows the direct conversion of ammonia from waste to hydrogen, while thermal cracking of ammonia requires pure ammonia.

Table 3
Process Efficiency Comparison for Ammonia Cracking and Ammonia Electrolysis

Methods	Conditions	Efficiency
Ammonia cracking and ammonia electrolysis	Thermodynamic at standard conditions.	112.1%
Ammonia cracking	Pure ammonia (excludes energy required to purify the hydrogen).	102.1%
Ammonia electrolysis	Glass cell, one compartment with 1 M NH_3.[3]	80 %
	Single cell electrolyzer (with separator), 1 M NH_3.[43]	104.5%
	9-cell stack, 1 M NH_3.[44]	98.2%
	Low concentration electrolyzer (20 mM NH_3): glass cell, one compartment.[45]	95.1%

Electrolysis of ammonia has evolved in recent years and its improvement is solely based on advances made in anode electrode and cell design. A detailed description of the advances made towards improving ammonia electrolysis is discussed in the following Sections. The initial study on using a glass cell (one compartment) for ammonia electrolysis with Pt-Ir anode electrode resulted in an efficiency rate of 80%.[3] The best recorded cell voltage for NH_3 electrolysis in a single cell electrolyzer is 0.35 V, which translates to 104.5% process efficiency.[43] In a recent study, a 9-cell stack was developed as an ammonia electrolyzer, which had a cell operating voltage of 0.63 V at room temperature and the efficiency was 98.2%.[44] One of the objectives of ammonia electrolysis is to clean the wastewater. Normally the concentration of ammonia in livestock wastewater is low (excluding the urea which can be transformed into ammonia to make up to 0.7 M NH_3), in the order of 20-mM NH_3, and electrolysis of low ammonia concentration had a cell voltage of 0.78 V at 25 mA cm^{-2}.[45] Low concentration electrolysis of NH_3 did yield 95.1% efficiency in converting NH_3 to H_2 gas.

Thermal cracking of ammonia like electrolysis does not take energy consumption for ammonia preparation from the Haber-Bosch process into consideration for the calculation of process efficiencies. But ammonia cracking has also skipped the energy requirement in cleaning the H_2 gas along with the thermal losses during the gas cooling step. On the other hand, ammonia electrolysis can produce pure hydrogen, without using the purification step from either commercially available ammonia or from ammonia waste, which would ultimately help clean the environment as well as producing inexpensive power. When comparing Tables 2 and 3, it is evident that ammonia wastewater electrolysis has the highest efficiency for the production of pure hydrogen.

IV. DEVELOPMENT OF COMPONENTS

The key components in an electrolyzer are: anode, cathode, and electrolyte. These components play important roles in the development of an ammonia electrolysis technology. In this Section, we concentrate on the materials used to build an ammonia electrolyzer with the research and development for the anode, the cathode, and the electrolyte.

1. Anode Electrode

The anode is the electrode where the oxidation reaction takes place. In the case of ammonia electrolysis, electro-oxidation of ammonia to nitrogen gas (reaction 9) is observed at the anode. The electrode is made of catalyst particles and a supporting material. The role of the supporting material is not only to provide physical support to catalysts but also to help in transferring current or electrons for the reaction. In order to obtain high efficiency in electrolysis, the anode electrode has to perform at its best. Various researchers around the world are working to improve the performance of electrodes to oxidize ammonia. Development of the anode electrode for ammonia electrolysis means developing both the substrate and the electro-catalyst for the alkaline solution.

The catalyst for ammonia oxidation has been widely studied mainly with the intention of removing ammonia from wastewaters and industrial effluents.[25,46-50] The mechanism for ammonia electro-oxidation over platinum in alkaline medium was proposed by Gerischer and Mauerer:[51]

$$NH_3(aq) \rightarrow NH_{3,ads} \tag{17}$$

$$NH_{3,ads} + OH^- \rightarrow NH_{2,ads} + H_2O + e^- \tag{18}$$

$$NH_{2,ads} + OH^- \rightarrow NH_{ads} + H_2O + e^- \tag{19}$$

$$NH_{x,ads} + NH_{y,ads} \rightarrow N_2H_{x+y,ads} \tag{20}$$

$$N_2H_{x+y,ads} + (x+y)\,OH^- \rightarrow N_2 + (x+y)\,H_2O + (x+y)\,e^- \tag{21}$$

$$NH_{ads} + OH^- \rightarrow N_{ads} + H_2O + e^- \tag{22}$$

$$N_{ads} + N_{ads} \rightarrow N_{2,ads} \tag{23}$$

$$N_{2,ads} \rightarrow N_2(aq) \tag{24}$$

with x, y = 1 or 2. The authors considered adsorbed nitrogen as a poison for the electrooxidation of the ammonia reaction. At potentials above 0.6 V vs. reversible hydrogen electrode (RHE), the platinum electrode deactivates and the main adsorbed species was reported to be adsorbed nitrogen. The adsorption of NH_3 is fast at potentials above 0.45 V, which suggests the kinetic order for NH_3 concentration is zero.

On the contrary to the above mechanism, the co-adsorption kinetics of OH^- ion with NH_3 over Pt surface was considered by Botte[52] in her study to formulate the reaction rate for electro-oxidation of NH_3 in an alkaline medium. Botte found the results from rotating disk electrode experiments to disagree with the reaction mechanism proposed by Gerischer and Mauerer.[51] She developed a reaction model for the experimental results based on the mechanism proposed by Tilak, et al.[53] using electroadsorption of OH^- ion over Pt:

$$Pt_4 + 2\,OH^- \leftrightarrows 2\,(Pt_2OH) + 2\,e^- \tag{25}$$

$$i = i_0 \theta \left(\frac{C_{OH}}{C_{OH,ref}}\right)^2 exp\left(\frac{2\alpha_a F \eta}{RT}\right) \tag{26}$$

where i is current density (mA cm^{-2}), i_0 is exchange current density (mA cm^{-2}), C_{OH} is OH^- ion concentration in the solution (mol cm^{-3}), $C_{OH,ref}$ is the reference concentration of OH^- ion in the solution (mol cm^{-3}), θ is surface coverage (dimensionless value), α_a is anodic transfer coefficient (dimensionless), F is Faraday's constant (96485 C eq-mol^{-1}), η is the overpotential, R represents universal gas constant (8.314 J mol^{-1} K^{-1}) and T is the temperature of the reaction (K). The surface coverage (θ) for the Pt electrode during the adsorption of OH^- ion was modeled to follow this equation:

$$\theta = 1 - ac_1 \, exp\left(\frac{ac_2 F \eta}{RT}\right)\left(\frac{C_{OH}}{C_{OH,ref}}\right)^{ac_3} \tag{27}$$

where ac_1, ac_2, and ac_3 are dimensionless constants.

Sasaki and Hisatomi described the cyclic voltammogram of ammonia over platinized platinum.[54] The ammonia oxidation peak appeared around 0.7 V vs. RHE, which was before the formation of the platinum surface oxide. The height of the peak was proportional to the square root of the sweep rate implying that the kinetics is diffusion limited. Wasmus, et al. used a newly developed technique—differential electrochemical mass spectroscopy (DEMS)—to determine the chemical composition of the products formed from the reaction of ammonia oxidation on Pt-black.[55] The authors reported that nitrogen oxides were observed at potentials higher than 0.8 V vs. RHE, while at the maximum of the oxidation peak (0.75 V vs. RHE), the selectivity toward N_2 formation is 100%. Gootzen, et al. studied the coverage of the surface by ammonia adsorbates during the electro-oxidation of ammonia by using DEMS.[56] According to this study, the selective formation of N_2 at lower potential may be explained by a lowering of the energy of adsorption of N_{ads} with decreasing potential.

De Vooys, et al. investigated the use of several transition and coinage metal catalysts for oxidation of ammonia to nitrogen gas.[57] The metal catalysts studied by the authors were platinum (Pt), palladium (Pd), rhodium (Rh), ruthenium (Ru), iridium (Ir), copper (Cu), silver (Ag) and gold (Au). Coinage metals (Cu, Ag, and Au) were unable to selectively oxidize ammonia to nitrogen gas because of metal ion— ammonia complex formation by the coinage metal catalyst leading to their electrodissolution. During the oxidation of NH_3 to N_2, the nitrogen intermediate species were adsorbed on the catalyst materials based on their affinity. Among the transitional metal catalysts, Ru, Rh and Pd showed high affinity towards the nitrogen intermediate species, which was the reason for no generation of N_2 gas by these catalysts. Both Pt and Ir displayed low affinity for the nitrogen intermediates and were able to form N_2 gas from the oxidation of ammonia.

The importance of ammonia oxidation lead to several investigations about metal catalysts like the use of platinized TiO_2 as an electrode for photocatalytic oxidation of NH_3 to N_2 [48] and a dimensionally stable anode (DSA) made of IrO_2 over Ti substrate for removal of NH_3 from wastewater to form N_2.[58] Recently, Yao and Cheng have used a binary alloy (Ni-Pt) as an anode for electro-oxidation of ammonia.[59] But the study reveals that the key material acting as catalyst in the anode is Pt and not Ni, so this anode–Ni-Pt—is basically made of micron size Pt catalyst particles dispersed over a Ni support. Lopez de Mishima, et al. studied the electrocatalysis of ammonia and its application to ammonia sensors.[60] Cyclic voltammetry studies were performed on Pt, Ir, and Pt-Ir (75:25 and 50:50 by weight) electrodes, and these electrodes displayed an apparent diffusion control mechanism in the kinetics of ammonia oxidation. Vidal-Iglesias, et al. observed that the oxidation of ammonia on platinum is a structure-sensitive reaction that takes place almost exclusively on Pt (100) sites.[49] This means that the geometric structure of the deposited species is as important as their composition.

Use of Pt based noble metals as the catalyst for ammonia oxidation was further extended by Endo, et al. in their quest to understand metal binary alloys Pt-Me (Me = Ir, Ru, Ni and Cu).[61,62] These authors prepared their alloy catalysts on glassy carbon electrode by thermal decomposition process and performed the experiments in ammonia containing a KOH solution. The starting potential for ammonia oxidation curve, obtained with electrodes having $Pt_{1-x}Ir_x$ ($0 \leq x \leq 1$) and $Pt_{1-x}Ru_x$ ($0 \leq x \leq 0.6$) as catalyst, was observed at –0.6 V (vs. Ag/AgCl), which was 0.1 V lower than the potential observed for Pt metal. Both Ir and Ru displayed high compatibility with Pt as the saturation current density for the electrodes at a high oxidation potential (NH_3 oxidation) was higher for $Pt_{1-x}Ir_x$ ($x \leq 0.8$) and $Pt_{1-x}Ru_x$ ($x \leq 0.4$) than pure Pt metal. But both Ni and Cu based alloy catalysts, $Pt_{1-x}Ni_x$ ($0 \leq x \leq 0.7$) and $Pt_{1-x}Cu_x$ ($x = 0.33$ and 0.5), were not able to lower the potential for ammonia oxidation and their current densities were lower than other alloy catalysts. Endo and his co-workers concluded that use of Ir and Ru along with Pt will enhance the oxidation of ammonia in the KOH solution.

Research at Ohio University has focused on using the electro-catalysts for electrolysis of ammonia to produce hydrogen. The electro-catalyst development found in the literature have used methods other than electrodeposition or electroplating of the metals over a substrate. But at EERL, electroplating is the approach to prepare both the anode and the cathode electrode dispersed with catalyst for ammonia electrolysis. Cooper and Botte studied electrodeposition of Raney nickel over titanium (Ti) gauze for application as an anode for ammonia electrolysis.[63] Raney Ni was deposited along with Ni particles over Ti grid. Authors found that in order to achieve at least 50% of Raney Ni in the deposit, a current density of 100 mA cm^{-2} has to be maintained.

The use of Raney Ni as catalyst prompted Cooper and Botte to investigate Raney Ni as a substrate of the noble metal alloy catalysts.[41] The metal alloy catalyst studied by the authors consisted of Rh and Pt. The experiments were designed to optimize Rh loading and followed with Pt loading optimization. Rh was electrodeposited over Raney Ni foil in loadings of 0.5, 1, 2, 3 and 4 mg cm^{-2}. These electrodes were further deposited with 1 mg cm^{-2} of Pt for ammonia electrolysis testing. The electrolyte used for ammonia electrolysis was 1 M NH_3 in 1 M KOH (ammonia solution). Cyclic voltammetry of these electrodes in an ammonia solution is shown in Fig. 4, where the current density is normalized to Rh loading for each electrode.

Each electrode displayed an oxidation peak in the range of –0.7 V to –0.57 V (vs. Hg/HgO), which has been related to the oxidation of OH^- ion by the electrochemical activity of Rh catalyst. Botte has postulated a mechanism to explain the surface blockage of the active sites during the oxidation of ammonia.[52] The surface blockage is considered to be the adsorption of OH^- ions rather than N_{ads} or its intermediates as widely believed in the literature.[51,55-57] In Fig. 4, Rh catalyst has proved the existence of OH^- ion adsorption by displaying the oxidation peak near –0.7 V. Among the different loadings for Rh, 1 mg cm^{-2} was chosen by the authors because it had the most catalytic activity per mass of Rh. During ammonia electrolysis, the cell voltage was 0.64 V for a current

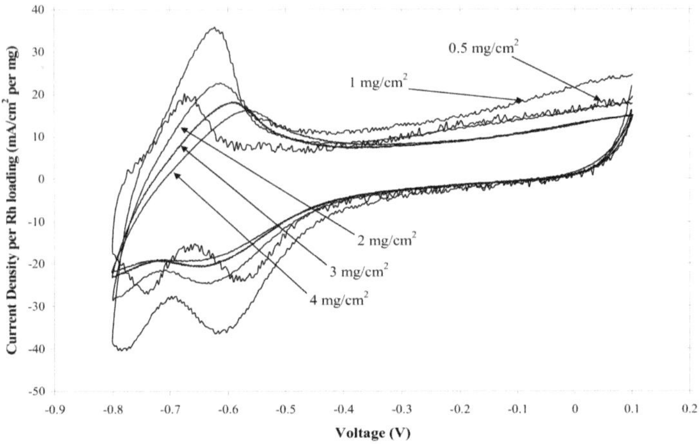

Figure 4. Comparison of various loadings of Rh with 1-mg cm^{-2} Pt in 1 M NH_3 in 1 M KOH solution at 25°C with 10 mV s^{-1} sweep rate. Reproduced from Ref. 41, Copyright (2006) by permission of The Electrochemical Society.

density of 0.25 mA cm^{-2} with the electrode carrying 1 mg cm^{-2} of Rh and Pt catalysts.

The authors proceeded to optimize Pt loading by comparing with 5 and 10 mg cm^{-2} of Pt deposition over 1 mg cm^{-2} of Rh loading. These two electrodes were tested in 1 M NH_3 in 5 M KOH solution so that higher current densities can be applied. The electrode with 5-mg cm^{-2} loading of Pt was able to hold only 2.5 mA cm^{-2} current density for a long duration, whereas 10-mg cm^{-2} Pt loading electrode sustained a maximum current density of 5 mA cm^{-2}. Morphological analysis of these electrodes is shown in Fig. 5 with SEM images.

The electrolysis of ammonia with 5 mg cm^{-2} Pt and 1 mg cm^{-2} Rh catalyst at 2.5 mA cm^{-2} current density resulted in a steady cell voltage of 0.633 V for 11 hrs. This voltage corresponds to 1.58 mW cm^{-2} of power requirement for electrolysis. The authors assumed 100% efficiency for ammonia electrolysis to produce 1.04 mL H_2 h^{-1} per cm^2 of catalyst or equivalently to generate 16.83 Wh per gram of H_2 from the hydrogen produced. On comparing, 10-mg cm^{-2} Pt and 1-mg cm^{-2} Rh electrode at 2.5-mA cm^{-2} current density had a constant cell voltage of 0.547 V with 1.37 mW cm^{-2}

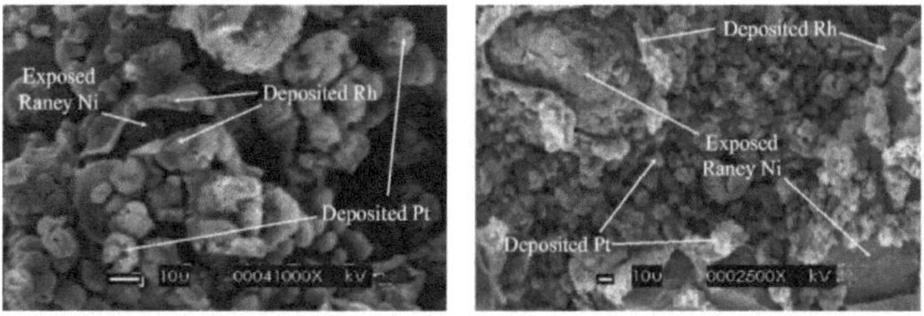

(a) (b)

Figure 5. SEM images of (a) 5-mg cm^{-2} Pt + 1-mg cm^{-2} Rh and (b) 10-mg cm^{-2} Pt + 1-mg cm^{-2} Rh. Reproduced from Ref. 41, Copyright (2006) by permission of The Electrochemical Society.

power consumption which accounted for 1.04 mL H_2 h^{-1} per cm^2 of catalyst or equivalently 14.54 Wh per gram of H_2 (13.6% power reduction). At the maximum current density of 5 mA cm^{-2}, the 10-mg cm^{-2} Pt and 1-mg cm^{-2} Rh electrode had a steady voltage of 0.599 V and produced 2.09 mL H_2 h^{-1} per cm^2 of catalyst at 15.93 Wh per gram of H_2 by consuming 3 mW cm^{-2} of power.

The initial studies conducted at Ohio University for demonstration of ammonia electrolysis involved work by Vitse, et al.,[3] they compared Pt metal and its alloys as catalyst for electro-oxidation of ammonia. The catalysts used were Pt black, Pt-Ir (10% Ir) and Pt-Ru (13% Ru) and these were electrodeposited over Pt foil substrate with 2.5 mg cm^{-2} loading. Polarization experiments were performed at 60°C with stirred 1 M NH_3 in 5 M KOH solution, using Pt-Ru as a counter electrode. A three-electrode 1-L glass cell, which can function as either a one compartment or a two compartment cell, was used in this study for batch experiments and the reference electrode was Hg/HgO. In a two-compartment situation a polypropylene membrane was used. A closer look at different anodes during the polarization experiments revealed that Pt-Ir electrode has 0.36 V overpotential at 70 mA cm^{-2} current density and 0.56 V overpotential for Pt black. Pt-Ru electrode overpotential was placed between 0.36 and 0.56 V. The polarization curve for ammonia electrolytic cell is shown in Fig. 6. In order to produce hydrogen gas, the ammonia electrolytic cell required a minimum of 0.4 V and if Pt-Ir was used as the cathode, then the cell voltage could be as low as 0.3 V.

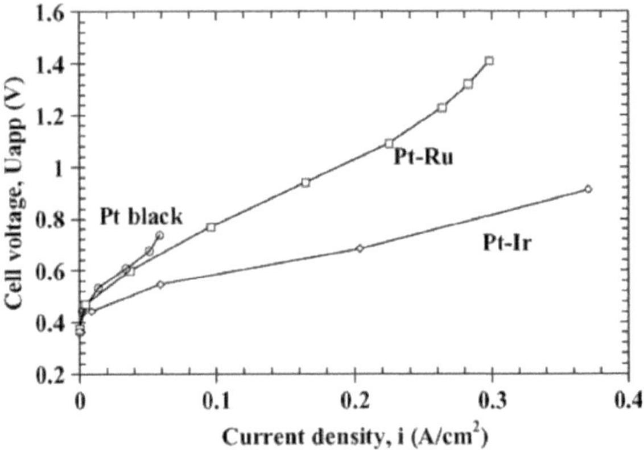

Figure 6. Polarization curves for ammonia electrolytic cell. Reprinted from Ref. 3, with permission from Elsevier.

Using the information on cell voltage, Vitse, et al. calculated the efficiency curves for each catalyst material used in the anode (Fig. 7). Efficiency for the ammonia electrolytic cell can be defined as the amount of energy that can be gained from burning H_2 gas divided by the total energy spent to obtain H_2 (energy from

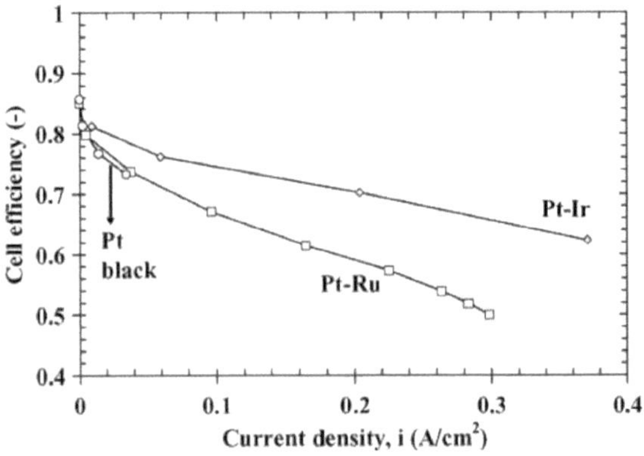

Figure 7. Efficiency curves for ammonia electrolytic cell with different anode catalyst. Reprinted from Ref. 3, with permission from Elsevier.

NH_3 plus electrical energy for electrolysis). The ammonia electrolytic cell efficiency was calculated using the following equation

$$\varepsilon = \frac{3 \times \Delta H_{H_2}}{2 \times \Delta H_{NH_3} + 6 \times F \times \Delta E} \tag{28}$$

where ΔH_{H_2} is the lower heating value of H_2 (242.7 kJ mol^{-1}), ΔH_{NH_3} is the lower heating value of NH_3 (320.1 kJ mol^{-1}), F is Faraday's constant (26.8 Ah mol^{-1}) and ΔE is the cell voltage. The efficiency for an ammonia electrolytic cell using Pt-Ir as catalyst in the anode varies from 80% at 10 mA cm^{-2} to 60% at 400 mA cm^{-2}.

The gases produced during the experiments were collected in two separate columns for nitrogen and hydrogen when the two compartment cell was used. Analysis of the collected gas was performed using gas chromatography and no CO_x, NO_x, or NH_3 was detected in both gas streams. Further, the hydrogen gas stream was passed through HPLC water at room temperature so that any NH_3 escaping through H_2 gas would increase the pH of HPLC water. The authors found that the pH of the HPLC water increased with the start of gas bubbling and settled at pH 8.4 after 5 hours. Based on the vapor-liquid equilibrium, the concentration of NH_3 in the gas phase of the bubbler was calculated to be less than 1 ppm. This analysis of gas streams produced by ammonia electrolysis has proved that high-purity hydrogen can be easily produced using this method.

In other experiments performed at the EERL, the substrate was changed from Pt foil to Ti foil and four different noble metals (Pt, Ir, Ru and Rh) were studied as monometallic, bi-metallic, and tri-metallic catalysts for ammonia oxidation.[64] Vitse, et al. found that among mono-metallic catalyst Ir and Ru had lower overpotential of 0.3 V, whereas Pt had 0.4 V as overpotential and higher current density for ammonia oxidation. In the case of bi-metallic catalysts, Pt-Rh and Pt-Ir displayed promising qualities. The overpotential for Pt-Rh catalyst was 0.27 V and it also exhibited a reduction peak characteristic of a Rh catalyst, related to reduction of dehydrogenated intermediates formed during electro-oxidation of NH_3. Pt-Ir bi-metallic catalyst combined the low overpotential from Ir and stable catalytic activity with higher current density from Pt. These results led to the formation of tri-metallic catalyst (Pt-Ir-Rh), which was able to achieve high current density at low overpotential of 0.27 V.

Next, the development in the substrate for the electro-catalyst was performed at EERL by wrapping polyacrylonitrile (PAN) carbon fibers (Celion G30-500) from BASF over Ti gauze. A bundle of carbon fiber consisting of 6000 filaments with an average length of 35 cm was wrapped around Ti gauze. Electro-catalysts Pt, Ir and Rh were preferentially electroplated over the carbon fibers with the intention of increasing their surface area to enhance catalytic activity. One of the objectives of the study conducted by Bonnin, et al. was to compare carbon fiber substrate with Raney nickel.[45] The authors electroplated the Pt-Ir alloy catalyst on a carbon fiber with a loading of 4.6 mg cm^{-1} as compared to 8 mg cm^{-1} on the Raney nickel substrate. The cyclic voltammetry of these two electrodes in 1 M NH_3 in 1 M KOH solution at 25°C is displayed in Fig. 8. An oxidation peak is observed for Pt-Ir on Raney nickel electrode at -0.2 V vs. Hg/HgO, which is electro-oxidation of NH_3. The slope of the curve at the beginning of the oxidation peak was identical for both Raney nickel and carbon fiber electrodes. But in the case of Raney nickel, the current decreased after the polarization potential, which the authors suggest is due to the blockage of active sites on Pt-Ir alloy by OH^- ion adsorption. The authors tested the stability of the Raney nickel electrode by exposing the electrode to air every week before cyclic voltammetry experiment. A decrease of 200 mA for the oxidation of NH_3 was noticed in a period of four weeks, which indicated the authors that carbon fiber substrate is better suited for Pt-Ir catalysts to perform electro-oxidation of NH_3.

A recent advancement in substrate for electrodes has been the use of carbon paper, especially from TORAY®. Carbon paper is a matte of carbon fibers pressed together with polymer binders to form a sheet. The carbon paper is used in the membrane electrode assembly (MEA) for fuel cells as a gas diffusion layer (GDL), primarily for diffusion of hydrogen and oxygen gas to catalyst particles. The anode electrode for electro-oxidation of NH_3 is prepared by sandwiching Ti gauze with two TORAY carbon paper (2 cm x 2 cm) and the papers were held together by Ti foil. Electroplating of the Pt-Ir alloy over the carbon paper electrode was studied using different compositions of Pt and Ir salt. The cyclic vol-

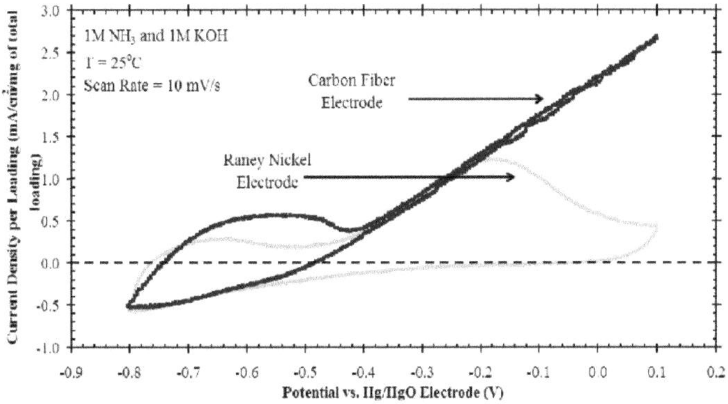

Figure 8. Cyclic voltammetry of Raney nickel and carbon fiber electrode with Pt-Ir catalyst in ammonia solution (1 M NH_3 in 1 M KOH). This article was published in Ref. 45, Copyright (2008) Elsevier.

tammetry of different Pt-Ir compositions on carbon paper electrodes (average loading: 1.25 mg cm^{-2}) in ammonia solution (1 M NH_4OH in 5 M KOH) is shown in Fig. 9.

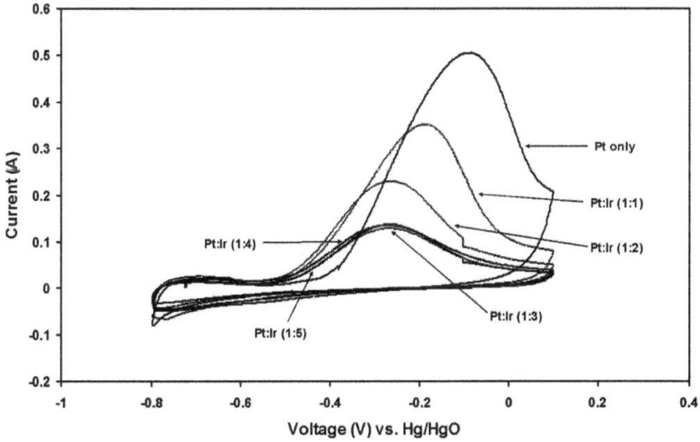

Figure 9. Cyclic voltammetry of carbon paper electrodes with Pt-Ir catalyst from different plating solution composition.

The carbon paper electrode has shown a better performance in ammonia electrolysis and electroplating Pt-Ir alloy over carbon paper has been easier compared to carbon fiber electrodes. Studies using carbon paper as an electrode are presently being pursued and the results will be available for future publications.

2. Cathode Electrode

Production of hydrogen gas occurs at the cathode electrode by reduction of water. This reaction is common for both alkaline water and ammonia electrolysis, so the catalyst and support materials for the cathode electrode can be same for both processes. The cathode material has to possess low overpotential for hydrogen evolution so that less energy will be consumed for electrolysis. Metals have different hydrogen overpotentials but among them Pt is considered to have the lowest H_2 overpotential. Pt is a very expensive metal so industrial water electrolyzers have used mild steel, Ni and Ni-based alloys as cathodes.

A study using Ni ultra fine particles (UFP) as a catalyst for the cathode was performed by Ezaki, et al.[65] The authors concluded that the performance of Ni UFP alloyed with Mo UFP (Ni-Mo UFP) had the lowest overpotential for hydrogen. Hu, et al. developed a novel multilayer cathode for water electrolysis to overcome the problem of losing the catalytic activity of the electrode if the electrolyzer was shut down for more than two weeks.[66] The authors' recommendation was to use a Mm-based hydride alloy, which will sever as the hydrogen absorbing alloy, and coat it with a layer of Ni-Mo alloy catalyst. The cathode material was made of porous Ni substrate, which had $MmNi_{3.6}Co_{0.75}Mn_{0.42}Al_{0.27}$ alloy coated with Ni-Mo alloy catalyst. H_2 overpotential for the cathode in 30% KOH solution with a current density of 200 mA cm^{-2} at 70°C, was 88 mV. This cathode material was able to maintain the same overpotential for a continuous operation of 4300 hours, even after 3000 hours of intermittent electrolysis. A ternary alloy like Ni-Fe-Mo was investigated for use as a cathode for water electrolysis.[67] The overpotential for hydrogen evolution with Ni-Fe-Mo (64:24:12) (wt. %) alloy in 30% KOH solution was 148 mV at 30°C for 100 mA cm^{-2}, which decreased to 104 mV at 70°C for the same current density.

The Ni-Fe-Mo alloy has demonstrated good catalytic activity, but adding Zn to this mixture will lead to large surface area structures as well. This was obtained by Crnkovic, et al. by co-deposition of Ni, Fe, Mo and Zn over mild steel substrate and leaching Zn in 28% KOH solution at 80°C.[68] This cathode displayed an adherent Ni-Fe-Mo-(Zn) alloy layer over mild steel. The performance of the cathode in 6 M KOH solution at a current density of 135 mA cm^{-2} was 83.1 mV H_2 overpotential at 80°C. But the cathode could not maintain the lowest reported hydrogen overpotential for a long time as the overpotential rose to 157 mV in 440 hours. In a recent study, different kinds of stainless steel (304, 316 and 430) were tested as cathodes for water electrolysis and it was found that SS 316 had a better cathodic performance since it has more Ni content.[69] In other investigations, sulfur is being introduced into the Ni based alloys to enhance the catalytic activity for cathode material either as Ni-S alloy coating on nickel foam substrate [70] or Ni-S-Mn alloy on different Ni substrates.[71]

In the case of ammonia electrolysis, generally the anode and the cathode were made of the same electrode material—Pt-Ir catalyst over carbon paper. All the studies conducted at EERL utilized Pt-Ir on carbon paper as cathode electrode. Presently, new projects are underway to identify the best cathode material for ammonia electrolysis. The overpotential for the oxidation of ammonia is lower than for water oxidation; therefore, depending on the application a noble metal based cathode will be more convenient. Catalysts being considered for the cathode electrode are Pt, Ir, Rh, Ni and Co or its alloys over carbon paper material.

3. Electrolyte

Composition of the electrolyte solution used for ammonia electrolysis is very important, as it will determine the cell voltage and current density for ammonia oxidation, which in turn dictates the energy consumption for hydrogen production. Oxidation of ammonia by an electrochemical method has been applied in an amperometric sensor for ammonia gas.[60] In this study, the electrolyte used to investigate the electro-catalyst for the anode had a varying concentration (0.4, 1.6, and 2.5 mM) of ammonium sulfate $[(NH_4)_2SO_4]$ in 1 M KOH. During sensor analysis, ammonia was introduced to the 1 M KOH solution as ammonium sulfate ranging from 10 to 200 ppm to evaluate the performance of the ammonia sensor.

Ammonium hydroxide (NH_4OH) or ammonia in water has been widely used as an electrolyte solution for electrochemical oxidation of ammonia. The electrolyte concentration used in the investigation of catalyst materials for the anode electrode has been ranging from 1 mM to 1 M NH_3 in 1 M KOH.[57,59,61,62] Other chemicals containing ammonia have also been used for electro-oxidation studies. In an investigation to identify the possibility of using dimensionally stable anode (DSA), namely IrO_2, for the treatment of wastewater, the electrolyte composition was 0.5 M $(NH_4)_2SO_4$ or 1 M NH_4Cl in 0.1 M Na_2SO_4 solutions.[58] The authors studied the effect of pH for the decomposition of NH_3 to N_2; they concluded that in the basic solution NH_3 mainly oxidized to nitrogen gas whereas the ammonium ion in the neutral and acidic solution was partly decomposed to N_2 by OH radicals generated under the oxygen evolution. It was noted that the decomposition rate for NH_3 to N_2 was fastest in the basic solution. A similar finding was observed in a recent study by Halseid et al.,[72] where the authors' investigation was focused on understanding ammonium oxidation in acidic solution. Halseid, et al. used perchloric acid ($HClO_4$) and sulfuric acid (H_2SO_4) as supporting electrolyte for ammonium hydroxide solution. They observed slow oxidation of ammonium ions and the presence of ammonium ion affected the formation and reduction of Pt oxide at higher potentials involving the formation of very stable nitrogen or nitrogen-oxygen species.

At EERL, the electrolyte composition has been a basic solution with ammonium hydroxide in KOH. Vitse et al. added 1 M NH_3 to 5 M KOH solution for electro-oxidation of ammonia at high temperature of 60°C.[3] On the other hand, Cooper and Botte investigated Pt and Rh catalyst over Raney Ni substrate using 1 M NH_3 in 1 M KOH solution at 25°C.[41] In a recent article, Bonnin, et al. have performed ammonia electrolysis in solutions having low concentration of ammonia.[45] Investigation on electro-oxidation of ammonia at low concentrations has indicated the possibility of cleaning ammonia from wastewater. Anode electrodes were made of carbon fiber wound around Ti gauze and different electrocatalyst with varying loadings were used in the study. The anode electrodes used are labeled (Fig. 10): Electrode 1 (Rh: 5 mg cm^{-1}), Electrode 2 (Pt: 5.3 mg cm^{-1}), Electrode 3 (Pt-Rh: 5.2 mg cm^{-1}),

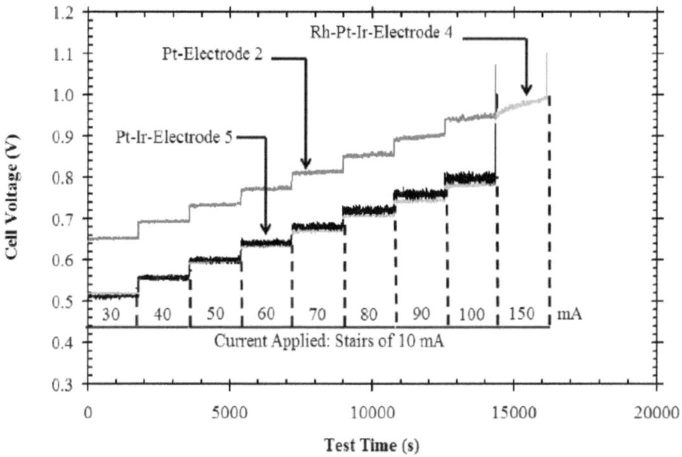

Figure 10. Galvanostatic performance of the electrodes in 20 mM NH_4OH in 0.2 M KOH. This article was published in Ref. 45, Copyright (2008) Elsevier.

Electrode 4 (Pt-Ir-Rh: 5.1 mg cm^{-1}), and Electrode 5 (Pt-Ir: 4.6 mg cm^{-1}). Figure 10 displays the galvanostatic performance of each anode with cathode being Pt-Ir-Rh catalyst over carbon fiber in a solution of 20 mM NH_4OH in 0.2 M KOH at 25°C. In a galvanostatic experiment, each anode electrode is subjected to a current staircase starting from 30 mA and increased to 150 mA with a 10 mA step, and at each step the current is held for 30 minutes. Among the electrodes, Electrodes 4 (Pt-Ir-Rh) and 5 (Pt-Ir) had similar cell voltage for each current step, but Electrode 4 was able to attain higher current of 150 mA before reaching a cell voltage of 1 V. On comparing with Electrode 2 (Pt), the cell voltage for Electrodes 4 and 5 was at least 100 mV less than Electrode 2.

Authors concluded that Pt-Ir-Rh on carbon fiber was the anode electrode for electrolysis of ammonia at low concentrations. They also calculated both ammonia conversion and Faradic efficiency for this anode. Electrode 4 (Pt-Ir-Rh) was able to oxidize ammonia from 21.5 mM to 1.83 mM in 13.8 hours at 100 mA with a conversion efficiency of 91.49% and the Faradic efficiency for the system was 91.81%. These high efficiency values demonstrate that ammonia electrolysis can also remove ammonia from low concentration solutions, which is significant towards wastewater treatment for ammonia so as to meet EPA standards of 0.08 mM NH_3.[73]

4. Separator Membrane

The electrolysis process consists of anode and cathode reactions. If both of these reactions take place in one container then it is known as an open cell system and if the reactions happen in two separate compartments then it is known as a divided cell system. In a divided cell system, ionic conductivity between anode and cathode compartment is established by a diaphragm or membrane. The function of the membrane is to avoid mixing of gases produced and electrolyte between the two compartments but allow the transport of ions for conductivity.

The diaphragm or membrane most commonly used in the alkaline water electrolyzer (industrial model) is asbestos. But asbestos is not the best membrane material available as it has limitations like low corrosion resistance to alkaline solution, cost and toxicity (carcinogenic). Literature on membrane research suggests that there are other available materials to act as diaphragm for water electrolysis. Mostly polymer compounds are considered for membranes like poly (phenylene sulfide) (PPS), poly (tetra fluroethylene) (PTFE) or commonly known as Teflon, and polysulfone (PSF).[74] Other materials that have been investigated as membranes are poly (ether sulfone),[75] Zirfon® is a porous composite membrane made of polysulfone matrix with ZrO_2 powder,[76] and homogenous mixture of poly (ether sulfone) and poly (vinylpyrrolidone) (PES-PVP).[77] It is generally known that Nafion membranes are used in an acidic environment, but Kim, et al. have experimented with Nafion 424 (cation exchange membrane) as a diaphragm to study ammonia electrolysis with IrO_2 as the anode and Ti as the cathode.[58] At EERL, Teflon has been used as membrane to separate both gas and electrolyte solution from mixing. It is expected that most membranes used in alkaline water electrolyzers can potentially be employed in an ammonia electrolyzer.

V. AMMONIA ELECTROLYZER: A PROTOTYPE

Demonstration of ammonia electrolysis technology requires the continuous functioning of an electrolyzer. Using the developments made in the area of anode electrode and utilizing the same material as cathode electrode, a continuous ammonia electrolytic cell was first constructed and operated by Botte and Benedetti.[43] The electrolytic cell was separated with a membrane and ammonia solution was flown through both anode and cathode compartments, which housed Pt-Ir on carbon fiber as electrodes (4 cm^2). A statistical analysis of the variables affecting the performance of ammonia electrolytic cell was studied using MINITAB software. Authors varied different parameters – concentration of KOH (0.5 and 7 M) and NH_3 (0.5 and 5 M), current (100 and 300 mA) and temperature of the electrolytic cell (25 and 50°C), to assess the electrochemical performance of the cell. The lowest energy consumption was found to be 8.6 Wh per gram of H_2 (that is 0.33 V operating voltage), for a system with 5 M NH_3 in 7 M KOH solution operated at 50°C with a current of 100 mA. On the other hand, a solution of 0.5 M NH_3 in 7 M KOH functioning at 50°C with 100 mA current exhibited only 0.35 V as cell voltage (9.4 Wh per gram of H_2).

The development of an ammonia electrolyzer prototype has made a strong stride with recent success in designing and operating a multi-stack, continuous closed cell ammonia electrolytic cell (AEC), which can produce hydrogen enough to sustain up to 2.5 W fuel cell.[44] In order to supply a 2.5 W fuel cell, a 9-cell AEC stack was designed starting with electrode preparation, membrane electrode assembly (MEA), and testing the prototype under varying parameters. The stack configuration for AEC was scaled up from 2-cell to 5-cell and finally extended to 9-cell set-up. Schematic representation of an assembled 9-cell AEC stack is shown in Fig. 11.

Electrodes for the AEC were prepared by electroplating Pt-Ir alloy as electro-catalyst over TORAY carbon paper. End plate, electrode plate, and gasket (separator plates) were designed and constructed with materials compatible to ammonia and potassium hydroxide solution. The anode side of the AEC was supplied with 1 M NH_4OH dissolved in 1 M KOH solution, whereas the cathode side was filled with only 1 M KOH solution. The cathode and the anode compartments were separated by a proprietary Teflon mem-

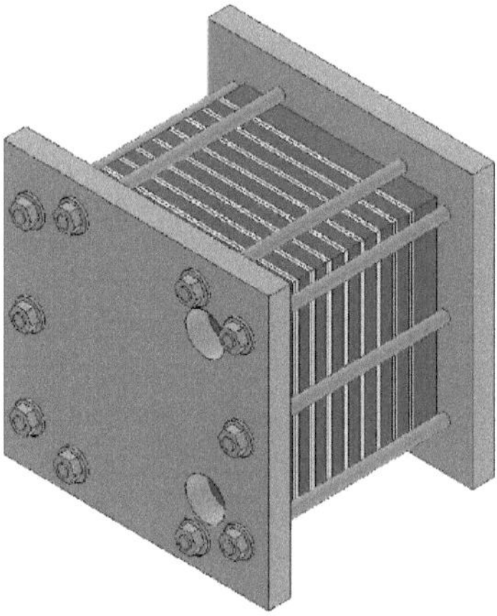

Figure 11. Schematic representation of 9-cell AEC stack.[44]

brane. The experimental set-up for 9-cell AEC with electrolytes flowing and gas collecting system is shown in Fig. 12.

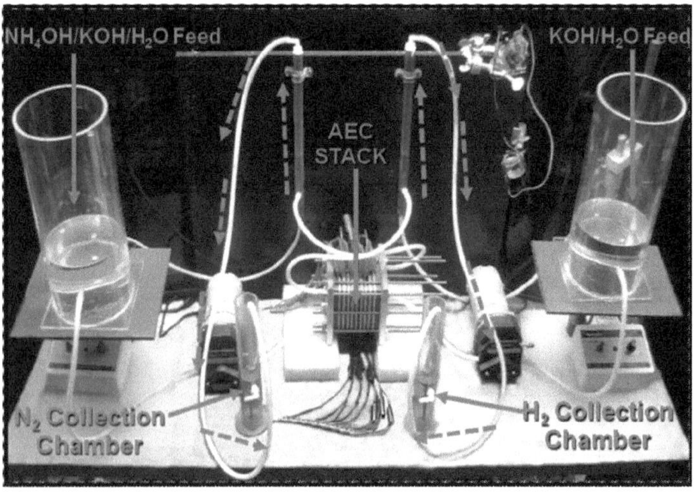

Figure 12. Experimental set-up for 9-cell AEC stack to produce hydrogen gas by ammonia electrolysis.[44]

Galvanostatic tests were performed on the AEC stack to evaluate the power consumption, H_2 gas generation and effect of temperature under a continuous flowing condition. Current applied for each cell was restricted to 500 mA so that high anode overpotential can be avoided, which means the maximum current passed through the 9-cell AEC stack was 4.5 A. At 25°C, the stack voltage was 0.633 V which was 73 mV higher than the voltage noticed at 55°C, that is, 0.56 V. Hydrogen gas was collected during the AEC stack testing. The efficiency for producing hydrogen gas was 97.55% at both temperatures, i.e., 0.164 g h^{-1} of H_2 was produced as compared to 0.168 g h^{-1} from Faraday's law. Net power consumed by the 9-cell AEC stack at 25°C was 2.85 W, which translates to 17.36 Wh per gram of hydrogen produced, but at higher temperature (55°C) the net power consumption was 2.52 W equivalent to 15.36 Wh per gram of H_2 produced.

The electrodes used in the AEC stack were tested in an open cell system and they had cell voltages similar to the ones reported by Botte and Benedetti.[43] This demonstrates that ammonia electrolysis is linearly scalable. But when these electrodes are placed in the AEC stack (9 cell closed system), their cell voltages were higher than open cell system. The voltage loss was 153 mV and 180 mV at 25°C and 55°C, respectively. This cell voltage increase in the electrolyzer is due to the ohmic loss from the present configuration of electrical connections for the AEC stack.

The 9-cell AEC stack was combined with a 4 W PEM fuel cell to assess the integration of ammonia electrolyzer with a fuel cell by calculating energy conversion efficiency. Energy conversion efficiency is calculated as the net useful energy available from the integration over the total energy consumed by AEC stack for ammonia electrolysis and the average energy efficiency for the fuel cell was assumed to be 65%. At 25°C, the energy conversion efficiency for AEC-PEMFC system was only 23.62% and 39.71% at 25°C and 55°C, respectively. This low efficiency is due to the significant ohmic loss mainly from the poor electrical connections to the electrodes in the AEC stack. Future improvement for the AEC stack is to eliminate the external electrical connection loss of 153 mV (25°C) and 180 mV (55°C). On removing the ohmic losses from AEC stack, the expected energy conversion efficiency for the AEC-PEMFC system would be as high as 62.87% (25°C) and

95.36% (55°C). This should enable ammonia electrolyzer to compete with any hydrogen producing technology.

VI. SUMMARY

The issue of hydrogen production, storage, and transportation can be solved with the use of ammonia as hydrogen carrier. Ammonia electrolysis demonstrates the simplicity in production of hydrogen gas and provides solution to the difficulty faced with hydrogen storage and transportation. The ammonia electrolyzer can be used to produce hydrogen on demand as shown on the experiments performed by Botte and Benedetti.[43] EERL group members have worked on a shoe-sized vehicle which proves the concept of on-board reforming of ammonia using the ammonia electrolyzer.[78] The ammonia electrolyzer in the shoe-sized car was able to produce 2.25 ml min^{-1} of H_2 by consuming energy equivalent to 13.1 Wh per gram of H_2. The PEMFC, connected to an ammonia electrolyzer, was able to generate 18.7 Wh energy for every gram of H_2 supplied, which was sufficient enough to power a 68 mW motor (5.6 Wh per gram of H_2) to drive the shoe-sized car and also use the remaining energy—13.1 Wh per gram of H_2—for the ammonia electrolyzer.

Furthermore, it has been established that ammonia electrolysis can be used as a remediation process coupled with hydrogen cogeneration (Bonnin, et al.).[45] Back of the envelope calculations indicate that hydrogen can be produced at less than $3 per kg using ammonia electrolysis. This cost could be significantly reduced if ammonia from waste is used for the electrolyzer. For example, this technology will be a beneficial process for farmers who could remediate their effluents or wastewater and produce electric power for their farms.

Further research and development is needed to make this technology more efficient. Current work at EERL has to do with the development of cell stacks, catalyst, and electrolyte. However, results from the stacks developed by the EERL group could be used to build large-scale demonstration units, which will facilitate early adoption of this technology for different applications including vehicle transportation, electricity for residential houses, back-up power, and ammonia remediation. Electrolysis of ammonia wastewater has the highest energy efficiency for the production of hydrogen among the different hydrogen production methods available.

REFERENCES

[1]J. Larminie and A. Dicks, Fuel Cell Systems Explained, 2nd ed., John Wiley & Sons Ltd, West Sussex, England, 2003.
[2]G. G. Botte, F. Vitse, and M. Cooper, U.S. Pending Patent, 2004.
[3]F. Vitse, M. Cooper, and G. G. Botte, *J. Power Sources* 142 (2005) 18.
[4]R. Farrauto, S. Hwang, L. Shore, W. Ruettinger, J. Lampert, T. Giroux, Y. Liu, and O. Ilinich, *Annu. Rev. Mater. Res.* **33** (2003) 1.
[5]Z. I. Onsan and Turk. *J. Chem.* **31** (2007) 531.
[6]D. R. Palo, R. A. Dagle, and J. D. Holladay, *Chem. Rev.* **107** (2007) 3992.
[7]N. T. Nguyen and S. H. Chan, *J. Micromech. Microeng.* **16** (2006) R1.
[8]S. Wasmus and A. Kuver, *J. Electroanal. Chem.* **461** (1999) 14.
[9]M. Boder and R. Dittmeyer, *J. Power Sources* **155** (2006) 13.
[10]S. H. Clarke, A. L. Dicks, K. Pointon, T. A. Smith, and A. Swann, *Catal. Today* **38** (1997) 411.
[11]A. L. Dicks, *J. Power Sources* **71** (1998) 111.
[12]J. M. Ogden, *Annu. Rev. Energ. Env.* **24** (1999) 227.
[13]R. M. Navarro, M. A. Pena, and J. L. G. Fierro, *Chem. Rev.* **107** (2007) 3952.
[14]J. P. Longwell, E. S. Rubin, and J. Wilson, *Prog. Energ. Combust.* **21** (1995) 269.
[15]S. P. S. Badwal, S. Giddey, and F. T. Ciacchi, *Ionics* **12** (2006) 7.
[16]G. Schiller, R. Henne, P. Mohr, and V. Peinecke, *Int. J. Hydrogen Energy* **23** (1998) 761.
[17]E. Varkaraki, N. Lymberopoulos, E. Zoulias, D. Guichardot, and G. Poli, *Int. J. Hydrogen Energy* **32** (2007) 1589.
[18]F. Barbir, *Sol. Energy* **78** (2005) 661.
[19]S. A. Grigoriev, V. I. Porembsky, and V. N. Fateev, *Int. J. Hydrogen Energy* **31** (2006) 171.
[20]A. Marshall, B. Borresen, G. Hagen, M. Tsypkin, and R. Tunold, *Energy* **32** (2007) 431.
[21]The *Hydrogen Economy: Opportunities, Costs, Barriers, and R&D Needs,* National Research Council and National Academy of Engineering, Washington, D.C., 2004.
[22]J. Phillips, *Control and Pollution Prevention Options for Ammonia Emissions*, ViGYAN Incorporated, Research Triangle Park, NC, 1995, EPA-456/R-95-002.
[23]T. Lipman and N. Shah, *Ammonia as an alternative energy storage medium for hydrogen fuel cells: Scientific and technical review for near-term stationary power demonstration projects*, University of California - Berkeley, Berkeley, CA, 2007, UCB-ITS-TSRC-RR-2007-5.
[24]G. Thomas and G. Parks, Pot*ential Roles of Ammonia in a Hydrogen Economy*, US Department of Energy (DOE), 2006.
[25]S. F. Yin, B. Q. Xu, X. P. Zhou, and C. T. Au, *Appl. Catal., A* **277** (2004) 1.
[26]R. A. Wynveen, *Fuel Cells* **2** (1963) 153.

[27]N. Maffei, L. Pelletier, J. P. Charland, and A. McFarlan, *J. Power Sources* **140** (2005) 264.
[28]A. F. Bouwman, D. S. Lee, W. A. H. Asman, F. J. Dentener, K. W. VanderHoek, and J. G. J. Olivier, *Global Biogeochem. Cycles* **11** (1997) 561.
[29]S. G. Sommer and N. J. Hutchings, *Eur. J. Agron.* **15** (2001) 1.
[30]D. Misenheimer, T. Warn, and S. Zelmanowitz, *Ammonia emission factors for the NAPAP emission inventory,* Alliance Technologies Corporation, 1987, EPA-600/7-87/001.
[31]G. E. Mansell, A*n improved ammonia inventory for the WRAP Domain*, ENVIRON International Corporation, 2005.
[32]Y. Lu, H. Wang, Y. Liu, Q. S. Xue, L. Chen, and M. Y. He, *Lab Chip 7* (2007) 133.
[33]R. Metkemeijer and P. Achard, *J. Power Sources* **49** (1994) 271.
[34]R. Metkemeijer and P. Achard, *Int. J. Hydrogen Energy* **19** (1994) 535.
[35]Y. Liu, H. Wang, J. Li, Y. Lu, H. Wu, Q. Xue, and L. Chen, *Appl. Catal.*, **A 328** (2007) 77.
[36]W. H. Chen, I. Ermanoski, and T. E. Madey, *J. Am. Chem. Soc.* **127** (2005) 5014.
[37]P. F. Ng, L. Li, S. B. Wang, Z. H. Zhu, G. Q. Lu, and Z. F. Yan, *Environ. Sci. Technol.* **41** (2007) 3758.
[38]V. Hacker and K. Kordesch, in: *Handbook of Fuel Cells - Fundamentals, Technology and Applications*, W. Vielstich et al. (Eds.), Vol. 3, Wiley, Chichester, 2003, p. 121.
[39]K. Kordesch, V. Hacker, R. Fankhauser, and G. Faleschini, *Apollo Energy Systems*, Inc. US 6,936,363 B2.
[40]K. Kordesch, V. Hacker, J. Gsellmann, M. Cifrain, G. Faleschini, P. Enzinger, R. Fankhauser, M. Ortner, M. Muhr, and R. R. Aronson, *J. Power Sources* **86** (2000) 162.
[41]M. Cooper and G. G. Botte, *J. Electrochem. Soc.* **153** (2006) A1894.
[42]R. Paur, Army Research Office, personal communication, 2004.
[43]G. G. Botte and L. Benedetti, *under preparation for publication*.
[44]M. Biradar, M. Muthuvel, G. G. Botte, *under preparation for publication.*
[45]E. P. Bonnin, E. J. Biddinger, and G. G. Botte, *J. Power Sources* **182** (2008) 284.
[46]C. G. Alfafara, T. Kawamori, N. Nomura, M. Kiuchi, and M. Matsumura, *J. Chem. Technol. Biotechnol.* **79** (2004) 291.
[47]K. W. Kim, I. T. Kim, G. I. Park, and E. H. Lee, *J. Appl. Electrochem.* **36** (2006) 1415.
[48]J. L. H. Park and W. Y. Choi, E*nviron. Sci. Technol.* **36** (2002) 5462.
[49]F. J. Vidal-Iglesias, N. Garcia-Araez, V. Montiel, J. M. Feliu, and A. Aldaz, *Electrochem. Commun.* **5** (2003) 22.
[50]L. Marincic and F. B. Leitz, *J. Appl. Electrochem.* **8** (1978) 333.
[51]H. Gerischer and A. Mauerer, *J. Electroanal. Chem.* **25** (1970) 421.
[52]G. G. Botte, *Electrode Processes VII*, 206th Meeting of The Electrochemical Society, Hawaii, USA, October 3-8, 2004.
[53]B. V. Tilak, B. E. Conway, and H. Angerstein-Kozlowska, *J. Electroanal. Chem.* **48** (1973) 1.
[54]K. Sasaki and Y. Hisatomi, *J. Electrochem. Soc.* **117** (1970) 758
[55]S. Wasmus, E. J. Vasini, M. Krausa, H. T. Mishima, and W. Vielstich, *Electrochim. Acta* **39** (1994) 23.
[56]J. F. E. Gootzen, A. H. Wonders, W. Visscher, R. A. van Santen, and J. A. R. van Veen, *Electrochim. Acta* **43** (1998) 1851.

[57]A. C. A. de Vooys, M. T. M. Koper, R. A. van Santen, and J. A. R. van Veen, *J. Electroanal. Chem.* **506** (2001) 127.
[58]K. W. Kim, Y. J. Kim, I. T. Kim, G. I. Park, and E. H. Lee, *Electrochim. Acta* **50** (2005) 4356.
[59]K. Yao and Y. F. Cheng, *J. Power Sources* **173** (2007) 96.
[60]B. A. L. de Mishima, D. Lescano, T. M. Holgado, and H. T. Mishima, *Electrochim. Acta* **43** (1998) 395.
[61]K. Endo, Y. Katayama, and T. Miura, *Electrochim. Acta* **49** (2004) 1635.
[62]K. Endo, K. Nakamura, Y. Katayama, and T. Miura, *Electrochim. Acta* **49** (2004) 2503.
[63]M. Cooper and G. G. Botte, *J. Mater. Sci.* **41** (2006) 5608.
[64]F. Vitse, J. Gonzales, and G. G. Botte, 205th Meeting of The Electrochemical Society, San Antonio, Texas, US, 2004.
[65]H. Ezaki, T. Nambu, M. Morinaga, M. Udaka, and K. Kawasaki, *Int. J. Hydrogen Energy* **21** (1996) 877.
[66]W. K. Hu, X. J. Cao, F. P. Wang, and Y. S. Zhang, *Int. J. Hydrogen Energy* **22** (1997) 621.
[67]L. Ramesh, B. S. Sheshadri, and S. M. Mayanna, *Int. J. Energ. Res.* **23** (1999) 919.
[68]F. C. Crnkovic, S. A. S. Machado, and L. A. Avaca, *Int. J. Hydrogen Energy* **29** (2004) 249.
[69]J. M. Olivares-Ramirez, M. L. Campos-Cornelio, J. U. Godinez, E. Borja-Arco, and R. H. Castellanos, *Int. J. Hydrogen Energy* **32** (2007) 3170.
[70]H. W. He, H. J. Liu, F. Liu, and K. C. Zhou, *Surf. Coat. Technol.* **201** (2006) 958.
[71]Z. Q. Shan, Y. J. Liu, Z. Chen, G. Warrender, and J. H. Tian, *Int. J. Hydrogen Energy* **33** (2008) 28.
[72]R. Halseid, J. S. Wainright, R. F. Savinell, and R. Tunold, *J. Electrochem. Soc.* **154** (2007) B263.
[73]EPA, *Water Quality Standards*, Columbus OH. OAC Chapter 3745-1.
[74]V. M. Rosa, M. B. F. Santos, and E. P. Dasilva, *Int. J. Hydrogen Energy* **20** (1995) 697.
[75]J. Kerres, G. Eigenberger, S. Reichle, V. Schramm, K. Hetzel, and W. Schnurnberger, I. Seybold, *Desalination* **104** (1996) 47.
[76]P. Vermeiren, W. Adriansens, J. P. Moreels, and R. Leysen, *Int. J. Hydrogen Energy* **23** (1998) 321.
[77]S. Lu, L. Zhuang, and J. Lu, *J. Membr. Sci.* **300** (2007) 205.
[78]B. K. Boggs, A. Weber, G. G. Botte, 211th Meeting of the Electrochemical Society, Chicago, Illinois, US, 2007.

5

Applications of Synchrotron X-Ray Scattering for the Investigation of the Electrochemical Interphase†

Zoltán Nagy** and Hoydoo You

Materials Science Division, Argonne National Laboratory, Argonne, Illinois 60439, USA
***Present address: Department of Chemistry, The University of North Carolina at Chapel Hill*

I. INTRODUCTION

The central phenomenon of surface electrochemistry, the transfer of charge between an electronically conducting phase and an ionically conducting phase, always occurs at a phase boundary—an interface—between the two phases. The region of interest, however, is usually wider than a simple two-dimensional interface. The atomic-level structure of both phases at the interface can be

†The submitted manuscript has been created by the University of Chicago as Operator of Argonne National Laboratory ("Argonne") under contract No. W-31-109-ENG-38 with the U.S. Department of Energy. The U.S. Government retains for itself, and others acting on its behalf, a paid-up, nonexclusive, irrevocable worldwide license in said article to reproduce, prepare derivative works, distribute copies to the public, and perform publicly and display publicly, by or on behalf of the Government.

R.E. White (ed.), *Modern Aspects of Electrochemistry, No. 45*, Modern Aspects of Electrochemistry 45, DOI 10.1007/978-1-4419-0655-7_5,

drastically different from those of the bulk structures. These *special regions* can penetrate from a few Å to a few thousand Å in both phases. To emphasize the three-dimensional nature of this region of interest, it is often called the *interphase.* The most common combination is a solid metal/aqueous solution interphase, although numerous other possibilities also exist. The important effect of the state of the metal surface on the rate and mechanism of charge-transfer reactions was recognized at the very earliest times of electrochemical research. Special *active surface sites* were proposed to explain many experimental observations, and great care was taken to use *reproducible surface preparation* techniques. However, real understanding of these *surface effects* was very much hampered by the virtual absence of experimental techniques for the determination of the *atomic- and molecular-level* structure of the interphase, both in the morphological and in the chemical sense. The situation has changed considerably in recent decades with the development of UHV surface science, modern spectroscopic techniques, and, more recently, the discovery of scanning tunneling microscopy and related techniques. Numerous approaches have been tried for the investigation of electrochemical interphases both *ex situ* and *in situ*. In the past, the ex-situ techniques were used more often than the in-situ techniques, but this tendency is slowly being reversed. While much useful information can be obtained using the ex-situ techniques, there always remains a nagging doubt about the effect that the loss of potential control and the changing environment from metal/solution to metal/vacuum may have had on the surface conditions. In contrast, the in-situ techniques, while experimentally more difficult, overcome these disadvantages. Namely, they permit the continuous electrochemical control of the interphase during the structural/chemical examination (permitting also dynamic measurements under changing electrochemical conditions), and they retain the aqueous condition during the measurement.

The experimental difficulties of in-situ techniques stem from the special nature of the electrochemical interphase, namely, that it is a *buried interface* and the probe used for the investigation must be able to penetrate either the electronically conducting phase or the ionically conducting phase. Consequently, the probe must fulfill certain criteria:

(*a*) the probe's interaction with the atoms in the interphase under investigation, at the probe's incident flux, must be sufficiently strong to be surface/interface sensitive, and

(*b*) the interaction of the probe with at least one of the phases must be sufficiently weak for penetration to and from the interface without significant intensity loss. Among the atomic-level structural probes—neutrons, electrons, and x rays— only synchrotron x rays meet both criteria for the in-situ investigation of electrochemical interphases.

While there are a number of synchrotron-x-ray techniques that are used for examining electrochemical interphases, in this chapter, applications of the *in-situ* synchrotron surface-x-ray-scattering techniques to electrochemistry problems are reviewed ranging from submonolayer level phenomena, through nanometer size phenomena, to submicron size phenomena; that is, covering the full range of the *interphase* at an electrode surface. The purpose of the chapter is to demonstrate the usefulness of synchrotron surface-x-ray scattering to a very wide range of electrode phenomena rather than to give a complete review of all published work. Only a few specific applications will be described in some detail for each electrochemical phenomena, for which the techniques are used, as examples to show the capabilities of the techniques, although a comprehensive listing of all research carried out will be attempted in the references. A number of reviews have been published in the past, but these are usually restricted to some specific problem area or to the work of one research group.[1-15] Description of other synchrotron-x-ray techniques used in electrochemistry research have also been published. These are EXAFS/XANES,[16,17] the standing-wave technique,[18] and the utilization of infrared radiation produced at synchrotrons.[19]

After a brief review of the physics of x-ray scattering, a number of applications of the technique will be described for the investigation of a variety of electrochemical phenomena.

II. THEORY AND PRACTICE OF X-RAY SCATTERING

1. General Description of Surface-X-Ray Scattering (SXS)

An electron density distribution representing the condensed matter is needed for a discussion of the interactions of x rays with matter. The electron density of an atom, f_n, can be approximated by a distribution centered at the position of the nucleus of the nth atom. Then, the total electron density in real space is:

$$n(\vec{r}) = \sum_{n}^{N} f_n(\vec{r} - \vec{r}_n) \tag{1}$$

Applying Fourier transformation and denoting it by $S(\vec{Q})$ gives:

$$\begin{aligned} S(\vec{Q}) &= \int n(\vec{r})\, e^{i\vec{Q}\cdot\vec{r}}\, d\vec{r} \\ &= \sum_{n}^{N} \int f_n(\vec{r} - \vec{r}_n) e^{i\vec{Q}\cdot\vec{r}}\, d\vec{r} \\ &= \sum_{n=1}^{N} F_n(Q) e^{i\vec{Q}\cdot\vec{r}_n} \end{aligned} \tag{2}$$

Within the first-order Born-Oppenheimer approximation, this is essentially an approximation for x-ray-scattering cross section, besides the proportionality factor, known as Thomson scattering length or classical radius of a free electron ($e^2/m_ec^2 = 2.83\text{x}10^{-13}$ cm). For a crystalline structure, Eq. (2) can be greatly simplified using the symmetry and periodicity of the structure.

Once periodicity is assumed, it is sufficient to consider the scattering only from the repeating element (the unit cell) to calculate the scattering amplitude. The scattering from the unit cell is called the *structure factor*, and it is usually denoted by S in analogy to the total scattering amplitude in Eq. (2), which, in fact, represents the structure factor for a giant unit cell incorporating every atom of the sample. Using $S_{\vec{G}}$ for the structure factor instead of $S(\vec{Q})$, for the sake of clarity, Eq. (2) can be rewritten as:

$$S(\vec{Q}) = \sum_{k=1}^{M} e^{i\vec{Q}\cdot\vec{R}_k}\left[\sum_{n=1}^{L} F_n(Q)e^{i\vec{Q}\cdot(\vec{r}_n-\vec{R}_k)}\right] = \sum_{\{\vec{G}\}} S_{\vec{G}}\delta(\vec{Q}-\vec{G}) \tag{3}$$

where L is the number of atoms in a unit cell and M is the total number of unit cells ($N = ML$). The summation is performed over all possible values allowed by the reciprocal lattice. The δ-function in Eq. (3) specifies the reciprocal lattice positions. The reciprocal lattice depends only on the lattice constants and the symmetry of the lattice, while the structure factor depends only on the atomic arrangement inside the unit cell. By an inverse Fourier transformation of Eq. (3), the electron density can be rewritten as a Bloch-wave expansion:[20]

$$n(\vec{r}) = \sum_{\{\vec{G}\}} S_{\vec{G}} e^{i\vec{G}\cdot\vec{r}} \tag{4}$$

From this equation, the structure factor can be identified as the corresponding Fourier component of the electron density. For the special case where the unit cell consists of a single atom, the form factor and the structure factor are equivalent.

A structure factor can, in principle, be inverted to a real electron density of the materials or their interfaces. However, only the absolute values of the structure factor are measurable, while the phase of the structure factor is generally lost. Furthermore, the range of experimentally accessible Q is limited to Ewald space[21] and it is often limited also by the geometry of the sample. Because of the limited information obtainable on structure factors, the usefulness of direct Fourier transformation using Eq. (4) to obtain the electron density is limited to very special cases. In most cases, a trial-and-error/curve-fitting procedure is used. Physically reasonable models are assumed for the system, the structure factors are calculated for the models, and the model parameters are obtained by curve fitting.

When the structure is periodic and has a true long range order, the summation, $\sum_{\{\vec{G}\}}$ must be carried out over three dimensions. When the structure is aperiodic in any one direction, the summation should be truncated or replaced by an integral for that direction. For example, for the case of a half-infinite single crystal, the structure is no longer periodic in the surface-normal direction and should be replaced by $\sum_{\{\vec{G}_{\|}\}} \int d\,\vec{Q}_{\perp}$ or truncated at the surface.

For a medium without any structure (a liquid is close to this limit for length scales larger than atomic distances), the Fourier transformation yields a three-dimensional delta function:

$$\begin{aligned} S(\vec{Q}) &= n \int d\vec{r} e^{i\vec{Q}\cdot\vec{r}} \\ &= n\delta(\vec{Q}) \end{aligned} \tag{5}$$

where n is an average electron density of this medium. The physical meaning of the three-dimensional delta function is that scattering is allowed only when the momentum transfer, $\vec{Q}$, is zero. In reciprocal space, there exists only one *spot*—the origin. That is, the x-ray beam propagates only to the forward direction. This provides an opportunity to study the atomic scale deviation from a continuum model of liquids by measuring non-forward scattering. When this medium is filled with a periodic array of electrons, the Fourier transformation yields a lattice of delta functions:

$$\begin{aligned} S(\vec{Q}) &= n \int d\vec{r}\, e^{i\vec{Q}\cdot\vec{r}} \sum_{n}^{N} \delta(\vec{r} - \vec{r}_n) \\ &= n \sum_{n}^{N} e^{i\vec{Q}\cdot\vec{r}_n} \\ &= n \sum_{\{\vec{G}\}} \delta(\vec{Q} - \vec{G}) \end{aligned} \tag{6}$$

where $\vec{G}$ is a reciprocal lattice vector. Assuming another medium that is essentially a sheet of electrons without any periodicity at $z = 0$, the equation for $S(\overrightarrow{Q})$ can be rewritten as:

$$\begin{aligned} S(\vec{Q}) &= n \int d\vec{r}\, \delta(z) e^{i\vec{Q}\cdot\vec{r}} \\ &= n\delta(\vec{Q}_{\parallel}) \end{aligned} \tag{7}$$

The result is a two-dimensional delta function and only one *line* of scattering exists that passes through the origin. On the other hand, when the sheet of electrons form a two-dimensional lattice, the Fourier transformation yields a set of lines:

$$\begin{aligned} S(\vec{Q}) &= n \int d\vec{r}\, e^{i\vec{Q}\cdot\vec{r}} \delta(z) \sum_{n}^{N} \delta(\vec{r} - \vec{r}_n) \\ &= n \sum_{\vec{G}_{\parallel}} \delta(\vec{Q}_{\parallel} - \vec{G}_{\parallel}) \end{aligned} \tag{8}$$

The real surface has both three-dimensional and two-dimensional characteristics because, even though the surface sometimes can be regarded as two dimensional, it generally has a finite thickness, and it is also part of a three-dimensional sample. Consequently, the two-dimensional and three-dimensional characteristics cannot be isolated. Therefore, the scattering results in a set of rods in reciprocal space, but the intensity along the rods varies as a result of convolution with the scattering from the three-dimensional system. These lines are often called *crystal truncation rods* (CTR), and are discussed in detail below.

2. X-Ray Reflectivity and Crystal Truncation Rods

From the point view of classical electromagnetism, the x-ray-scattering process is the response of a medium to high-frequency electromagnetic waves. A simple model for the dielectric constant, using the equation of motion for an electron bound by a harmonic force and acted upon by an electromagnetic wave field, gives:[22]

$$\epsilon(\omega) = 1 + \frac{4\pi N e^2}{m} \sum_{j}^{Z} \left(\omega_j^2 - \omega^2 - i\omega\Gamma_j\right)^{-1} \tag{9}$$

where NZ is the total number of electrons involved in the scattering process, ω_j is the resonance frequency, and Γ_j is the damping constant. The damping constants are generally small compared with the *resonant frequencies*, and the dielectric constant rises sharply at $\omega = \omega_j$, which is a phenomenon known as anomalous dispersion.

The complex dielectric constant is directly related to the complex scattering factor of an atom as:

$$\begin{aligned} \epsilon(\omega) &= 1 - \frac{4\pi N e^2}{m\omega^2} \sum_{j}^{Z} \omega^2 \frac{\omega^2 - \omega_j^2 - i\omega\Gamma_j}{\left(\omega^2 - \omega_j^2\right)^2 + \left(\omega\Gamma_j\right)^2} \\ &= 1 - \frac{4\pi N e^2}{m\omega^2} F \end{aligned} \tag{10}$$

where F $(= F_1 + i\,F_2)$ is a form factor. This expression can simply be extended to the multi-atom unit-cell case by substituting F by S, a structure factor, regarding N as the number of unit cells, and summing over the all the electrons in the unit cell. When the x-ray energy is significantly higher than the binding energies of most electrons in the medium, within an approximation the dielectric constant takes the simple form of:

$$\begin{aligned} \epsilon(\omega) &\approx 1 - \frac{\omega_p^2}{\omega^2} \\ \omega_p^2 &= \frac{4\pi N F_1 e^2}{m_e} \end{aligned} \tag{11}$$

and the frequency ω_p is called the plasma frequency of the medium.

For the case of an ideally sharp interface, the scattering can be described by the Fresnel reflectivity equation of classical optics,[22] since x rays are electromagnetic waves. The Fresnel equation can be expressed with the perpendicular components of momentum-transfers using the complex dielectric constant as follows. The equation can be written in terms of the angle of incidence with respect to the interface:

$$R(Q) = \frac{\sqrt{\varepsilon}cos(\pi/2 - \theta) - \sqrt{\varepsilon' - \varepsilon\, sin^2(\pi/2 - \theta)}}{\sqrt{\varepsilon}cos(\pi/2 - \theta) + \sqrt{\varepsilon' - \varepsilon\, sin^2(\pi/2 - \theta)}} \tag{12}$$

where the angle θ is measured from the surface and ε and ε' are the dielectric constants of the media across the interface. By defining the critical angle of the interface as $\cos^2\theta_c = \varepsilon'/\varepsilon$ from Snell's law and by defining momentum transfer as $Q = 4\pi\sin\theta/\lambda$, the reflectivity equation becomes

$$R(Q) = \frac{Q_\perp - Q'_\perp}{Q_\perp + Q'_\perp} \approx \left(\frac{Q_c}{2Q_\perp}\right)^2 \tag{13}$$

where $Q'_\perp = \sqrt{Q_\perp^2 - Q_c^2}$. The critical momentum transfer of a surface can also be written using the plasma frequency for the limiting case as:

$$\begin{aligned} Q_c &= \frac{4\pi}{\lambda}\frac{\omega_p}{\omega} \\ &= \frac{2\omega_p}{c} \\ &= 0.029\sqrt{\rho Z/A} \end{aligned} \tag{14}$$

where ρ is the mass density, and A is the atomic number of the sample. For a perpendicular component of the momentum transfer that is smaller than the critical momentum transfer, the x-ray beam undergoes a total reflection and a complete reflection of the incident x ray occurs.

When there is more than one interface, the total reflectivity can be obtained simply by adding the reflections from each interface. More rigorously, multiple scattering has to be taken into account and the total reflectivity can be obtained by appropriate summation of reflection amplitudes.[23]

The total external reflection is a dynamical process; consequently, very little information can be obtained easily about the nature of an interface near or below the critical momentum transfer. On the other hand, the scattering process is nearly kinematic in the large momentum-transfer limit. Consequently, the reflectivity can be obtained from the Fourier transformation of a step function $\Theta[g(\vec{r})]$ where $g(\vec{r}) = 0$ defines the surface:

$$R(Q) = \frac{i0.029^2 \rho Z/A}{4Q_\perp} \int e^{i\vec{Q}\cdot\vec{r}}\, \Theta[g(\vec{r})] d\vec{r} \tag{15}$$

For a single-crystal case, it is convenient to start with the density of an infinite single crystal in the form of a Bloch-wave expansion.[20] Subsequently, the electron density of an infinite single crystal can be expressed as the product of a step function and the sum of Bloch waves as in Eq. (4):

$$n(\vec{r}) = \Theta[g(\vec{r})] \sum_{\vec{G}} S_{\vec{G}}\, e^{-i\vec{G}\cdot\vec{r}} \tag{16}$$

where $S_{\vec{G}}$ is the structure factor of the single crystal. Therefore, the scattering amplitude in the kinematic limit is:

$$R(\vec{Q}) = \frac{i0.029^2 \rho/A}{4Q_\perp} \sum_{\vec{G}} S(\vec{G}) \int e^{i(\vec{Q}-\vec{G})\cdot\vec{r}}\, \Theta[g(\vec{r})] d\vec{r} \tag{17}$$

$$= \frac{0.029^2 \rho/A}{4Q_\perp} \sum_{\vec{G}} \frac{S_{\vec{G}}}{\vec{Q}_\perp - \vec{G}_\perp} \int_S e^{i(\vec{Q}-\vec{G})\cdot\vec{r}} d\vec{S}_\perp \tag{18}$$

Equation (18) is analogous to Eq. (15). Equation (18) can be obtained from Eq. (17) by using Green's theorem. A comparison of Eqs. (15) and (18) reveals that CTR is the sum of the reflected amplitudes from the surface with the momentum transfer reduced by the crystal momentum transfer. The reflections emanate from each reciprocal lattice vector position in the direction perpendicular to the surface or surfaces for facetted samples. When the surface is parallel to a crystallographic plane, reflectivities from the aligned reciprocal lattice points may interfere and become very sensitive to the interfacial structure of the sample. In this case, the reflectivity can be written as a simple one-dimensional sum of scattering from each plane:

$$R(\vec{Q}) = \frac{i0.029^2\rho/A}{4Q_\perp}\sum_{\vec{G}_\parallel}\delta(\vec{Q}_\parallel - \vec{G}_\parallel) \times \sum_{n}^{N} V_n e^{i(Q_\perp r_{n\perp} + \vec{G}_\parallel \cdot \vec{r}_{n\parallel})} \tag{19}$$

where $\boldsymbol{Q}_\parallel$ and $\boldsymbol{Q}_\perp$ denote parallel and perpendicular components of a vector $\boldsymbol{Q}$, respectively, and V_n is the scattering amplitude of the nth layer.

For a three-dimensional single crystal with one of the dimensions finite, as for a two-dimensional crystal, thin film, or truncated surface, only $\vec{G}_\parallel$ is discrete. Then, the $R(\vec{Q})$forms an array of rods in the reciprocal space and the intensity variations along the rods reflect the electron density profile along the direction of the finite dimension.

The mathematical representations of the x-ray-scattering processes are pictorially summarized in Fig. 1. Following customary naming convention, surface-x-ray scattering (SXS) in general can be categorized into three types of surface scattering, namely, x-ray reflectivity (specular and off-specular), glancing angle in-plane x-ray diffraction, and crystal truncation rod (CTR) measurements:

(a) The specular and off-specular x-ray-reflectivity techniques are used to study surface morphology. They are not typically sensitive to crystalline structures of the surface but sensitive to nano- to micro-meter length scales.
(b) Glancing angle in-plane diffraction, measured parallel to the surface with incident and diffracted x rays at glancing angles, is extremely sensitive to the two-dimensional structure and reconstruction.
(c) Crystal truncation rods, measured normal to the surface, are extremely sensitive to the surface coverage and relaxation.

Numerous examples of applications of these techniques to electrochemical problems will be given in the rest of this chapter. To point out just the first few examples:

(a) the use of the reflectivity measurement is illustrated in Section III.2(*ii*) (cf. Fig. 9);

(b) that of the glancing angle diffraction in Section III.1(*ii*) (Fig. 3); and
(c) that of the CRT measurements in Sections III.1(*i*) and (*iii*) (Figs. 2 and 4).

The electrochemical aspects of the synchrotron-x-ray-scattering measurements, such as detailed descriptions of several x-ray-electrochemical cell designs and other experimental details, will be given in the Appendix.

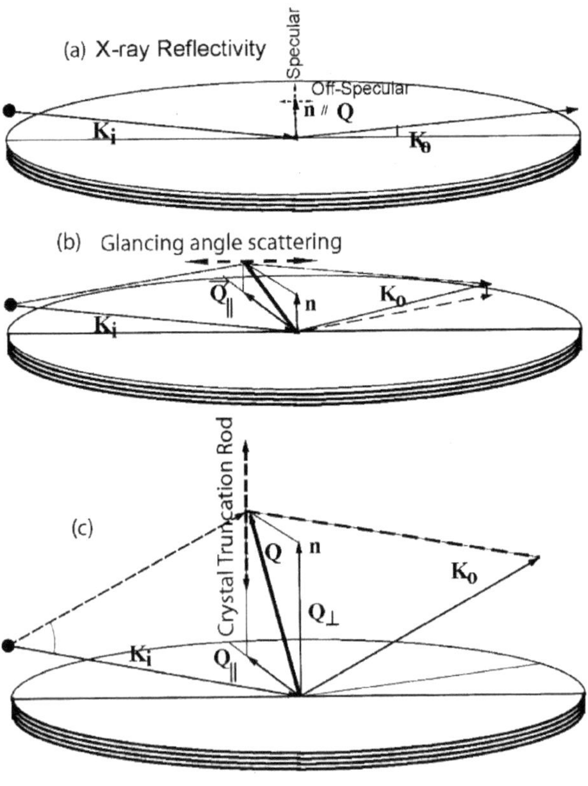

Figure 1. Experimental geometries of surface-x-ray scattering.[24]

3. Resonance Anomalous Surface X-Ray Scattering

So far, the discussion was limited to an approximation given in Eq. (11) where the energy of x rays is far above the absorption edges of the atoms involved in the scattering processes. However, it is well known that, at least in bulk x-ray scattering techniques, x-ray energies can be tuned through the absorption edges for resonance scattering and the anomalous dispersions at the edges can deliver elemental and chemical information about the atoms probed. A similar concept can be used for surface scattering processes for elemental and chemical sensitivity of surface atoms involved in surface scattering processes.

To understand the resonance anomalous scattering processes, it is necessary to examine Eq. (9). It is obvious that the scattering factor, F, can have anomalous dispersions whenever $\omega \approx \omega_j$. By tuning the x-ray energy through the anomalous scattering energies of surface atoms, one can examine the chemical states of the atoms at the buried interfaces at which ordinary surface-science probes cannot be used.[25]

The quantum mechanical version of the numerator of Eq. (9) exhibits explicit polarization dependence. The polarization dependence was recently used to study the structures and chemical states of light elements, even when the elements do not have edges in the hard x-ray energies, by tuning the x-ray energy to the binding energy of *substrate* atoms.[26]

A detailed discussion of resonance anomalous surface x-ray scattering is beyond the scope of this chapter. However, in-depth discussion of the technique and examples of applications can be found elsewhere.[24] One example of the application of this technique to electrochemistry is briefly discussed in Section VIII.2.

III. METAL SURFACE PREPARATION, RESTRUCTURING, AND ROUGHENING

An initial experiment is often the investigation of the structure of a well prepared, clean single-crystal surface under an electrolyte, because such a surface will eventually serve as the stage of electrochemical reactions. The preparation of single crystal surfaces for electrochemical investigations has been somewhat of an *art* for quite some time, and the cleanliness and well defined morphology of the surface often were judged by the results of an electrochemical experiment (e.g., a cyclic voltammogram) rather than by direct observation. Synchrotron-x-ray techniques provide a very direct way for this observation. This was utilized to investigate the different preparation techniques for Cu(111) film,[27] gold[28-32] and platinum[33] single crystals, and for the investigation of nanostructured platinum sufaces.[34,35]

1. Relaxation/Reconstruction of Metal Surfaces

The surfaces of single crystals are often structurally different from the structure of the bulk crystal. Surface restructuring (relaxation and/or reconstruction) is a well-known and much investigated phenomenon in UHV studies, but the presence of an electrolyte solution and the application of an electrical potential often strongly influence the reconstruction. Much work has been carried out in this area as shown in Table 1.

(*i*) *Relaxation of Platinum Surfaces*

The relaxation of Pt(hkl) surface structures was described in a series of papers.[33,51,76,78,79,81,83] It was found that close to the hydrogen evolution potential there was an increase in the separation of

Table 1
Surface Relaxation and Reconstruction

System	References
Ag(111)	36-40
Au(100)	28,29,32,41-54
Au(110)	48,51,54,55
Au(111)	7,48,51,54,56-71
Pd on Pt(100)	72
Pd on Pt(111)	73-75
Pt(100)	51,76-78
Pt(110)	33,51,78-80
Pt(111)	51,78,81-83
Pt(111)/CO	84
$Pt_3Sn(111)$	85,86
$RuO_2(100)$	87
$RuO_2(110)$	87,88

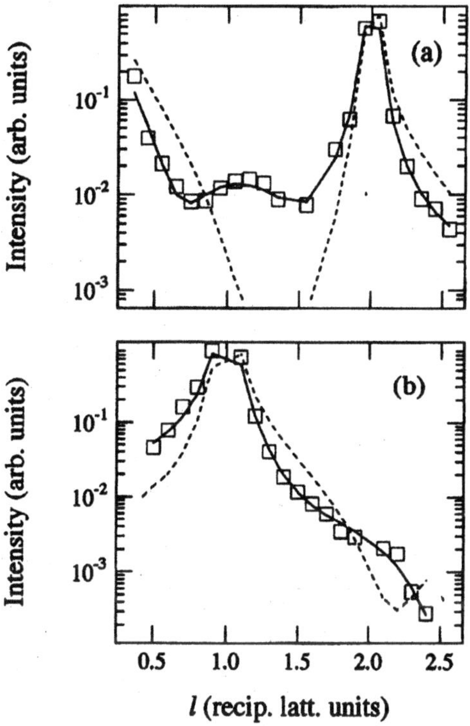

Figure 2. The measured x-ray intensity along (a) (0, 0, *l*) and (b) (0, 1, *l*) of the Pt(110) surface in 0.1 M NaOH at an electrode potential of 0.1 V. The solid lines are calculated fits to the data according to the relaxation model, and the dashed lines are the calculated intensity for a contracted (1 × 2) missing-row model of the surface. (Reprinted with permission from reference 79. Copyright (1996) by the American Physical Society.)

the top surface layer from the second layer, and that this expansion of surface atoms increased in the sequence of Pt(111) < Pt(100) < Pt(110) (in the amounts of 1.5%, 2.5%, and 25%, respectively) in both acidic and basic solutions. It was suggested that the expansion is connected to the adsorption of hydrogen on the surface and the results can be explained by the differences in the adsorption sites and the strength of the hydrogen-metal bonding. Fig. 2 shows the good fit of the CTR data to the surface relaxation model for the Pt(110) surface.

(*ii*) *Reconstruction of Platinum Surfaces*

The reconstruction of platinum surfaces under aqueous solutions was also investigated. The Pt(100)[76] and the Pt(111)[81] surfaces were found to be stable and unreconstructed at all potentials investigated, while the Pt(110) was found to exhibit either a (1 × 1) or a (1 × 2) structure depending on the method of preparation (Fig. 3).[33,79] A quick quenching of the flame-annealed crystal freezes the unreconstructed surface, while a slow cooling allows the reconstruction to occur below the critical temperature. Both surfaces were found to be stable under solution within the potential range of hydrogen adsorption and the onset of oxide formation.

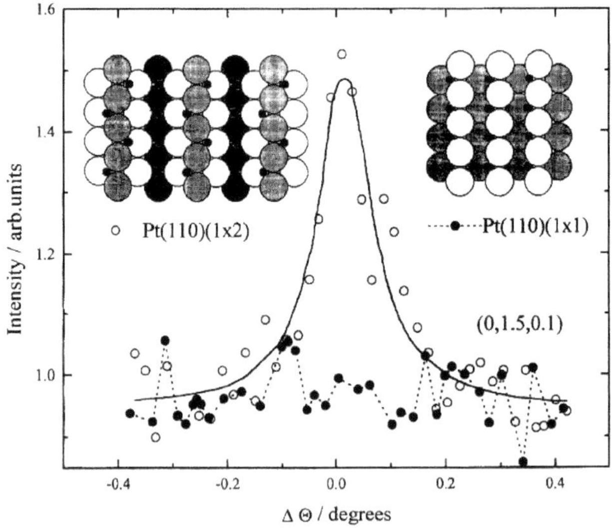

Figure 3. Top: ideal models for the (1 × 2) (right) and (1 × 1) (left) surfaces; solid dots represent adsorbed hydrogen atoms. Bottom: the measured x-ray intensity at (0,1.5,0.1) along the [100] direction for a Pt(110)-(1 × 2) (solid line) crystal, and for a Pt(110)-(1 × 1) crystal (dashed line—note the absence of the 1 × 2 peak). (Reprinted from reference 33. Copyright (1997), with permission from Elsevier.)

(iii) Reconstruction of Gold Surfaces

The Au(hkl) surfaces behave quite differently from those of Pt. All three major faces were found to be reconstructed at negative potentials, but the reconstructions are lifted at more positive potentials, as reported in a series of papers.[41,55-57] The Au(100) exhibited a (5 × 20) reconstructed surface at negative potentials, similar to the surface observed in UHV, which, in perchloric acid solution, was lifted at more positive potentials with the formation of a partial top layer containing the excess atoms from the earlier reconstructed layer (Fig. 4).[41] The Au(110) exhibited a (1 × 3) reconstructed surface at negative potentials in a number of halide solutions, in contrast with the (1 × 2) reconstruction observed in

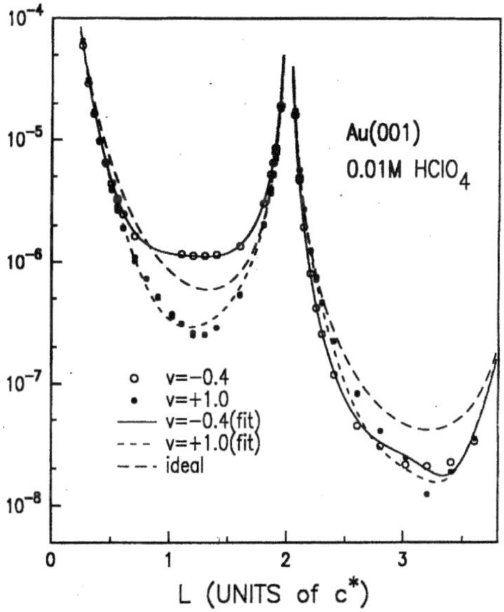

Figure 4. Absolute reflectivity data of Au(100) for the (0,0,L) rod at -0.4 V (open circles) and 1.0 V (solid circles). The long-dashed line is for ideal termination with no rms displacement amplitude. The solid line is fit for the reconstructed surface and the short-dashed line is fit for the unreconstructed surface with a partial top layer. (Reprinted with permission from reference 41. Copyright (1990) by the American Physical Society.)

Table 2
Surface Roughness

System	References
Ag(polycrystalline)	89
Ag(111)	40,90
Au(100)	28,29,31
Au(111)	29,30,64
Cu (polycrystalline), Cu/oxide	23
$Cu_3Au(111)$	91
GaAs(001)	92
Nb(polycrystalline), Nb/Nb_2O_5	93
Ta(polycrystalline), Ta/Ta_2O_5	93
Pd on Pt(111)	73,75
Pt(111)	4,10,94-97
$Pt_3Sn(111)$	86

UHV, this reconstruction was lifted at more positive potentials, presumably due to the adsorption of anions.[55] The Au(111) exhibited a $(23 \times \sqrt{3})$ surface at negative potentials in a number of halide solutions, similar to the surface observed in UHV. This reconstruction was also lifted at more positive potentials, presumably due to the adsorption of anions.[56]

2. Surface Roughness Measurements

The surface roughness of an otherwise clean and well ordered surface is also important in electrochemical investigations. X-ray-scattering techniques are very well suited to determine even atomic level roughness, as shown by the investigations collected in Table 2. Roughness at a much higher level can also be investigated, as will be described in Section X on *porous silicon*.

(*i*) *Platinum Oxidation/Reduction*

While the characterization of relatively thick (multiple atomic layers) oxide films is important both for science and technology, the most fundamental understanding of oxide film growth requires an examination of the oxide film at the *submonolayer* level and at the *incipient* stages of the formation process. The oxidation/reduction of the Pt(111) single-crystal surface was selected for such an investigation.[95,97] Experiments in 0.1 M CsF solution, have shown that the surface irreversibly roughens when the potential is cycled beyond 1.3 V (HE), while the initially flat surface can be completely recovered after electrochemical reduction of the oxide film formed at potentials more negative than 1.3 V. A series of normalized crystal-truncation-rod scans is shown in Fig. 5. Each scan was made at 0.8 V, after oxidation at the potential indicated and reduction at 0.0 V. All data are divided by the 0.8 V data to emphasize the surface contribution and each scan from the top is offset sequentially by an order of magnitude for display purposes.

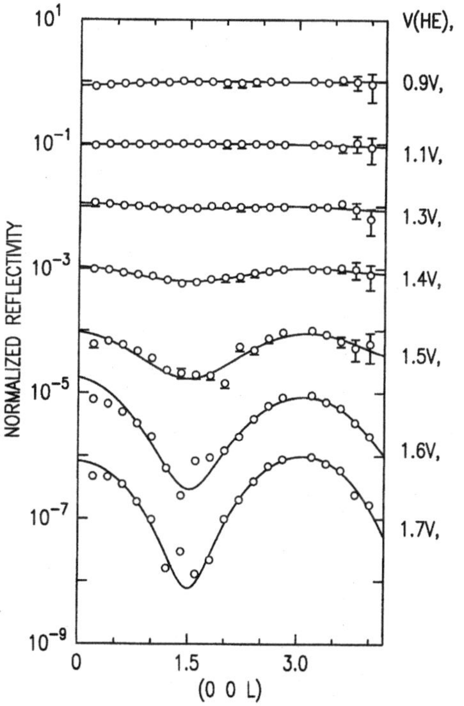

Figure 5. Crystal-truncation-rod scans of Pt(111) oxidized at potentials indicated. Open circles represent the experimental data, and the solid line were calculated with the three-step model.[4, 97]

The abscissa is the reciprocal lattice unit of platinum in the hcp unit cell. Electrochemical measurements (cyclic voltammetry) of the oxygen coverage of the platinum surface indicated that a full monolayer of oxide coverage is not achieved at potentials negative to approximately 1.4 V, suggesting that oxidation/reduction of less than a monolayer of oxide film will not destroy the originally flat surface, but the surface will roughen if more than a monolayer of oxide is formed/reduced.

The roughness of the surface atomic layers could be determined by comparing the experimental data to predictions of the scattering equation applied to some simple models of rough surfaces (Fig. 6). In these calculations only occupational disorder was considered. The experimental data are compared to curves calculated with the various models in Fig. 7. It is quite evident that only the three-step model fits the data. The calculated curves in Fig. 5 were all obtained with the three-step model.

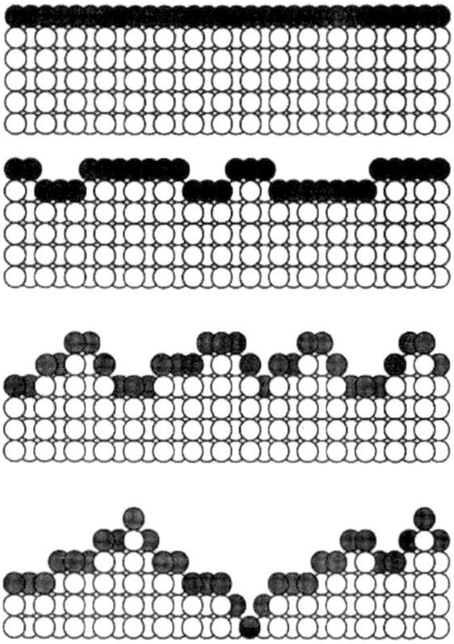

Figure 6. Surface roughness models. From top to bottom: ideal surface, two-step model, three-step model, and multiple-step model. The shaded circles represent platinum atoms exposed to the solution.[10, 97]

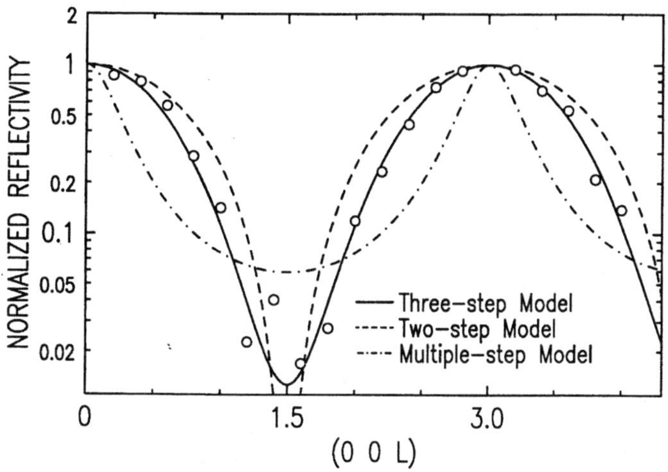

Figure 7. Normalized reflectivity data and comparison to models for a reduced surface obtained after oxidation to 1.7 V in CsF solution.[95]

Some conclusions can be drawn from these results about the mechanism of surface oxidation. The total number of platinum atoms is not conserved in the two-step model. The surface structure represented by the two-step model can evolve from an ideally flat surface only by the removal of some atoms from the first layer, thereby exposing some atoms of the second layer to the solution. One possibility is the chemical dissolution of platinum oxide. Another possibility arises if one considers that the surface is not ideally flat, rather it is a stepped surface with large flat terraces between steps. Then, a two-step surface can be produced between steps if some of the atoms removed from the first layer migrate on the surface until they find step edges and become incorporated at kink sites. The poor fit based on the two-step model is clear evidence against these mechanisms. This indicates that the mobility of platinum atoms at room temperature is not sufficiently large for the atoms to migrate thousands of angstroms to the step edges. It also indicates that the solubility of platinum oxide should be near zero or too small to be detectable in our measurements. This is further substantiated by the good fit of the data to the three-step model because the morphological transformation from the flat surface to the surface of the three-step model is a mass-conserving process (with the restriction that the step-up and step-down probabilities are the same, which is a consequence of the zero dissolution of the oxide). Atoms removed from the first layer *remain* on top of the first layer.

The good fit of the data to the three-step model identifies it as the one most likely describing correctly the morphology of the rough surface induced by oxidation. The fit of this model to the data indicates that the oxide formation involves mostly the platinum atoms at the first layer and only very few atoms in the second layer. Adding a small fraction of second-layer oxidation (that would result in a five-step model of the surface because of the restriction of equal probability of step-ups and step-downs) did not significantly improve the fit, and the fraction of the second-layer oxidation found in the fitting was small. Furthermore, although the morphological transformation from a flat surface to a multiple-step surface can also be achieved with mass conservation (by successive, single-height movements of the atoms) the poor fit resulting from the application of this model also substantiates the correctness of the three-step model.

(ii) Copper Passivation

The surface roughness of the passive film on copper was investigated using specular the x-ray-reflectivity technique.[23] Figure 8 shows the cyclic voltammogram of copper passivation/ depassi-

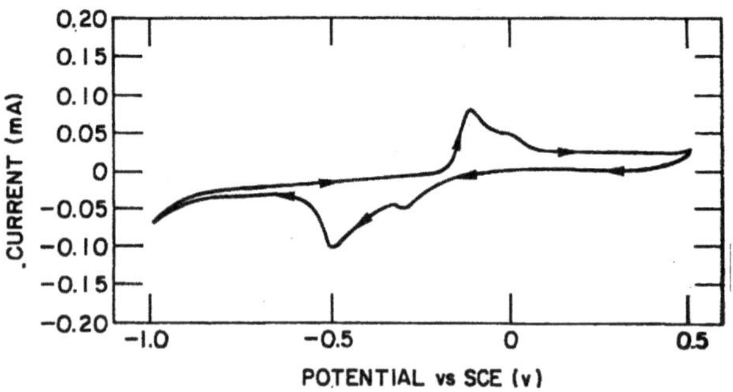

Figure 8. Cyclic voltammogram of the Cu/Si electrode in borate buffer solution (pH 8.4), scan rate = 10 mV/sec.[23]

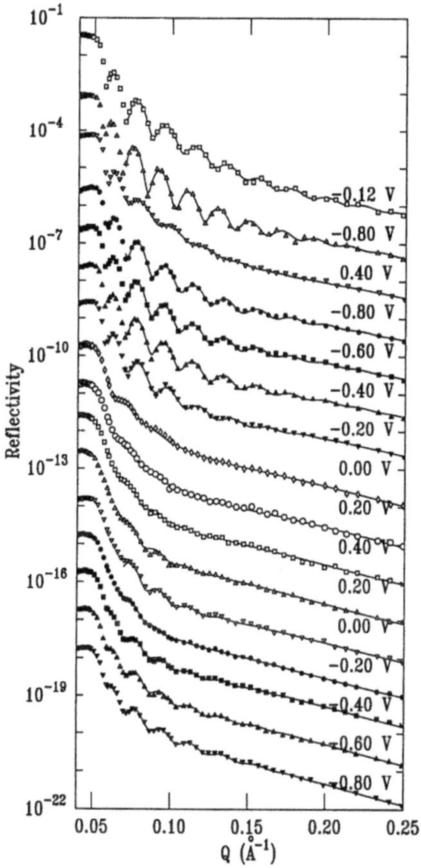

Figure 9. A series of reflectivity scans for the copper/silicon sample in borate buffer solution made at the potentials indicated. For display purpose each reflectivity scan was sequentially offset by one order of magnitude.[23]

vation, in borate buffer solution at pH of 8.4, indicating the formation and reduction of the passive oxide film. The electrode used in this work consisted of a thin film (~ 250 Å) of copper which was evaporated in vacuum onto a polished silicon substrate.

To follow this process, a series of x-ray-reflectivity scans (Fig. 9) were taken at several potentials between –0.8 V and 0.4 V (all potential changes were made at 10 mV/sec and the scans were taken after the current reached a steady state value of less than 1 μA). The changes in the reflectivity (the reversible changes in the peaks) correlate well with the oxidation/reduction peaks in the CV. The solid lines are fits to the data and only one out of every five data points are shown for clarity.

The scans in Fig. 9 were fit with a reflectivity equation developed for three interfaces, i.e., solution/copper oxide, copper oxide/copper, and copper/silicon interfaces; the fitting parameters were the thickness of the copper layer and that of the copper oxide layer, the roughness of the copper/copper-oxide interface, and that of the copper-oxide/solution interface, and the correlated roughness of the two interfaces. The most important results can be summarized as follows. The oxide film thickness was found to be about 25 Å (corresponding well with the measured charge transfer). The corresponding changes in the copper layer thickness are shown in Fig. 10 (less than 25 Å due to the higher density of copper metal). The hysteresis corresponds well with the peak positions in the CV (Fig. 8). The roughness of both interfaces increased from about 4 to about 9 Å during the course of the experiment. The correlated roughness was a function of potential (Fig. 10) with the same hysteresis as the copper thickness, except that the loop did not close. This was probably due to preferential roughening at grain boundaries. The significance of these experiments was to show the unique capability of the x-ray-scattering technique for the simultaneous, in-situ determination of all these parameters (the film thicknesses and the roughness of several interfaces including the correlated roughness of two interfaces) during metal passivation.

IV. DOUBLE LAYER STRUCTURE STUDIES

An important area of electrochemical research deals with the structure of the electrical double layer at the surface of electrodes. Surface-x-ray-scattering techniques have been utilized to investigate this phenomenon including

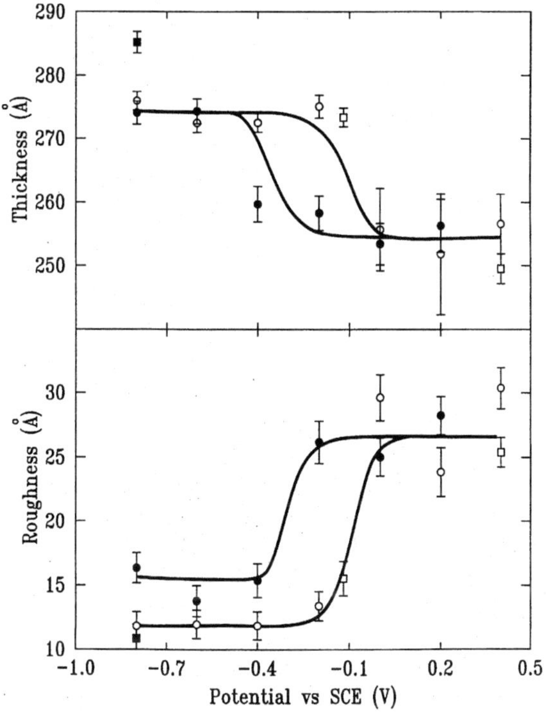

Figure 10. Plots of key fitting parameters from Fig. 9. (top) The thickness of the copper film on silicon. (bottom) The correlated roughness of the copper/oxide and oxide/water interfaces. The solid lines are guides to the eye. The open circles are the fitting parameters in the oxidation cycle and the closed circles are the fitting parameters in the reduction cycle. The electrochemical irreversibility in these two fitting parameters as a function of potential is apparent.[23]

(*a*) the observation of the structure of water in the interphase at a Ag(111),[36-39] a Au(111),[7,58] and a ruthenium dioxide electrode,[87,88] and

(*b*) the distribution of ions at a platinum electrode[10,97] and at a liquid/liquid interface.[98-101]

1. Water at the Silver Surface

The investigation of the water structure at the Ag(111) produced some unexpected results.[36-39] No *ice-like* structures were found and the main features concluded from the oxygen distribution (Fig. 11) were the potential dependent layering of the water molecules extending to about three molecular diameters with a reorientation of the water as a function of potential. What was unexpected was the density of the water molecules in the first layer: 1.1 (at –0.23 V) to 1.8 (at +0.52 V) per silver atoms. One would expect ~ 0.8 water molecules per silver atom. This high surface density of water was tentatively explained by the effect the large electric field (~ 10^7 V/cm) at the surface of the electrode.

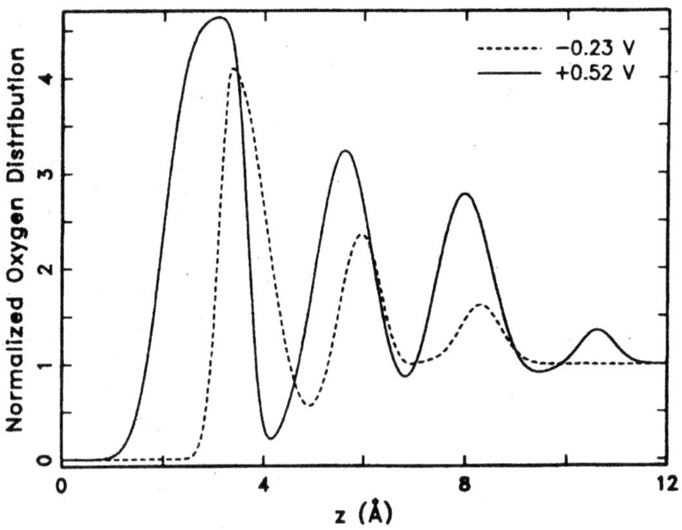

Figure 11. Best-fit of the CTR data for the normalized oxygen distribution of water near a Ag(111) electrode surface at -0.23 (dashed line) and +0.52 (solid line) V to the pzc as a function of the distance above the top silver atomic plane. (Reprinted from reference 38. Copyright (1995), with permission from Elsevier.)

2. Water at the Ruthenium Dioxide Surface

A different picture emerged from a study of ruthenium dioxide(110) single crystal electrodes.[87,88] There is a special characteristics of this facet that makes its study especially informative: certain crystal truncation rods are dominated by scattering from oxygen atoms only (called further as *oxygen-rods*) due to the fortuitous cancellation of the signal from the two Ru sublattices, while most other rods are typically dominated by the much heavier ruthenium atoms. Cyclic voltammogram and x-ray-intensity data (Fig. 12) indicated a complex behavior, including oxidation/reduction of the electrode and the beginning of oxygen evolution.

However, the most interesting feature of these changes was the formation of a commensurate water layer on the surface with a

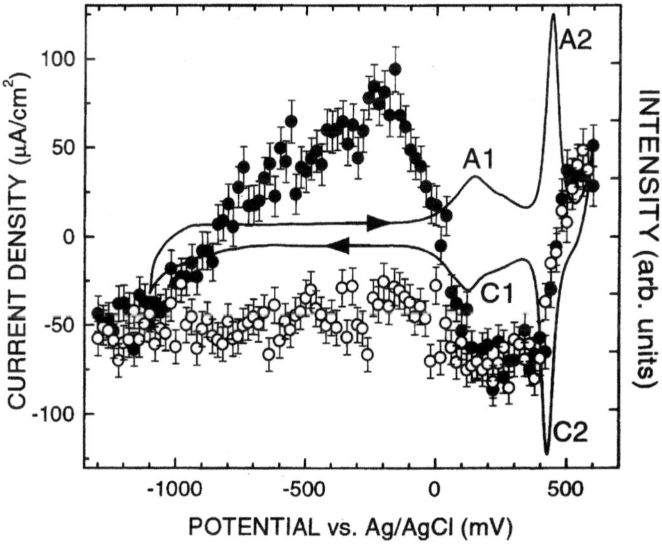

Figure 12. Cyclic voltammogram (solid line) and x-ray intensity (circles with error bars) at (014) measured for the $RuO_2(110)$ surface in 0.1 M NaOH. Filled and open circles are for cathodic and anodic sweeps, respectively. The arrows indicate the directions of the potential change.[88]

structure very closely resembling that of *ice X*. The x-ray intensity near the Bragg peaks was very strong in the truncation rods because the x-ray-scattering amplitudes from many layers are added in-phase, but this signal was insensitive to the surface structure. The intensity distribution around the anti-Bragg point (mid-point between two adjacent Bragg peaks) was typically weak and concave because x rays scattered from different layers were out of phase, but it was sensitive to the positions of surface atoms. For instance, if a new layer would form over an ideal termination at a height twice the bulk spacing, the scattered x-ray amplitude from the new layer would be added in-phase and the anti-Bragg point would change from a minimum to a local maximum. Therefore, the concave-to-convex change in the off-specular oxygen rods, seen in these measurement near the anti-Bragg point (e.g., (014) and (105) in Fig. 13), demonstrated the possibility of the formation of such an extraneous layer of oxy-species. Additionally, this proved the commensurate nature of the extraneous layer since an incommensurate or disordered layer would not affect the off-specular rods.

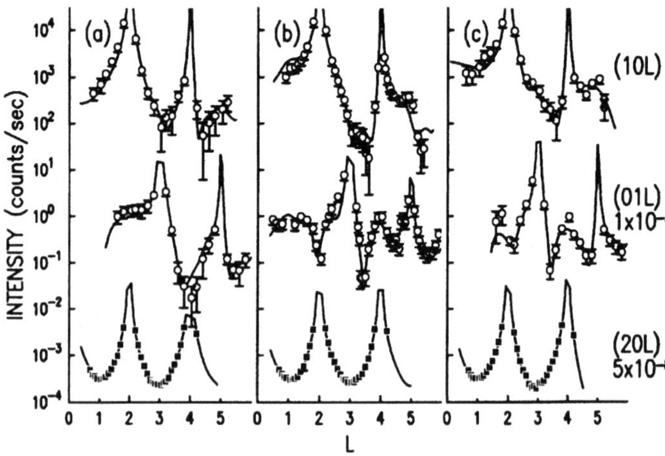

Figure 13. RuO_2(110) CTRs at (10L) and (01L) (oxygen rods) and (20L), measured at potentials of (a) 330 mV, (b) 500 mV, and (c) -200 mV. Solid lines are the results of the best curvefits.[88]

The assumed structure of this layer is shown in Fig. 14(b). The x-ray scan superimposed on the CV in Fig. 12 was one that was very sensitive to this layer. These data agreed with the model of the ice-layer formation as the surface is oxidized, however, there was an interesting hysteresis in the x-ray scattering not present in the CV. Apparently, as the surface was reduced there existed a *memory effect* and the ice-like water layer structure was retained, although it was not strongly bound to the surface any more. As the potential was swept further negative, this layer was slowly lifted to further and further distance from the surface, Fig. 14(c), till it finally disintegrated. On the other hand, on reversal of the sweep in the anodic direction, the ice-like layer could not reform till the surface was oxidized. This is the first instance of the demonstration of an ice-like water structure in the electrochemical double layer.

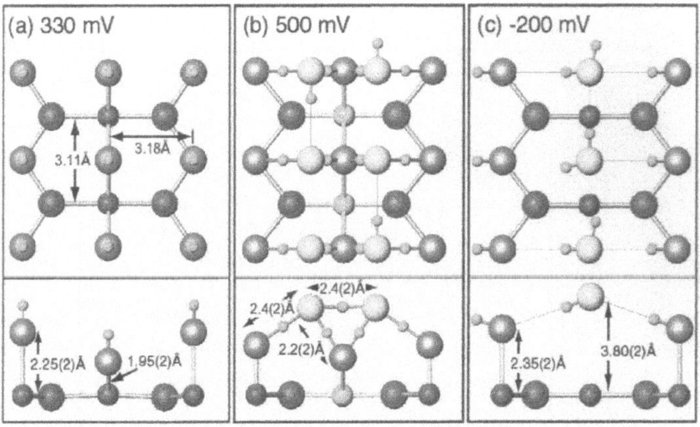

Figure 14. Ball-and-stick models for (a) 330-mV, (b) 500-mV, and (c) –200-mV structures. The medium-size balls represent the Ru atoms. The large balls represent oxygen atoms in the bulk, on the surface bonded to Ru, and in the water molecules, respectively. The small balls represent the hydrogen atoms, conjectured to show conceivable hydrogen bonds.[88]

3. Ionic Distribution in the Double Layer

Some preliminary measurements were also carried out to test the feasibility of detecting the ionic distribution changes in the solution side of the double layer as a function of potential.[10,97] These measurements were carried out with a Pt(111) electrode in a 0.1 M CsF solution; this solution gives a very large double-layer window, and it was expected that the results will be affected only by the concentration changes of the Cs^+ ion since it is much heavier than the fluoride, which is very similar to water as far as the x rays are concerned. A potential shift of 800 mV from the point of zero charge indeed produced a clearly measurable x-ray response (Fig. 15) attributable to the ionic concentration changes since no Faradaic reactions take place in this potential range. Ionic distributions were also examined recently at the nitrobenzene/aqueous solution electrified interface, with the preliminary conclusion that the Gouy-Chapman model does not correctly describes the system because it ignores the molecular-scale structure of the solutions.[98-101]

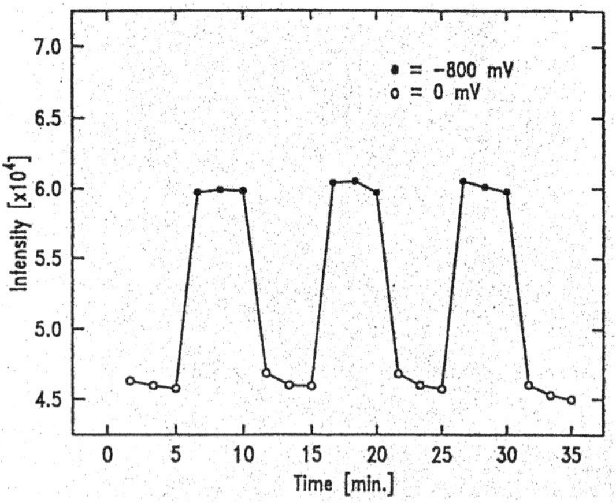

Figure 15. Reflectivity measured at the potential of zero charge and at a very negative potential, indicating a systematic difference that may be explained by the presence of excess cations in the double layer.[10]

V. ADSORPTION/ABSORPTION AT ELECTRODE SURFACES

While it could be considered part of double layer research, the contact adsorption of ionic, atomic, and molecular species is a large area of research in itself. Many systems have been investigated using synchrotron x-ray-scattering as listed in Table 3.

1. Adsorption of Bromide on Gold Surfaces

The adsorption of bromide on Au(100) and Au(111) electrodes have been described in a series of papers.[110,111,113,114] On the Au(111) electrode, electrodeposited Br forms an incommensurate, rotated-hexagonal phase (Fig. 16). The Br coverage increases with potential and the rotated-hexagonal diffraction pattern appears when the potential is increased positive of 0.42 V. Between 0.42 and 0.72 V the Br coverage normalized to that of the substrate increases from 0.465 to 0.515, while the interatomic Br-Br spacing decreases from 4.24 Å at 0.42 V to 4.03 Å at 0.72 V. This electrocompression of the adlayer proceeds monotonically.

Bromide adsorption on the Au(100) surface behaves quite differently (Fig. 16) and the potentials where new diffraction features appear are well correlated with the peaks in the cyclic voltammogram (Fig. 17). At potentials below P2, all diffraction features are associated with the square pattern from the underlying gold substrate. Positive of P2, additional reflections appear, which correspond to a rectangular, commensurate $c(\sqrt{2} \times 2\sqrt{2})R45^{\circ}$ bromide unit cell. Over the entire potential range between P2 and P3, there

Table 3
Adsorption at Electrodes

System	References
Acetate on UPD Hg/Au(111)	102
Br on Ag(100)	103-109
Br on Ag(111)	110
Br on Au(100)	50,53,106,109-112
Br on Au(111)	7,56-58,61,62,110,113-116
Br on Pt(100)	117-119
Br on Pt(111)	82,106,109,112,120-127

Table 3. Continuation

System	References
Br on UPD Cu/Pt(100)	128-131
Br on UPD Cu/Pt(111)	130,132
Br on UPD Pb/Pt(100)	130,131
Br on UPD Pb/Pt(111)	130,132
Cl on Au(111)	7,56-58,113
Cl on Pt(111)	82,120,122,133
CO on Au(100)	52-54
CO on Au(110) and (111)	54
CO on Pt(100)	51,77,80
CO on Pt(110)	33,51,80
CO on Pt(111)	24,51,67,80,83,84,127,134-137
CO on UPD Bi/Pt(100)	131
CO on UPD Bi/Pt/(111)/Bi	138
CO on UPD Cu/Pt(100)	130,131
CO on UPD Cu/Pt(111)	130,132
CO on UPD Pb/Pt(100)	130,131
CO on UPD Pb/Pt(111)	130,132
CO on Pt(100)/Pd	72
CO on $Pt_3Sn(111)$	85,86
CsI on Au(110)	139
Cu on Au(100)	140
CuBr on Pt(111)	122,124,125
CuCl on Pt(111)	120,122,133,141-144
Dodecanethiolate on Au(111)	145
I on Ag(111)	110
I on Au(111)	7,110
I and alkali metals on Au(110)	139
I on Au(111)	60,146
I on Pt(111)	82
NO on Pt(111)	147
O/OH on Cu(111)	148
OH on UPD Tl/Au(111)	149
S on Ag(111)	40
S on Au(111)	150,151
Silicotunstate on Ag(100)	152
SO_4 on Au(111)	153
SO_4 on UPD Hg/Au(111)	102,142,153
SO_4 on Pt(111)	24,81
SO_4 on UPD Co/Pt(111)	154
SO_4 on UPD Cu/Pt(111)	155
SO_4 on UPD Cu/Pt(111 × 110) stepped surfaces	142
SO_4 on UPD Tl/Pt(111)	156
SO_4 on Ru(001)	157,158

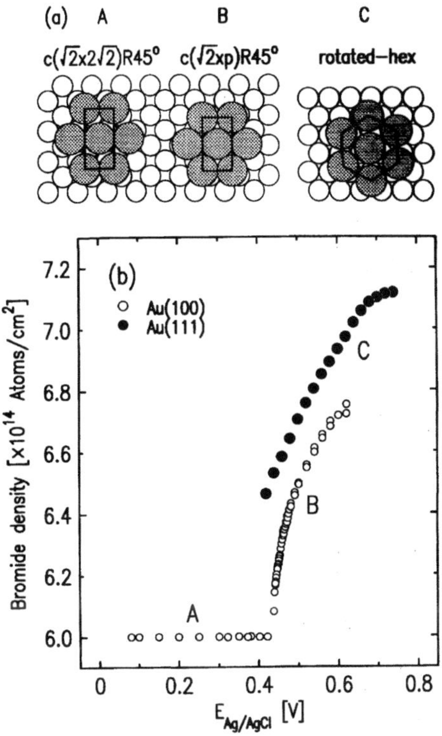

Figure 16. (a) The structures of Br on Au(100) and Au(111). They correspond to A: the commensurate c($\sqrt{2}$ × 2$\sqrt{2}$)R45° structure, B: the uniaxial-incommensurate c($\sqrt{2}$ × p)R45° structure, and C: the rotated-hexagonal structure. (b) The potential-dependent bromide coverages, determined from the in-plane diffraction, versus the applied potential. The Au(100) and Au(111) studies were carried out in 0.05 M NaBr and 0.10 M NaBr solutions, respectively. (Reprinted from reference 110. Copyright (1996), with permission from Elsevier.)

is no change in the symmetry or in the coverage of bromides. At potentials above P3, the bromide monolayer is monotonously compressed uniaxially into an incommensurate c($\sqrt{2}$ × p)R45° phase ($p < 2\sqrt{2}$). The phase transition is nearly reversible and appears to be second order.

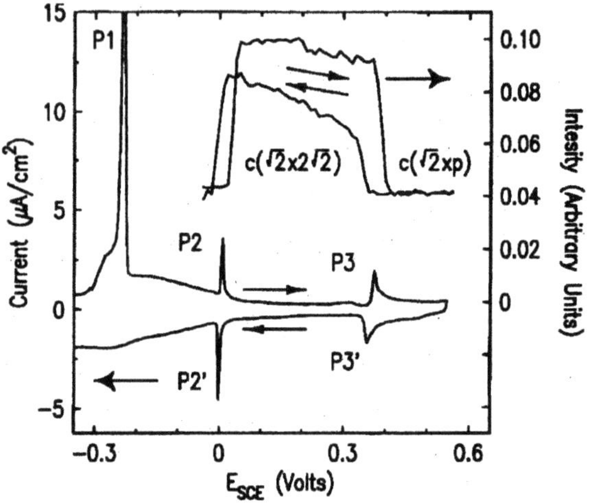

Figure 17. Cyclic voltammogram (10 mV/sec) and the corresponding (0,1) x-ray intensity vs. potential curve (1 mV/sec) for the Au(100) electrode. Peak P1 corresponds to the rearrangement of the top gold layers from a reconstructed-hexagonal to an ideally terminated surface. (Reprinted with permission from reference 111. Copyright (1996) by the American Physical Society.)

2. Adsorption of Carbon Monoxide on Platinum Surfaces

The adsorption of CO on the Pt(111) electrode surface recently became subject of intense research because of its direct relevance to the mechanism of poisoning in modern low-temperature fuel cells. Measurements using STM and IR spectroscopy suggested the existence of two types of structures: hexagonal close-packed (2 × 2)-3CO lattice with a coverage of 3/4 per surface Pt atom at potentials close to 0 V (RHE), that transforms to another hcp with a coverage of 13/19 at more positive potentials. The (2 × 2)-3CO structure was also confirmed by early synchrotron-x-ray-scattering investigations, but no additional structure was found.[80,83,127,134] Recent measurements have also confirmed the existence of the ($\sqrt{19} \times \sqrt{19}$)R23.4°-13CO lattice.[135] Potential-dependent x-ray-scattering measurements were performed of both the (2 × 2) and the ($\sqrt{19} \times \sqrt{19}$) superlattice reflections. The cyclic voltammograms of Pt(111) with and without preadsorbed CO are shown in

Fig. 18 in 0.1 M $HClO_4$ + 10 mM NaBr (bromide was added because it was reported to increase the size of CO domains by blocking OH adsorption[127]). Integrated scattering intensities as a function of potential in a CO-saturated solution are also shown in Fig. 18. The current peak at around 780 mV (RHE) in the anodic scan (solid line) is due to the oxidation of cc 9% of the preadsorbed CO, which accompanies transition from (2 × 2) to (√19 × √19). This transition occurs at a less positive potential (650 mV) in the x-ray measurements. This is due to the difference in CO pressure and in the time scale of the two experiments. Whereas the CV

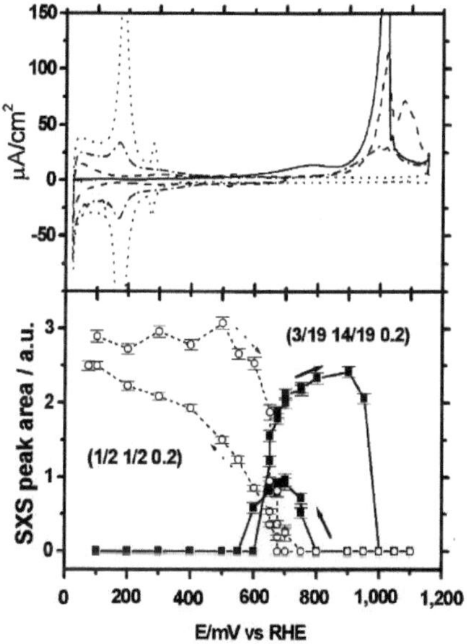

Figure 18. Top: CVs (first anodic scans) at 50 mV/s with CO preadsorbed at 50 mV (solid line), at 700 mV after 30 min of Ar purge (dashed), at 700 mV after 60 min of Ar purge (dash-dot), and without preadsorbed CO (dotted line). Bottom: Integrated SXS intensities (filled triangles) at (1/2 1/2 0.2) and (open triangles) at (3/19 14/19 0.2) as a function of potential in CO-saturated solution. Vertical size of the symbols corresponds to error bars (standard deviation) of the signal. Lines are drawn for visualization only.[135]

was acquired at 50 mV/s, the X-ray data were acquired stepwise with cc 4 min time delay. Upon further increase in potential, adsorbed CO is completely oxidized as manifested by the anodic CV peak at 1020 mV and disappearance of the ($\sqrt{19} \times \sqrt{19}$) reflection at 930 mV. The discrepancy between the values of the CO stripping potential can be again attributed to the different experimental conditions. Upon reversal of the potential scan direction, restoration of the ($\sqrt{19} \times \sqrt{19}$) structure is delayed to 720 mV, showing about 200 mV hysteresis between CO stripping and reordering. Because the CV in the presence of solution-phase CO shows a similar hysteresis between CO oxidation and adsorption peaks, it was concluded that CO does not adsorb on Pt(111) in the potential range between CO stripping and ($\sqrt{19} \times \sqrt{19}$) formation. This hysteresis appears to be an intrinsic phenomenon related to competition between CO and anions for Pt sites and it is not related to the slow ordering of the ($\sqrt{19} \times \sqrt{19}$) phase.

3. Absorption of Oxygen below Platinum Surfaces

Some species can not only be adsorbed on the surface of electrodes but they can penetrate under the surface layers in an *absorption* or *intercalation* process, such as hydrogen into palladium[72,73,159,160] and palladium/niobium[159,160] films and oxygen into platinum.[95,161-163]

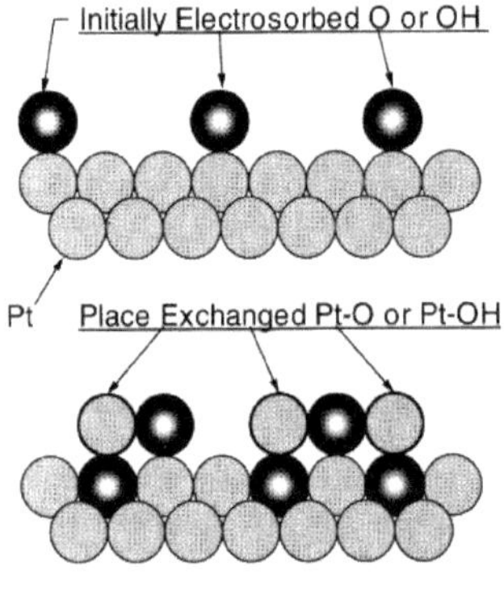

Figure 19. A schematic representation of the *place-exchange* model.[161]

The so called *dermasorption* of oxygen into platinum (Fig. 19) was investigated in a number of experiments.[95,161-163] These are described in more detail in Section VIII.2 with respect to oxide film formation on platinum.

VI. KINETICS/MECHANISM AND ELECTROCATALYSIS OF CHARGE TRANSFER REACTIONS

The kinetics and mechanism of electron-transfer reactions can be very strongly influenced by the atomistic structure of the electrode surface and the absence or presence of adsorbed/deposited species (and their structures) at the interface. The use of synchrotron-x-ray-scattering techniques coupled with electrochemical measurements and other surface techniques is a powerful combination to detect such effects (see Table 4).

1. Oxygen Reduction Reaction on Gold Surfaces

The oxygen reduction reaction (ORR) is of considerable interest both because of its rather complex mechanism and because of its practical importance in fuel cells. An example of bromide adsorption on Au(100) electrodes was already discussed in Section V.1,[111,110] and the same system was also investigated in the presence of dissolved oxygen.[106] Figure 20 shows the voltammetry curve and determination of the Br adlayer structure and stability during O_2 reduction in 0.1 M $HClO_4$ solution containing 10 mM KBr. The thin solid line in Fig. 20b shows that the diffraction intensity at (1/2, 1/2, 0.1) rises between 0.3 and 0.66 V in the absence of O_2, indicating the existence of the $c(\sqrt{2} \times 2\sqrt{2})R45^\circ$ adlayer in this potential region. In the presence of oxygen, the x-ray intensity (bold line) rises over the same potential region, although the intensity is slightly weaker. This indicates that the $c(\sqrt{2} \times 2\sqrt{2})R45^\circ$ Br adlayer is stable over the entire potential region even in the presence of oxygen. The electrochemical results with stationary Au(100) and Au(100)/Br electrodes are shown in Fig. 20c. The bromide adlayer causes a considerable inhibition of ORR, shifting its half-wave potential by approximately 0.3 V to more negative values. A comparison of Fig. 20b and 20c reveals that the ORR in the presence of Br^- is completely blocked in the potential region of existence of the c($\sqrt{2} \times 2\sqrt{2}$)R45° phase. The reaction takes place only at potentials negative to 0.3 V, where the ordered Br^- structure vanishes and the coverage decreases with decreasing potential. Therefore, at smaller coverages, a disordered bromide

Table 4
Reaction Kinetics/Mechanism and Electrocatalysis

System	References
Au(111) reconstruction	58
Bi deposition on Au(111)	164
Br adsorption on Ag(100)	108
Br adsorption on Au(100)	50
CO oxidation on Au(100)	49,52,54
CO oxidation on Au(110) and (111)	54
CO oxidation on Pt(100)	77,80
CO oxidation on Pt(110)	33,80
CO oxidation on Pt(111)	51,67,80,83,134
CO oxidation on Pt(111)/Bi	165
CO oxidation on Pt(111)/Pd	74
Cu deposition on Pt(111)	68
Cu UPD on Pt(111)	129,141,142,144
CuCl deposition on Pt(111)	141-144
H_2 evolution/oxidation on Pt(111)/Pd	73
H_2 evolution on RuO_2(100) and (110)	166
H_2 oxidation on Pt(100)	77
H_2 oxidation on Pt(111)	134
H_2 oxidation on Pt(111)/Bi	165
HCOOH oxidation on Pt(111)/Bi	165
H_2O_2 reduction on Au(100)	47
H_2O_2 reduction on Au(111)/Tl	167
Hg UPD on Au(111)	144
Formaldehyde oxidation on Au(100)	49
Glucose oxidation on Au(100)	49
NO oxidation on Pt(111)	147
O_2 reduction on Ag(100)/Br	105,106,109
O_2 reduction on Au(100)	47,52
O_2 reduction on Au(111)/Bi	168
O_2 reduction on Au(100)/Br	106,109,112
O_2 reduction on Au(111)/Tl	105,167,169,
O_2 reduction on Au(111)/TlBr	109
O_2 reduction on Pt(111)/(100) nanofacets	35
O_2 reduction on Pt(111)/Ag	109,123,170,171
O_2 reduction on Pt(100)/Bi	165
O_2 reduction on Pt(111)/Br	106,109,112,123
S adsorption on Au(111)	150,151

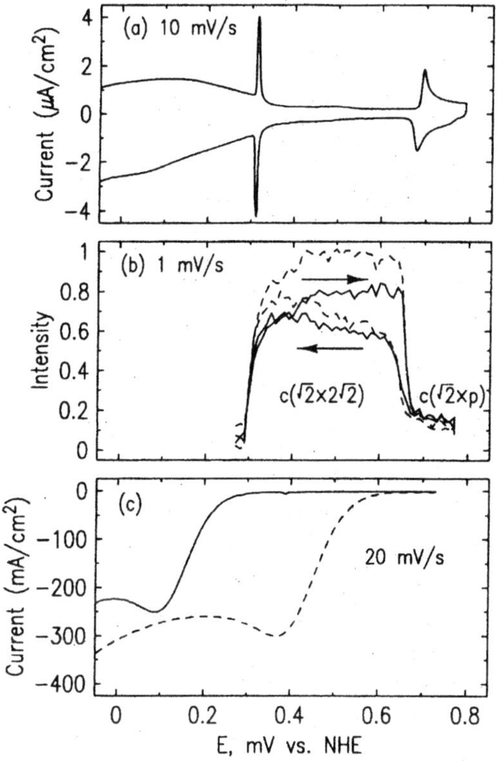

Figure 20. (a) Voltammetry curve of the Au(100) electrode in 0.1 M $HClO_4$ with 20 mM Br^-, sweep rate 20 mV/s. (b) X-ray intensity at (1/2, 1/2, 0.1) position as a function of potential for Br/Au(100) in the same solution in the absence (dashed line), and in the presence (full line) of oxygen. (c) O_2 reduction on Au(100) in 0.1 M $HClO_4$ in the absence (dashed line) and in the presence of 20 mM Br^- (full line). Sweep rate 20 mV/s. (Reprinted from reference 106. Copyright (2000), with permission from Elsevier.)

adlayer does not cause a pronounced inhibition in the given potential region of the O_2 reduction on Au(100) while a complete blocking is caused by the c($\sqrt{2} \times 2\sqrt{2}$)R45° phase.

2. Oxygen Reduction Reaction on Platinum Surfaces

Another example of the effect on the ORR kinetics and mechanism is the irreversible adsorption (underpotential deposition) of bismuth on the Pt(100) electrode.[165] In the absence of oxygen, a c(2 × 2) Bi structure was found that was stable over the full potential range investigated (Fig. 21). From a fit of a Lorentzian line shape to this data and to other c(2 × 2) reflections it was calculated that the coherent domain size in the Bi adlayer was in the range 25 to

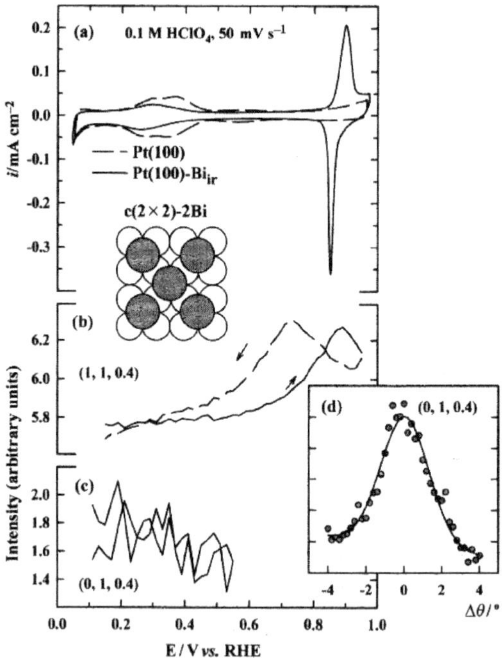

Figure 21. (a) Base voltammetry of Pt(100) and Pt(100)-Bi_{ir}(~ 4 ML) in 0.1 M $HClO_4$ (50 mV/s). (b) Potential dependence of the x-ray intensity at the (1, 1, 0.4) reciprocal space position. (c) Same at (0, 1, 0.4). (d) Rocking scan through the (0, 1, 0.4) lattice point, which gives rise to a c(2 × 2)Bi_{ir} structure and fit to Lorentzian lineshape. (From reference 165. Reproduced by permission of the PCCP Owner Societies.)

40 Å. It was concluded that Bi orders into small patches of c(2 × 2) structure on the Pt(100)Bi_{ir} surface. The fact that the calculated Bi coverage (~ 0.4 ML), obtained from the CTR results, is smaller than the ideal coverage of 0.5 ML implies that this structure does not fully cover the Pt(100) surface.

The electrochemical data on this electrode in the presence of oxygen is shown in Fig. 22. Several conclusions were drawn from these data. At positive potentials (> 0.3 V), the ORR on Pt(100)-Bi_{ir} and Pt(100) obeyed the same reaction mechanism, which was determined from Levich plots to be the four-electron reduction of oxygen to water. The activity on the former surface was reduced by adsorbed Bi_{ir} due to a reduction of the number of active surface sites, that is, a pure site-blocking effect was observed. On the other

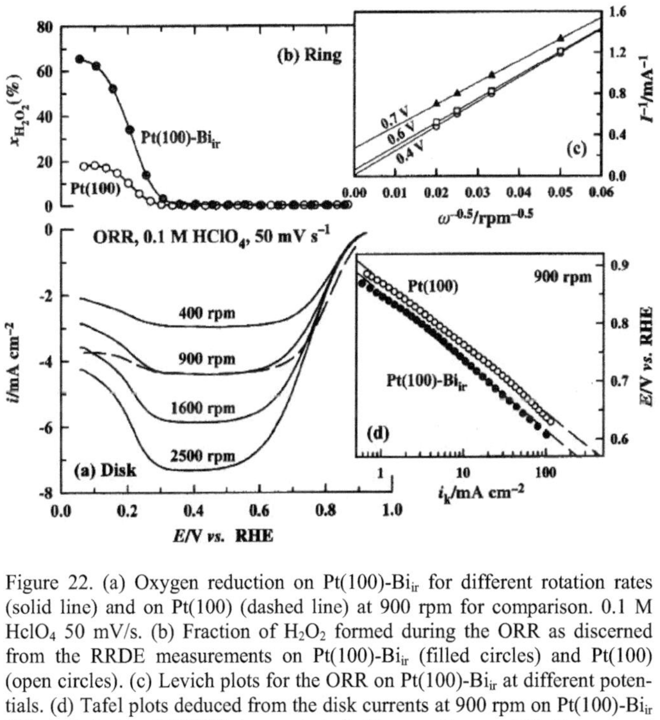

Figure 22. (a) Oxygen reduction on Pt(100)-Bi_{ir} for different rotation rates (solid line) and on Pt(100) (dashed line) at 900 rpm for comparison. 0.1 M $HclO_4$ 50 mV/s. (b) Fraction of H_2O_2 formed during the ORR as discerned from the RRDE measurements on Pt(100)-Bi_{ir} (filled circles) and Pt(100) (open circles). (c) Levich plots for the ORR on Pt(100)-Bi_{ir} at different potentials. (d) Tafel plots deduced from the disk currents at 900 rpm on Pt(100)-Bi_{ir} (filled circles) and Pt(100) (open circles). (From reference 165. Reproduced by permission of the PCCP Owner Societies.)

hand, in the more negative potential region (< 0.3 V), much more peroxide is formed on Pt(100)Bi_{ir} since the formation of pairs of Pt sites is effectively suppressed by the Bi_{ir}. Consequently, the adsorbed bismuth can not only passively block the ORR, but it can affect the mechanism by changing the ratio of the four-electron to two-electron paths.

VII. METAL DEPOSITION AT THE SUBMONOLAYER, MONOLOYER, AND MULTILAYER LEVEL

Some charge-transfer reactions will materially change the surface composition and structure of the electrode surfaces. One of these reaction groups is metal deposition either at the submonolayer level (underpotential deposition, Table 5) or at the multilayer level (bulk deposition, Table 6). The underpotential deposition (UPD) is probably the phenomenon most investigated by synchrotron-x-ray-scattering techniques. In addition of being rather important for basic electrochemistry, including electrocatalysis, these systems are ideally suited for these techniques because the layers are typically ordered and the metals of importance, having high atomic numbers, are excellent scatterers of x rays. Only four examples will be described briefly to show typical applications. For the first three systems, the investigations were carried out from UPD to bulk deposition.

1. Deposition of Lead on Silver Surfaces

The deposition of lead on Ag(111) is described in a number of reports.[184-190] The UPD layer of lead (Fig. 23) was found to have a close-packed-hexagonal structure, incommensurate with the underlying silver, and rotated 4.5° about the six-fold axis of the silver substrate.

An interesting feature of this layer was that it exhibited a two-dimensional compressibility as a function of potential (Fig. 24). The lead-lead interatomic distance decreased linearly with potential until the start of bulk deposition; at more negative potentials, the structure of the UPD layer remained unchanged with a 2.8% contraction from bulk lead. The rotational epitaxy degree did not change with potential. The compression of the structure with

Table 5
Underpotential Metal Deposition

System	References
Ag on Au(111)	172,173
Ag on Pt(111)	109,123,170,171,174,175
Bi on Ag(111)	36,176,177
Bi on Au(111)	7,60,164,168,178
Bi on Pt(100)	131,165
Bi on Pt(111)	138,165
Cd on Au(111)	179
Co on Pt(111)	154
Cu on Au(100)	140,180,181
Cu on Au(111)	39,115,116,182
Cu on Pt(100)	128-131
Cu on Pt(111)	120,122,124,125,130,132,133,141,142,144,155
Cu on Pt(111 × 110) stepped surfaces	142
Hg on Au(111)	102,142,144,153,183
Pb on Ag(111)	184-190
Pb on Au(100)	45,46
Pb on Au(111)	186,173, 190
Pb on Cu(111)	148,191
Pb on Pt(100)	118,119,130,131,192
Pb on Pt(110)	192
Pb on Pt(111)	67,126,130,132,192
Pb-Br on Au(111)	193
Pd on Au(100)	194
Pd on Au(111)	69,194
Pd on Pt(100)	72
Pd on Pt(111)	73-75
Tl on Ag(100)	195
Tl on Ag(111)	90,190,196-198
Tl on Au(100)	199
Tl on Au(111)	62,105,109,149,167,169,190
Tl on Pt(111)	156
Tl-Br on Au(111)	105,109,200,201
Tl-Cl on Au(111)	193
Tl-I on Au(111)	193

potential was tentatively explained by a thermodynamic driving force due to an *ionic overpressure*. This would arise because the ionic concentration of lead in the solution (which is in equilibrium with the UPD layer at the potential of its formation) is unchanged, but at the more negative potentials, a lower concentration would be

Table 6
Bulk Metal Deposition

System	References
Ag on UPD Pb/Au(111)	173
Au(100) on itself	202
Au(111) on itself	71
Au on Si(111)	203
Co on Cu(100)	204,205
Co on Si(111)	91
Cu on Au(100)	68,181,206
Cu on Au(111)	68
Cu on GaAs(001)	92,207
Cu on Si(111)	91
Ni and Ni/Fe alloy (polycrystalline)	208
Os on Pt(111)	209
Pb on Ag(111)	184,188,189
Pb on Au(100)	45,46
Pb on Si(111)	203,210
Pd on Pt(100)	72
Pd on Pt(111)	75
Tl on Ag(100)	195
Tl on Au(100)	199

needed for equilibrium. This driving force was assumed to exist even under nonequilibrium conditions.

Lead deposited over the UPD layer was not found to be epitaxial with the UPD layer and the crystallites must have been randomly oriented as no reflections were observed from them. After about five *equivalent monolayer* deposition, the signal completely disappeared, indicating that the UPD layer itself became restructured. The lead signal reappeared after about 100 *equivalent monolayer* deposition, and the signal could be explained by (111) fiber-structured islands which were somewhat randomly oriented in the plane of the substrate.

2. Deposition of Palladium on Platinum Surfaces

The growth of palladium films on Pt(111) from the submonolayer to the multilayer level has been described in a number of papers.[73-75] It was found that Pd films deposited electrochemically onto Pt(111) exhibited a pseudomorphic growth, whereby a full pseu-

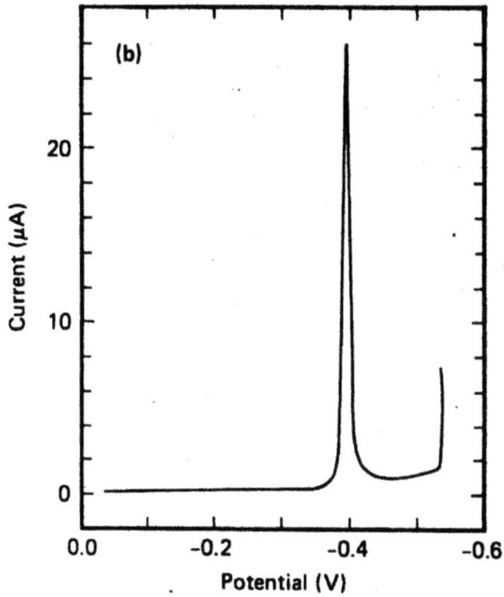

Figure 23. Voltammogram for the deposition of lead on silver(111). Potentials are measured relative to Ag/AgCl. Scan rate 20 mV/s, 0.005 M lead acetate, 0.1 M sodium-acetate, 0.1 M acetic acid. Shows UPD at about -0.4 V and the beginning of bulk deposition at about -0.55 V. (Reprinted with permission from reference 184. Copyright (1988) by the American Physical Society.)

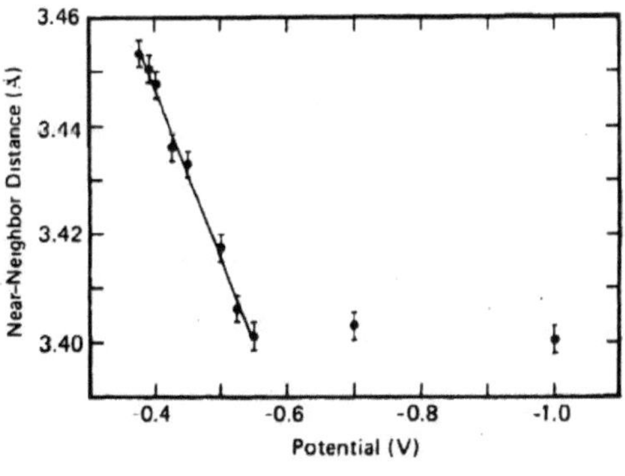

Figure 24. Lead-lead near-neighbor distance vs. electrode potential. (Reprinted with permission from reference 184. Copyright (1988) by the American Physical Society.)

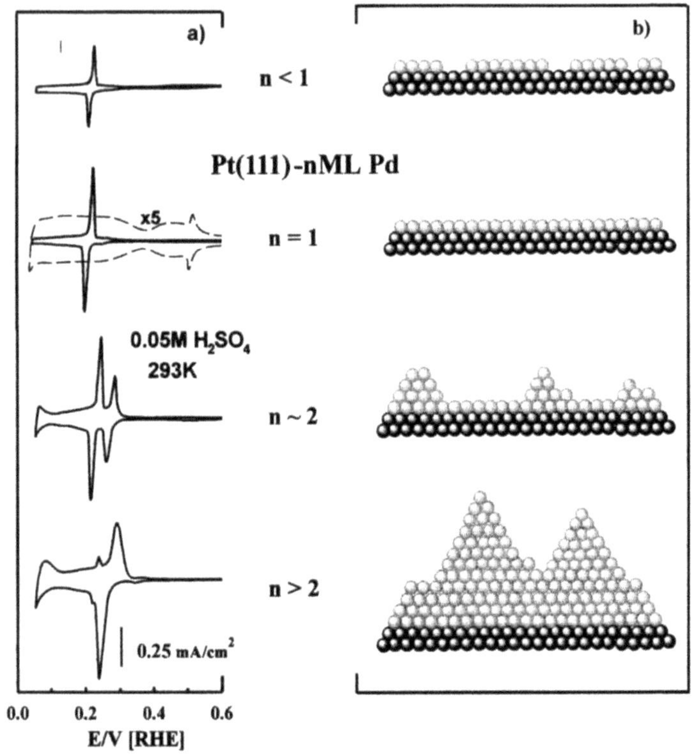

Figure 25. (a) CV from the Pt(111)–n ML Pd surface at differing levels of Pd coverage. For $n < 1$ (the SUB-ML regime), the CV shows the emergence of a peak at ~ 0.21 V due to Pd surface adsorption sites; for $n = 1$ (the ML regime), the ~ 0.21 V peak has saturated (the dashed line shows the CV of the clean Pt(111) surface); for $n = 2$ (the 2-ML regime) the emergence of a second peak at ~ 0.3 V due to Pd step sites concurrent with a reduction of the peak due to Pd surface terrace sites is evident; for n > 2 (the N-ML regime) it is evident that the second peak now dominates indicating a much greater number of step sites than terrace sites. (b) A schematic of the surface in each case as inferred from the Pd coverages obtained from the SXS results, the darker circles are Pt atoms and the lighter circles are Pd atoms. (Reprinted from reference 75. Copyright (2002), with permission from Elsevier.)

domorphic monolayer was completed before the onset of three-dimensional island formation (Fig. 25). The Pd atoms were found to be located at threefold (fcc) hollow sites continuing the ABC stacking of the bulk lattice.

Up to two monolayers, the CTR data were fitted to a model with the following structural parameters: the relaxation of the topmost Pt atomic layer from the bulk position, the coverage of Pd per Pt atom, the Pt–Pd surface normal spacing, and a Debye-Waller type rms roughness of the topmost Pt atomic layer and of the Pd adlayer. Good fits were obtained with the pseudomorphic model with a slight expansion of the topmost Pt layer and the first Pt-Pd distance. With two monolayer equivalent deposition, the best fit indicated that a full second layer was not formed, rather islands of up to three layer heights were formed, continuing the ABC stacking. Similarly, for the N-layer cases, island growth was found. A good fit to the data was obtained (Fig. 26) using a statistical model of the layer occupancy with Lorentzian distribution. The electrocatalytic properties of these films were also investigated for the hydrogen evolution/oxidation reactions[73] and the oxidation of carbon monoxide.[74]

3. Deposition of Thallium on Silver Surfaces

Thallium was found to form two UPD layers on the Ag(100) surface before bulk deposition.[195] The cyclic voltammogram (Fig. 27) showed a broad cathodic peak at –0.225 V that is the initial UPD of Tl. It was followed by the sharp peak at –0.335 V which, according to the x-ray-scattering measurements, corresponded to the formation of a close-packed Tl monolayer. At –0.460 V, the large peak signaled the deposition of the second T1 layer which occurred at a potential about 10 mV positive of the bulk deposition. Below the reversible Nernst potential of –0.472 V (dashed line), the onset of bulk deposition gave rise to a sharp increase of the cathodic current.

X-ray-scattering results (Fig. 28) indicated that thallium formed a $c(p \times 2)$ close-packed monolayer which compresses uniaxially (p decreasing from 1.185 to 1.768) with decreasing potential. Upon deposition of the second layer, the first layer expands slightly along the incommensurate direction and both layers form a $c(1.2 \times 2)$ structure. After deposition of a few hundred nominal atomic layers, the x-ray results (Fig. 29) indicated that several very sharp and intense peaks appeared in the two surface rods. These equally spaced spikes along the L direction were indicative of the formation of well-ordered three-dimensional crystallites. Despite the presence of these spikes, the intensity at all other positions are nearly the same as the ones measured at –0.47 V (circles) for the Tl bilayer. A tentative explanation for the constant intensity in the bilayer rods was suggested: the bulk crystallites only covered a small region of the crystal.

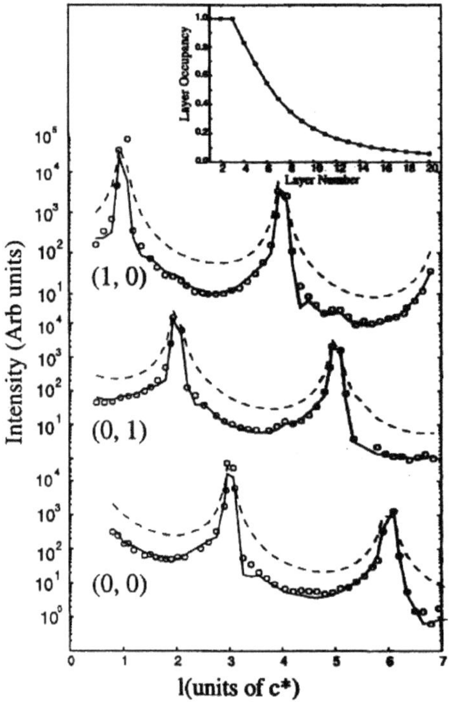

Figure 26. Specular and first order non-specular CTRs for the N-ML surface in 0.05 M H_2SO_4, taken at an electrode potential of 0.05 V. The circles represent the data, the solid line is a best fit to the data using a structural model involving a half-Lorentzian distribution of layer occupation for pseudomorphic layers of palladiums. The dashed lines are calculated for an ideally terminated, clean Pt(111) surface. Inset: plot of the Pd layer occupancy derived from the fit to the CTR data. (Reprinted from reference 75. Copyright (2002), with permission from Elsevier.)

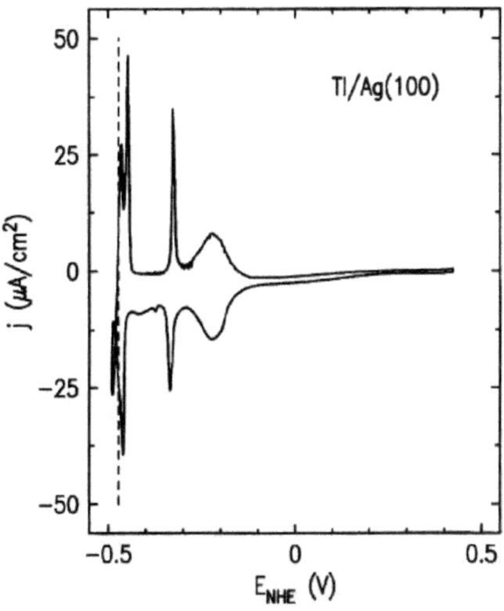

Figure 27. Linear sweep voltammogram for UPD of Tl on Ag(100) in 0.1 M $HClO_4$, with 2.5 mM Tl_2SO_4 obtained in a conventional electrochemical cell. The Nernst potential for T1 bulk deposition is indicated by the dashed line. Sweep rate 5 mV/s. (Reprinted from reference 195. Copyright (1995), with permission from Elsevier.)

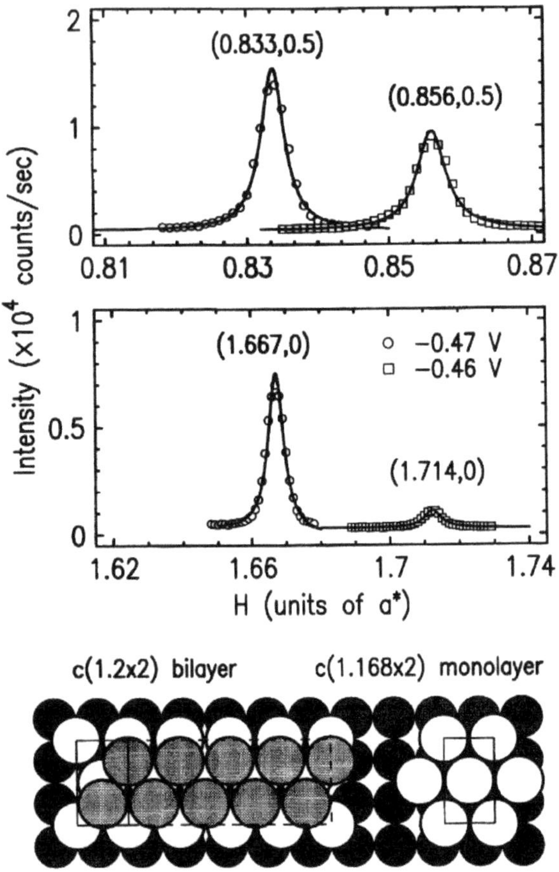

Figure 28. Upper panels: diffraction scans along the H direction through two low-order diffractions of the c(1.2 × 2) Tl bilayer (circles) at –0.47 V and of the c(p × 2) monolayer (squares) at -0.46 V. The solid lines are the fits to a Lorentzian line shape. Lower panel: the real space models for the commensurate c(1.2 × 2) bilayer and the c(p × 2) monolayer. The diameter ratio of the open (the first Tl layer) and lightly shaded (the second Tl layer) circles to the filled (Ag) circles is 3.456/2.889 = 1.196, which equals to the ratio of bulk Tl separation in the hexagonal plane (3.456 Å) over the Ag-Ag separation (2.889 Å). (Reprinted from reference 195. Copyright (1995), with permission from Elsevier.)

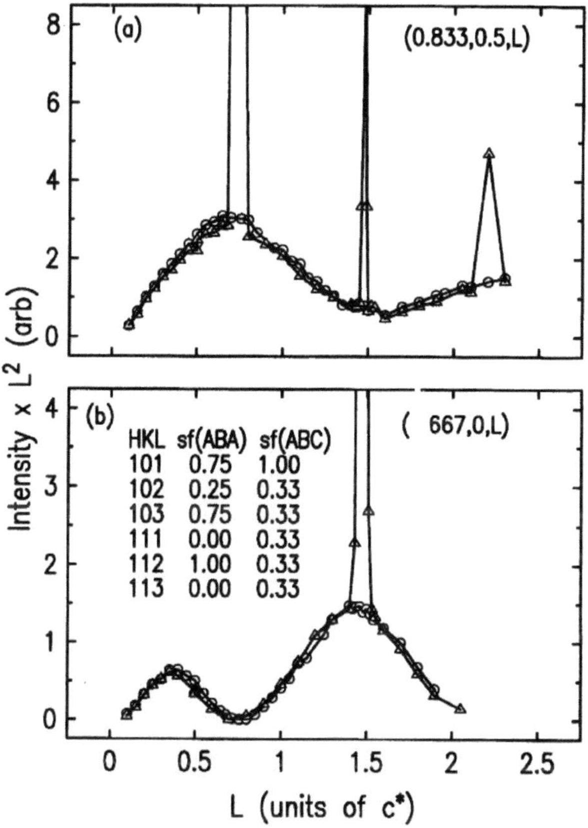

Figure 29. Surface rods measured at –0.47 V For the c(1.2 × 2) bilayer (circles) and at –0.53 V for the c(1.2 × 2) bilayer plus hcp multilayers (triangles). The solid lines are for guiding the eyes. The (10) and (11) Tl rods are measured at (0.833, 0.5) and (1.667, 0) in-plane positions of the substrate coordinate, respectively. The inset of (b) lists the calculated structure factors for AB and ABC stacking sequences of Tl multilayers at six Bragg positions. (Reprinted from reference 195. Copyright (1995), with permission from Elsevier.)

4. Deposition of Copper on Gold Surfaces

Bulk deposition of copper on Au(111) was found to form pseudomorphic epilayers that were fully relaxed after the deposition of the first monolayer.[68] A monotonic increase of the x-ray signal was observed, indicating a homogeneous growth rate (Fig. 30). An interesting feature of these results was the observation that the copper layer thickness decreased in time after the cessation of deposition, this was explained as the dissolution of copper by the free radicals formed through the radiolysis of the solution. This phenomenon will be discussed in more detail in the Appendix.

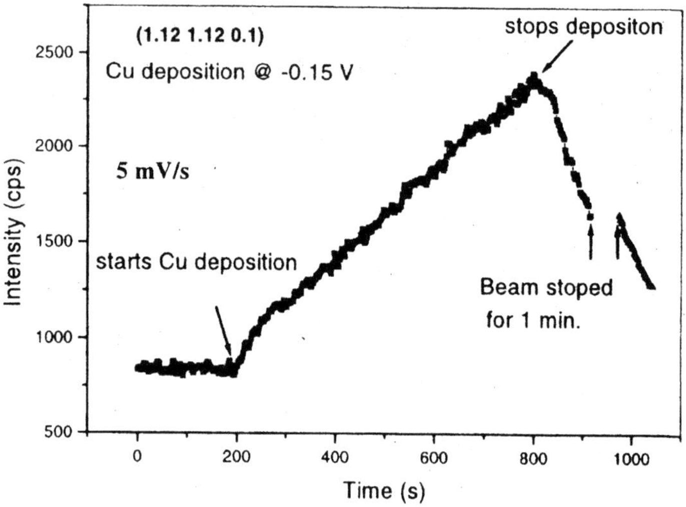

Figure 30. Evolution of the fully relaxed copper peak as a function of the deposition time. Scan starts with the potential at 0.8 V (no Cu adsorbed), the potential is then reduced at 5 mV/s until the deposition potential of –0.15 V is reached. (Reprinted from reference 68. Copyright (1999), with permission from Elsevier.)

VIII. OXIDE FILM FORMATION ON METALS AND PASSIVATION

Another charge-transfer reaction that changes the composition and structure of metal electrodes is oxidation and the resulting oxide film formation (Table 7). Some of these films may be *passive* films providing corrosion protection; consequently there is some overlap between this and the following Section IX on corrosion. The formation of passive oxide film on polycrystalline copper was already discussed in Section III.2(ii). Oxidation/reduction of electrode surfaces other than metals have also been investigated, such as RuO_2.[87,211]

1. Oxidation of Ruthenium Surfaces

The electrochemical surface oxidation of Ru(001) in 1 M H_2SO_4 was found to be limited to a one-electron process resulting in one monolayer oxygen uptake at potentials below the onset of bulk oxidation.[157,158] At a potential negative to surface oxidation, a bisulfate coverage of approximately 1/3 was found with an oxygen adlayer of about the same coverage, while the Ru(001) surface remained nearly ideally terminated. At a potential positive to the surface oxidation, the oxygen coverage increased to one while the bisulfate coverage decreased to 0.08 (Fig. 31 and 32). However, no

Table 7
Oxide Films

System	References
Ag(polycrystalline)	89,212
Cu(polycrystalline)	23,163,212-214
Cu(111)	148,215
Fe(polycrystalline) passivation	216
Fe(100) and (110) passivation	217,218
Nb(polycrystalline)	93
Ni(111) passivation	219-221
Pt(111)	4,10,24,25,95,96,161-163,222
Ru(001)	157,158
Si(100)	214,223
Stainless steel passivation (polycrystalline)	216,224
Ta(polycrystalline)	93
Tl UPD on Au(111)	149

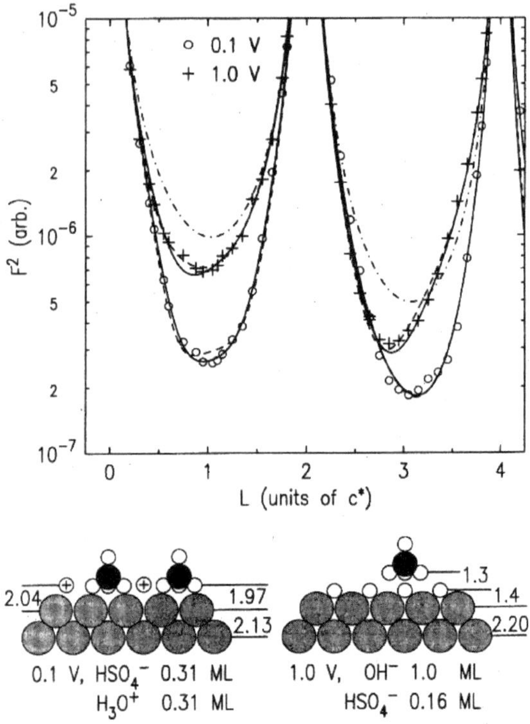

Figure 31. Upper panel: Specular reflectivity (structure factor squared) measured for Ru(001) at 0.1 V (circles) and 1.0 V (plus signs) in 1 M H_2SO_4 with the dot-dash line showing the calculated curve for an ideally terminated Ru(001). The dashed and solid lines are the fits to the models shown in the lower panel, dashed: sulfate only, solid: sulfate and oxygen. Lower panel: Proposed structural models where the O, S, and Ru atoms are represented by the open, heavily shaded, and lightly shaded circles, respectively. The layer spacings are given in Å and coverages are given in monolayer (ML). (Reprinted with permission from reference 157. Copyright (2001), from American Chemical Society.)

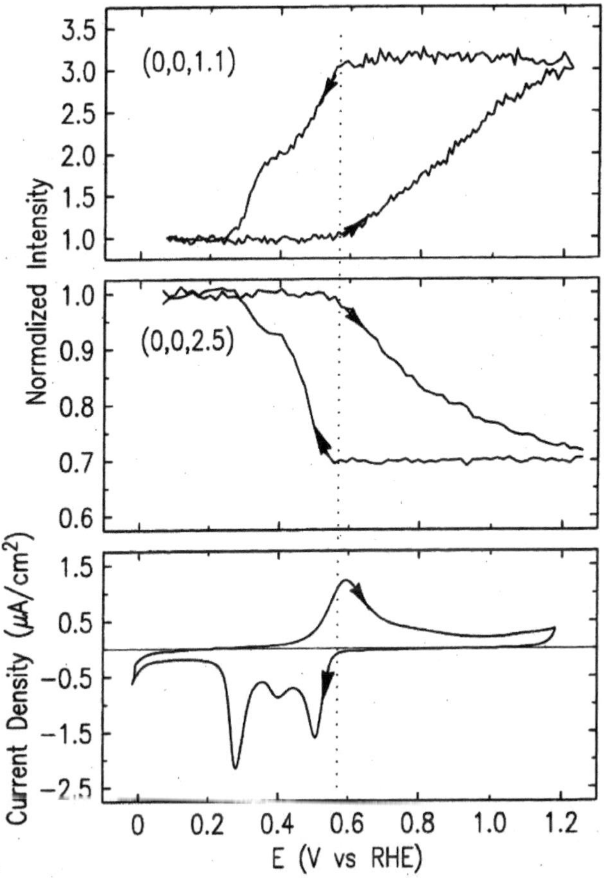

Figure 32. X-ray intensities as a function of potential and the corresponding voltammetry curve for Ru(001) in 1 M H_2SO_4. Sweep rate = 1 mV/s. The x-ray intensities measured at the (0,0,1.1) and (0,0,2.5) positions are shown after subtracting the diffuse scattering background and normalizing the intensities at the most negative potentials to unity. The vertical dotted line highlights the critical potential (0.57 V) where the onset of structural changes due to the surface oxidation/reduction occurs. (Reprinted with permission from reference 157. Copyright (2001), from American Chemical Society.)

in-plane diffraction features could be correlated to any ordered (sub)monolayer oxygen phases, presumably because the expected x-ray intensity from an oxygen adlayer would be very weak relative to the diffuse scattering background originating from the thin solution layer and the plastic film of the x-ray-electrochemical cell.

Further insight was obtained into these surface processes by monitoring the x-ray intensities at the (0,0,1.1) and (0,0,2.5) positions while sweeping the potential between 0 and 1.2 V at 1 mV/s. Figure 32 shows these potential-dependent intensities together with the voltammetry curve. Both the bisulfate desorption and a layer expansion of the Ru surface would be expected to contribute to the intensity increase at the (0,0,1.1) position at high potentials, but the latter has a dominant and opposite effect on the x-ray intensity at (0,0,2.5). The (0,0,1.1) intensity was found to be constant up to 0.57 V in the positive potential sweep suggesting that no significant change in either Ru surface or adlayer coverage occurred below this critical potential, even though the anodic current started to rise at a slightly more negative potential. Above 0.57 V, the (0,0,1.1) intensity continuously increased with increasing potential, which is accompanied by the intensity decrease at the (0,0,2.5) position. This confirmed that an electrooxidation-induced surface expansion occurred on Ru(001) above this critical potential. After the potential sweep reversal, both x-ray intensities remained constant over the entire oxide formation potential region down to 0.57 V. At this critical potential, the onset of oxide reduction occurred, which was followed immediately by changes in the (0,0,1.1) and (0,0,2.5) intensities. The steep slopes in the x-ray-intensity curves correlated well with the reduction current peaks.

2. Oxidation of Platinum Surfaces

The oxidation of the Pt(111) surface was already briefly discussed in Section III.2(*i*) with respect to surface roughening. X-ray scattering was also used to investigate many other aspects of the submonolayer/monolayer oxidation/reduction of that surace.[4,10,95,96,161-163] It was found (see Fig. 33 and 34) that there are three distinctly different behaviors that can be best separated in terms of the total amount of charge transferred during the surface oxidation (ex-

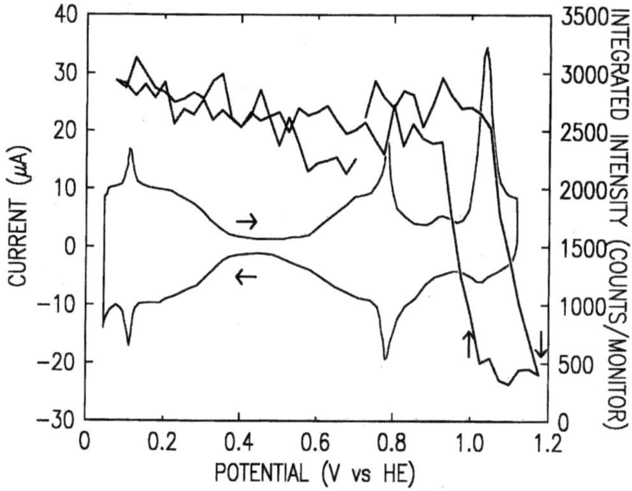

Figure 33. Voltammogram of Pt(111) in 0.1 M $HClO_4$ solution (bottom curve) and the change of scattered x-ray intensity at (0,0,1.3) versus potential (top curve). The potential scale is referenced to the potential of the beginning of hydrogen evolution in the CV.[10,162]

pressed in units of *number of electrons transferred per surface platinum atom*, e^-/Pt):

(*a*) Up to approximately 0.6 e^-/Pt, (~ 0.9 V) there was no change in the x-ray scattering in spite of the obvious electrochemical indication of surface oxidation.

(*b*) Between approximately 0.6 and 1.7 e^-/Pt (up to approximately 1.2 V), an increasing deviation was found from the scattering measured at oxide-free surfaces. However, upon electrochemical reduction of the surface, the original oxide-free surface scattering was always reproduced, indicating that the surface processes were completely reversible.

(*c*) Oxidation for > 1.7 e^-/Pt irreversibly changed the x-ray scattering, indicating an irreversible roughening of the surface (see Fig. 6 in Section III.2*i*).

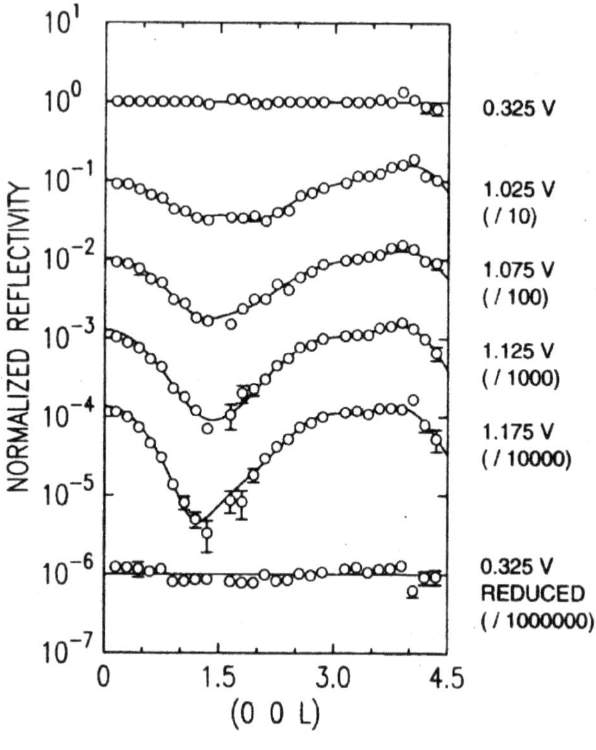

Figure 34. (00L) scans of Pt(111) in 0.1 M $HClO_4$ solution, normalized to the oxide-free scan. The scans are offset for clarity. The potential was changed to four consecutively more oxidizing potentials between 1.025 V and 1.175 V. Finally, the surface was reduced, and the final x-ray measurements indicated that the original smooth surface was recovered. The lines are fits with the place-exchange model.[95]

The following explanation of these observations was proposed. (With the caveat that only the positions of the Pt atoms can be considered in the modeling of the x-ray scattering because the electron densities of both the OH^- and O^{2-} species are negligible compared with that of Pt. Therefore, x rays cannot distinguish between the OH^- and O^{2-}, nor can they be used to determine directly the position of these species.) During the first phase of surface oxidation, either the OH^- or O^{2-} was chemically adsorbed onto the surface without changing the structure of the underlying platinum. These structures represent dipoles and, because the high repulsive dipole-dipole interaction energy, adsorption occurred in *nearest-neighbor-avoiding* configuration till at a critical point of coverage some of the dipole structures was flipping around resulting in *place exchange* (sometimes called *dermasorbed oxygen*). Modeling of the x-ray-scattering data indeed gave the best fit with the place-exchange model (Fig. 34 and 35). The fraction of place exchanged platinum atoms increased linearly from about 0.21 to 0.33 for the four scans shown in Fig. 34. At this phase of the oxidation, it was possible to essentially recover the original smooth platinum surface with cathodic reduction of the oxide layer. The proposed consecutive surface structures are shown in Fig. 36.

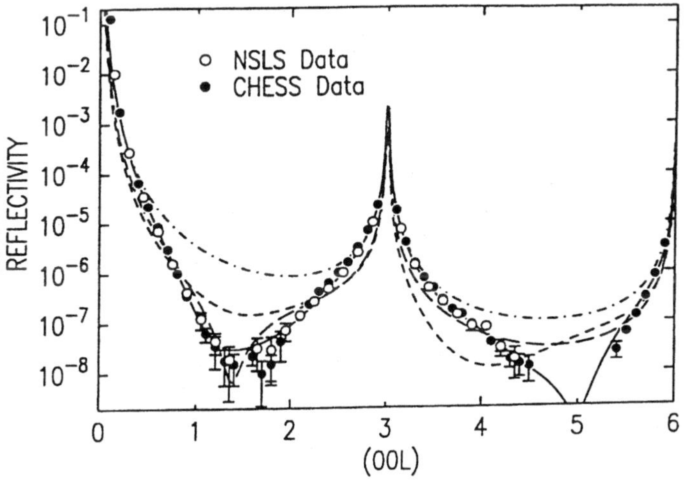

Figure 35. The (00L) scan at 1.175 V with fits based on several models. (The data reproducibility is indicated by the good agreement between datasets taken at two different synchrotrons.) Dot-dashed: ideal surface; long-dashed: buckled surface; short-dashed: ClO_4^- adsorption; solid: place exchange.[10,95]

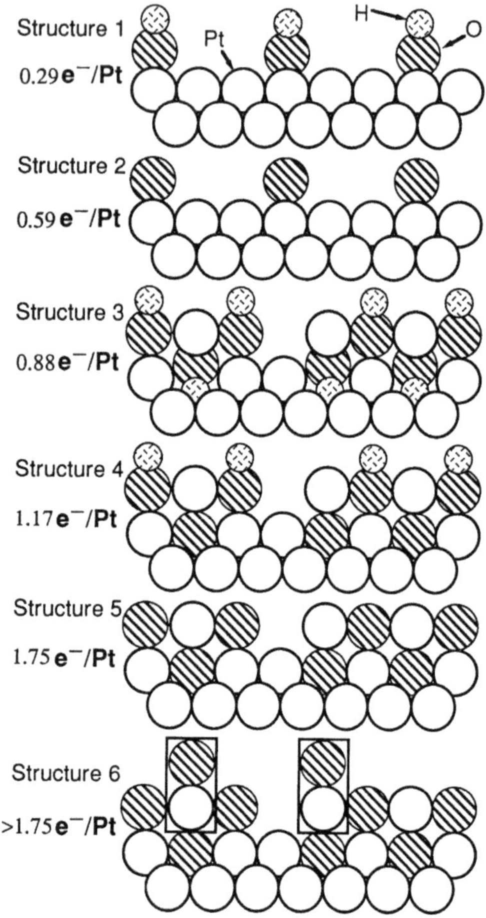

Figure 36. Six possible structures of the oxidized surface.[10,162]

Once a full monolayer of the place exchanged structure was completed (at the oxidation of 1/3 of the platinum atoms in the surface), upon further oxidation OH^- or O^{2-} reacted with the *place-exchanged* Pt atoms. Therefore, the formation of the Pt—O or Pt—OH (shown in the boxes in Fig. 36) was accompanied by a decrease of the metal-oxygen bond strength in the second layer of platinum atoms. Consequently, the newly formed Pt—O or Pt—OH would easily move away from their original positions by surface diffusion, and reduction of the oxide would not recover the original Pt surface structure, resulting in irreversible roughening of the surface.

While the above results permitted the examination of the physical morphology of the surface, the novel technique of *resonance surface scattering* (RASXS) permitted the examination of the surface *oxidation state.*[24,25,222] In structures 1 and 2 in Fig. 36, where oxyspecies chemisorbed on the surface without disrupting the top Pt layer, the RASXS measurements indicated no measurable chemical shift despite the expected chemisorption. It was concluded that in this case the chemical shift is too small to be measured with this technique. In Structures 3-4, it was found that the measured RASXS spectra were easily modeled by the oxide monolayer (1/3 Pt layer and 2/3 oxyspecies). The fit value for the resonance energy shift of the top 1/3 monolayer Pt atoms was a surprisingly large value of 9(±2) eV. This large shift was attributed to the change in the work function of Pt surface. The top Pt oxide layer becomes an insulating layer which has no longer a Fermi level. The resonance energy is the energy *difference* between the ground and the lowest unoccupied intermediate states.[25] In case of bulk platinum in metallic state, the lowest unoccupied intermediate state is at the Fermi level. In the case of oxide layer(s), the Fermi level falls at the insulating band gap, thus the lowest intermediate state available is at the vacuum level. Since the work function of platinum metal is ~ 6 eV, the actual binding energy shift of the oxidized Pt atoms is ~ 3 eV. This value is not unreasonable for the L_{III} core binding energy shift upon oxidation and it indicates the higher oxidation state of these structures. Similarly, it was possible to model Structure 6 with the resonance energy shift of the top *two* layers (1/3 first and 2/3 second layers). It was concluded that the top two layers of Pt and oxygen atoms form a bilayer oxide of PtO_2 when the electrochemical potential exceeds ~ 1.2 V or the charge transfer amount is larger than 1.75 e^-/Pt.

IX. CORROSION AND METAL DISSOLUTION

The opposite of metal deposition is the charge-transfer reaction involving metal dissolution. These reactions are also the central reactions in the phenomenon of metallic corrosion (see also the previous Section VIII on passive film formation). The corrosion/electrodissolution of As from GaAs(001),[225-227] $Cu_3Au(111)$,[91,228] polycrystalline copper,[229,230] and steel[231] have been reported.

1. Pitting Corrosion of Copper Surfaces

In one study, the electrochemically-induced pitting process on a copper surface in $NaHCO_3$ solution was monitored using in-situ x-ray off-specular reflectivity measurements.[229,230] The morphology and growth dynamics of the localized corrosion sites or pits were studied. Analysis of the experimental results indicated that early pitting proceeded in favor of nucleation of pit clusters over individual pit growth. It was found that the lateral distribution of the pits was not random but exhibited a short-range order as evidenced by the appearance of a side peak in the transverse off-specular reflectivity. Fitting of the experimental results employing an *island* model with the distribution of islands satisfying the *hard-sphere* model (Fig. 37) yielded the average size, nearest-neighbor distance (within any one of the clusters), and over-all density of the pits averaged over the entire surface (Fig. 38).

X. POROUS SILICON FORMATION

Roughness of surfaces can also be investigated at a much larger scale than atomic layer level (discussed in Section III.2), as it was carried out for the investigation of the anodic formation of *porous silicon* layers.[211,232-234]

The experiments were performed with p^+-type silicon with a resistance of 0.015 Ω/cm, and the (100) surface investigated had an initial roughness of 2 to 3 Å. The electrolyte solution was ~ 25% HF solution (50% aqueous HF diluted with an equal volume of ethanol). The two main characteristics of specular reflectivity utilized in this work were:

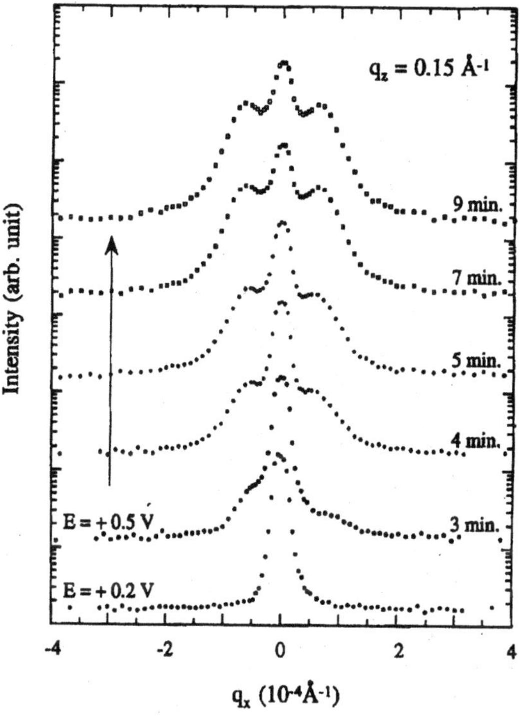

Figure 37. X-ray transverse off-specular reflectivity at the pitting potential of 0.5 V after various holding times. (Reprinted from reference 230. Copyright (1996), with permission from Elsevier.)

(*a*) the critical angle for total external reflection, and
(*b*) the Q dependence of the reflectivity.

Diffuse scattering was also used around the Bragg reflections to study the in-plane structure of the pores. One series of reflectivity scans obtained are shown in Fig. 39. The top two scans were carried out before any anodic dissolution. The subsequent scans were made after the application of a successively increasing number of dissolution pulses at 10 mA/cm^2 resulting in the dissolution of the

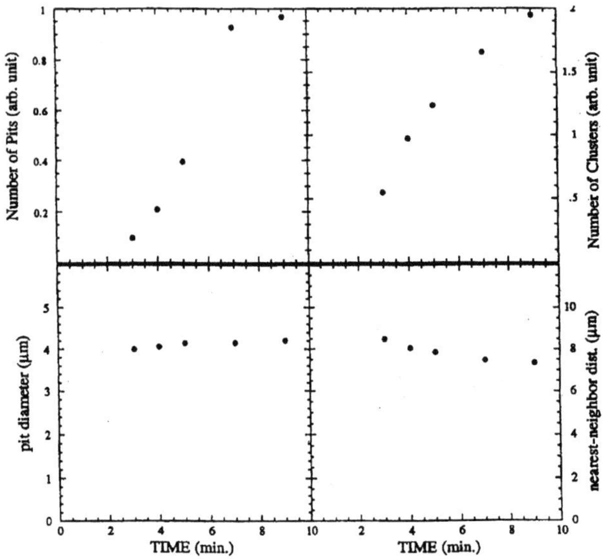

Figure 38. Time dependence of the average size, nearest-neighbor distance of pits within a cluster, the number of pits per cluster, and the number of clusters. (Reprinted from reference 230. Copyright (1996), with permission from Elsevier.)

equivalent of solid silicon layers ranging from approximately 135 to 2,500 Å, for a grand total of 24,000 Å. All the scans were linear for angles larger than the critical angle, and the slope of the reflectivity scans changed during the first few successive dissolutions but remained constant thereafter. In other words, the interface density profile reached a steady state functional form at a critical dissolution depth (~ 300 Å of solid silicon for this particular set of data). This observation indicated the existence of an interfacial region where the density gradually changed, followed by another layer where dissolution in excess of a critical depth did not significantly alter the density profile any more. Furthermore, this observation also indicated that the interface density profile should follow a power law. The critical angle was well defined for all the scans, and it decreased continuously as the total dissolution amount increased. The critical angle of the top two scans (zero porosity) was 0.025 Å^{-1}, and it decreased to 0.018 Å^{-1} for the last scan (indicated by the arrows in Fig. 39). The porosity of silicon could be estimated from the measured critical angles. For example, the porosity of the last scan was approximately 52%. This porosity was averaged over a depth of 1.4 μm, which was the penetration depth of x rays at the critical angle.

Two possible pore distribution models consistent with the power-law density profile are schematically shown in Fig. 40:

(a) tapered pores with relatively uniform length across the sample surface, and
(b) pores with uniform radius and a length distribution matching the overall power-law interfacial-density profile.

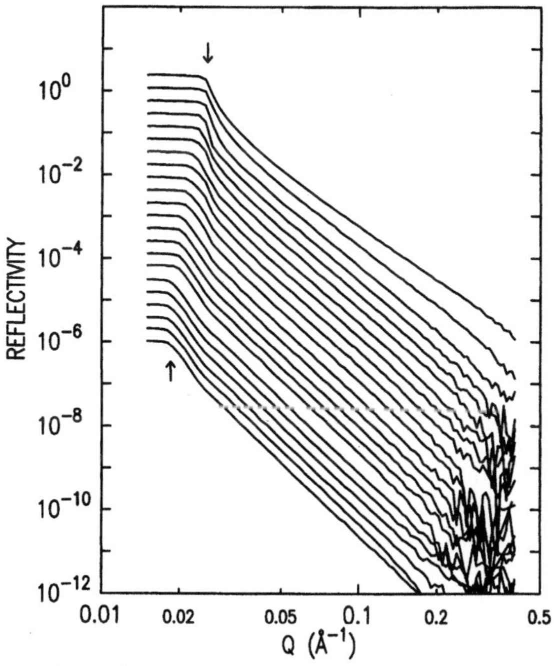

Figure 39. A series of reflectivity scans during anodic oxidation of silicon. The scans are offset from each other for display purposes.[10,232]

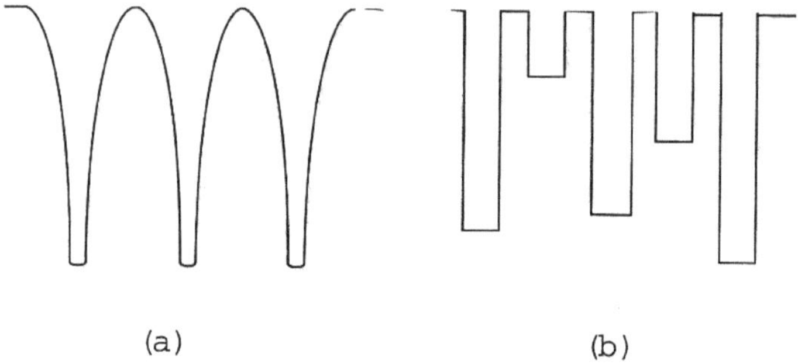

Figure 40. Two possible pore distributions consistent with the reflectivity scans.[10,232]

The reflectivity data were consistent with either possibility or with any combination of them. Therefore, additional measurements were made for diffuse scattering around the Bragg reflections to determine which model was closer to the real structure of the pores. It was found that a funnel-shaped (tapered) interfacial layer, followed by a layer of uniform porosity with relatively straight circular pores and increased branching with depth provided the best fit to the data. For the particular conditions, the porosity was around 50%, and the pore diameter was about 50 to 70 Å. It was also found that the space-charge regions became *insulating* layers between the pores, because the holes must overcome the potential barrier of the space-charge region for the dissolution process to continue. To illustrate this, schematic views of pore cross sections at a given depth are shown in Fig. 41. (a) Illustrates the mechanism of pore *repulsion.* The arrows represent growth rates, the white area is space-charge region, and the gray background is solid silicon. Holes cross freely in the direction of the long arrows, but fewer hole crossings are possible across the space-charge region in direction of the short arrows. Consequently, a pair of holes will grow apart, thereby, avoiding the overlap between nearest neighbors. Figure 41(b) illustrates the termination of lateral growth of the pores. The growth of the pores towards each other will be terminated when three pores meet. Only three pores are shown for illustration purposes, but lateral growth of pores will be similarly terminated when all pores are closely packed.

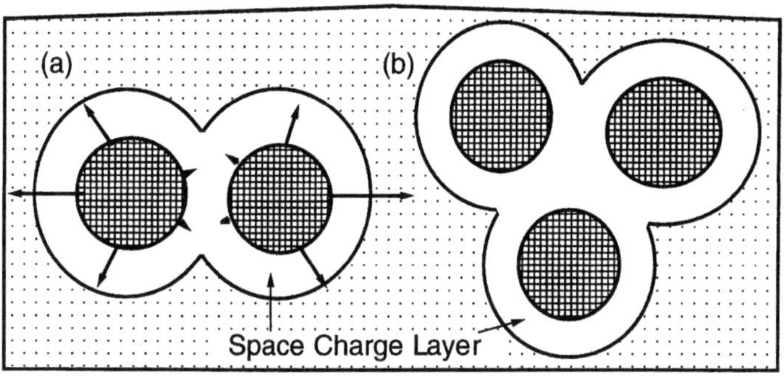

Figure 41 Schematic diagrams for pore formation: (a) pore *repulsion* (b) termination of lateral growth.[233]

XI. INVESTIGATION OF INTERPHASES OTHER THAN METAL/ELECTROLYTE

Synchrotron-x-ray-scattering techniques can be used also for the investigation of electrode surfaces other than solid metals and metal alloys under aqueous electrolytes. Electrode surfaces of liquid gallium and mercury,[235-237] semiconductors such as Si[203,210,211,214,223,232-234] and GaAs,[92,207,226,227] and bulk oxides such as RuO_2[87,88,166,211] were also investigated under aqueous electrolytes. Electrodeposition by electrochemical atomic layer epitaxy (ECALE), CdS deposition on Ag(111), was also investigated.[40] As was the Au(111)/polymer-electrolyte interface[66] and a liquid/liquid interface.[98-101] The ionic adsorption and the structure of the electrical double layer at mineral/electrolyte interfaces were also investigated without using potential control.[14] A number of these systems have been briefly discussed in previous sections.

1. Oxidation of Ruthenium Dioxide Surfaces

The surface structure and oxidation/reduction of $RuO_2(100)$ in H_2SO_4 solution was investigated and described in a number of papers.[87,88,211] The structure was found to display marked changes with potential. The cyclic voltammogram and the very closely corresponding x-ray-scattering response are shown in Fig. 42. A fitting analysis revealed that the Ru plane at the surface is significantly expanded (about 0.15 to 0.2 Å) when reduced. In addition to the surface expansion, the Ru—O bonds at the surface were found to be significantly distorted. The distortion of the surface structure is consistent with the reduction of RuO_2 to Ru(OH)O, in which a proton is inserted in the surface of the RuO_2 lattice to balance the change of the oxidation state of Ru from +4 to +3. This behavior is completely reversible over many cycles. Some results reported for the RuO_2(110) facet were already described in Section IV.2 in connection to the study of water structure in the double layer. The catalytic behavior of both these facets for hydrogen evolution reaction was also investigated.[166]

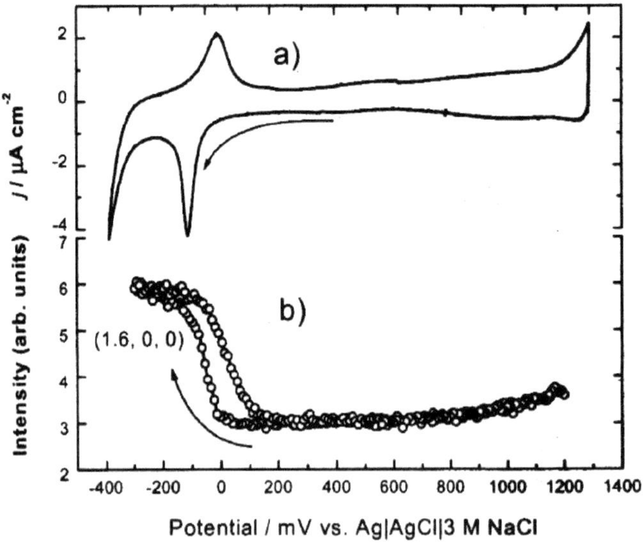

Figure 42. Cyclic voltammogram of RuO_2(100) in 0.5 M H_2SO_4 solution and the corresponding x-ray-scattering intensity.[87]

XII. CONCLUDING REMARKS

It was the intent of this chapter to demonstrate the usefulness of synchrotron-x-ray-scattering techniques for the *in-situ* investigation of a wide range of phenomena occurring in the electrochemical interphase. In the interphase that extends from a few Å to a few thousand Å in both the electronically and the ionically conducting phases. The phenomena that can be investigated include, among others, the morphological and chemical nature of the solid interface, adsorption and absorption occurring at the interface, the structure of the electrochemical double layer, and the deposition/dissolution and oxidation of metal interfaces. Many interphases other than those occurring between a solid metal and an aqueous solution can also be investigated. These in-situ measurements were made possible by the advent of synchrotrons, which provide x rays orders of magnitude more brilliant that prior sources. While experiments at synchrotron beamlines can be difficult, time consuming, and sometimes frustrating,[238] the results, which usually cannot be obtained with any other technique, are making them worthwhile.

APPENDIX

Typical schematics of an experimental setup at a synchrotron beamline using a transmission-geometry x-ray-electrochemical cell is shown in Fig. 43.

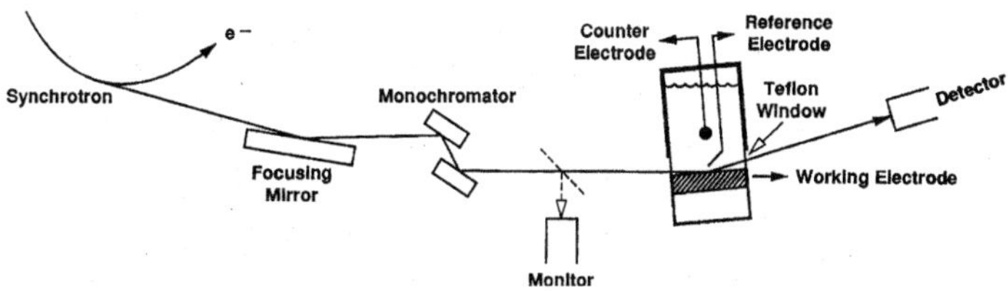

Figure 43. A schematic diagram of the experimental set-up of the electrochemical cell and the beamline at the synchrotron.[10]

Several x-ray-electrochemical cell designs have been developed based on different x-ray geometries. Some of the design features will be described in this Appendix, together with some special precautions needed in electrochemical experiments at synchrotron beamlines.

1. X-Ray-Electrochemical Cell Designs

There are two x-ray geometries used in cell designs, which are called: *reflection* (*Bragg*) and *transmission* (*Laue*) geometries. The reflection geometry is shown schematically in Fig. 44a. The window of the cell through which x rays pass is parallel to the interface of interest. It is analogous to Bragg geometry for the single-crystal scattering case where the interface of interest is considered as a Bragg plane. The other possibility is to perform the x-ray-scattering studies in a transmission geometry (Laue geometry) where the incoming x rays pass through a window perpendicular to the interface of interest and the outgoing x rays pass through a second perpendicular window. The cell geometry for the Laue case is shown schematically in Fig. 44b. The earliest reports for x-ray-

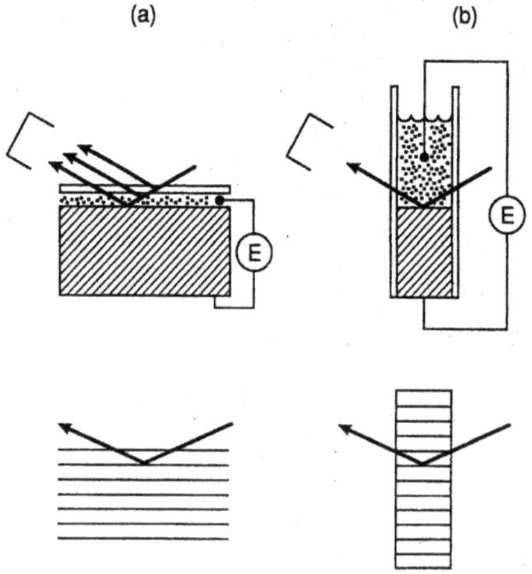

Figure 44. (a) Reflection (Bragg) geometry. (b) Transmission (Laue) geometry. The upper figures are for the solid/water (electrochemical) interface and the lower ones are for single crystals.[10]

electrochemical cell designs probably were those published in the mid 1980s for EXAFS experiments using the reflection geometry[239] and the transmission geometry,[240,241] and for surface x-ray diffraction experiments.[242]

(*i*) *Reflection-Geometry Cell Design*

The reflection-geometry cells are, by necessity, *thin-layer* cells (Fig. 45). A few micrometer thin plastic film (typically polypropylene) window covers and seals the cell with a thin capillary electrolyte film of a few micrometers between the working electrode surface and the plastic film. The counter and reference electrodes are positioned in the solution reservoir behind and surrounding the working-electrode holder. For electrochemical manipulations, the thin film is expanded by the application of a slight overpressure, followed by the application of a slight underpressure to collapse the film for x-ray measurements. An early design[185] is shown in Fig. 45, while several variations, using essentially the same design principles, were also reported.[41,92,194,204,243]

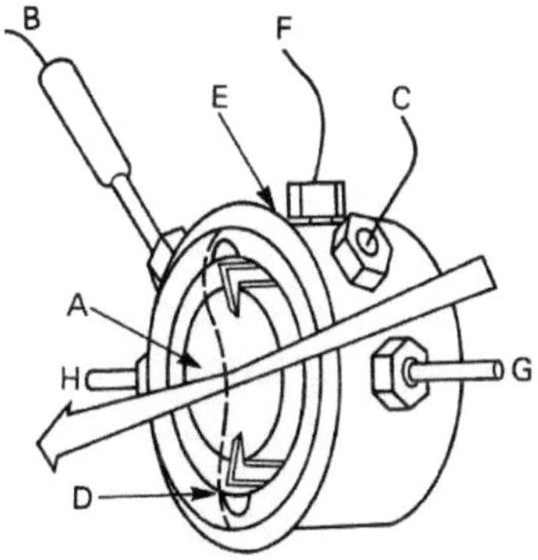

Figure 45. Thin-layer, reflection-geometry x-ray-electrochemical cell. (A) working electrode, (B) Ag/AgCl reference electrode, (C) counter electrode, (D) polypropylene window, (E) o-ring holding polypropylene film to cell, (F) external electrical connection to the working electrode, (G) solution inlet, and (H) solution outlet. (Reprinted from reference 185. Copyright (1988), with permission from Elsevier.)

(*ii*) *Transmission-Geometry Cell Design*

A transmission-geometry type cell is shown in Fig. 46.[10,89,244] The cell is a layer-type, sandwich configuration, consisting of a Teflon center section that houses the electrodes and shapes the electrolyte cavity, surrounded on both sides by thin plastic membranes; the membranes are pressed in place by a pair of Teflon inner frames, which, in turn, are supported by a pair of outer metal frames. The schematic view in Fig. 46 shows only the center section and one of the two sets of framing pieces of the cell. The pair of inner and outer frames is cut with appropriately shaped slits to permit the x rays to enter and exit the cell through the membrane cell windows in a range of 45°. The assembly is held together by bolts that compress the inner frames and the center section to provide a solution-tight seal around the edges. The outer frames are also equipped with appropriate mounting fixtures to attach the cell to a four-circle x-ray spectrometer.

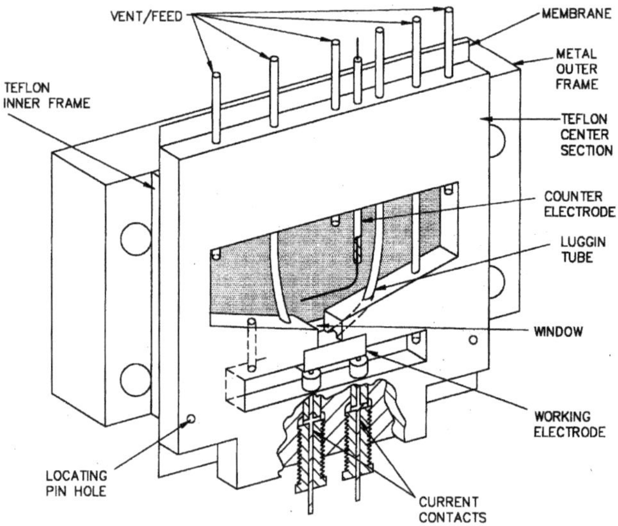

Figure 46. Transmission-geometry x-ray-electrochemical cell design. Schematic drawing, not to scale.[10]

The electrolyte cavity is so shaped as to accommodate rotation of the cell by 90° in either direction and still provide gas space above the electrolyte with the working and counter electrodes fully submerged. For the same reason, there are two vent holes, one in each upper corner of the electrolyte cavity. The cell is further equipped with solution-fill and gas-bubbler tubes, and connections to purge or fill the space below the working electrode. These input/output connections are made of Teflon tubings that are tightly fitted into holes drilled in the Teflon center section, their functions are generally interchangeable. Electrical contact is made at both ends of the working electrode using lead wires that are pressed against the bottom of the electrode. The lead wires are sealed in a Teflon sheath that forms a tightly fitting, sliding seal in the center section. The inert metal wire counter electrode is positioned above the working electrode surface, traversing its full length to provide a uniform current distribution. The reference electrode is positioned close to the working electrode surface to ensure a small uncompensated ir-drop in the solution. Either internal or external reference electrodes can be used. Internal reference can be, for example, chloride or oxide coated silver wire sheathed in Teflon tubing except for a few mm at the working end. In the case of an external reference electrode, a small Teflon tube can serve as both Luggin tip and solution bridge. This Teflon-tube Luggin, coupled with the feed tubing leading to the other end of the working electrode, can also be used to continuously flush the solution over the working electrode surface, if needed. Several other cells built on essentially the same design principles were also reported.[64,68,227,245] Transmission-geometry cells for use under controlled hydrodynamic conditions[40,208] and for use with liquid metal (mercury or gallium) working electrode[235-237] have also been reported.

(*iii*) *Hanging-Drop Transmission-Geometry Cell Design*

A common problem of both the reflection and transmission-geometry cell designs is that it is not easy, and sometimes it is impossible, to ensure that only the crystal face of interest is exposed to the solution. Seepage around the edges of the crystals is difficult to control. To overcome these problems, a *hanging-drop* configuration transmission-geometry x-ray-electrochemical cell has been designed (Fig. 47).[44,46] The hanging-drop geometry has very often been used in single-crystal electrochemical experiments. A small droplet of the test solution is extruded from a capillary tubing and it is touching only the crystal face to be examined. The whole assembly is surrounded by a cylindrical Kapton window and the atmosphere within the cylinder is flushed with pure nitrogen or other inert gas. Furthermore, this geometry provides a 360° in-plane accessibility for the x rays. The counter and reference electrodes are placed in the inlet/outlet tubing for the solution, resulting in rather high-resistance circuits; but the uncompensated ir-drop need not be large, being controlled by the diameter of the solution drop. Preliminary experiments with a similar design, with the modification to permit the circulation or continuous flushing of the solution in the drop (see the effect of radiolysis below) were also reported. It was shown that the drop is stable with flowing solution and withstands the movements of the x-ray goniometer.[10,246]

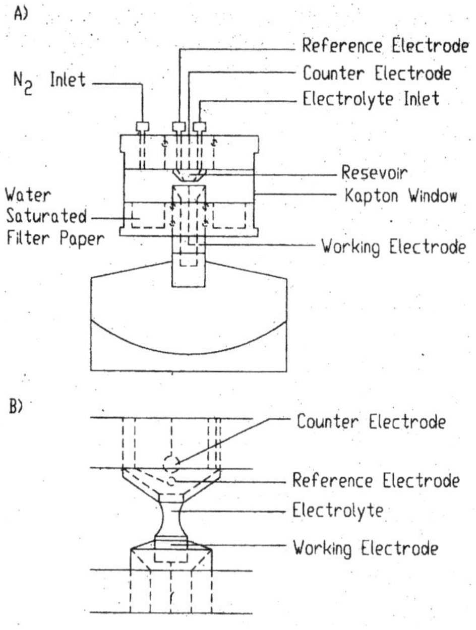

Figure 47. Hanging-drop, transmission-geometry x-ray-electrochemical cell. (Reprinted with permission from reference 44. Copyright (1993) by the American Institute of Physics.)

(*iv*) *Comparison of Cell Designs*

The transmission geometry has several advantages from the standpoint of desirable electrochemical practice. In the reflection geometry (Fig. 44a), the current distribution over the working electrode surface is very nonuniform and the uncompensated ir-drop is very large because of the geometrical arrangement of the electrodes and the large ohmic resistance in the electrolyte gap. It is common practice with this type of cell to expand the electrolyte gap by inflating the membrane (or withdrawing the electrode assembly) during electrochemical manipulations. However, during the measurements with x rays, the electrolyte gap is constricted to a layer of the order of μm, which could lead to the loss of meaningful potential control except at very low current densities. A cell designed with transmission geometry (Fig. 44b) alleviates these problems at the price of increased attenuation of the beam intensity, but choosing an optimum cell thickness can minimize that. A properly designed transmission-geometry cell also permits the continuous flushing of the solution at the electrode surface. This is important for removing radiolytic products of the electrolyte solution that would otherwise interfere with the surface electrochemistry (see below). With a reflection-geometry cell, the radiolysis products of the electrolyte and the membrane itself are trapped in a thin layer of electrolyte near the electrode surface and can cause unwanted contaminations.

The transmission geometry is also advantageous from the x-ray scattering standpoint:

(*a*) In the transmission geometry (Fig. 44b), x rays reflect to the detector from the electrochemical interface only, while in the reflection geometry (Fig. 44a), x rays reflect from the air/membrane and membrane/solution interfaces as well as the interface of interest. Therefore, the reflectivity data in transmission geometry can be interpreted directly, while the data obtained in reflection geometry require deconvolution or subtraction of the interference from air/membrane and membrane/solution interfaces for small angles of incidence. Furthermore, the nature of those additional interfaces may also vary as the experimental conditions change, which can compromise the correct interpretation of the data.

(*b*) Reflectivities measured in transmission geometry require little absorption correction because the path length of x rays is nearly constant over the angular range of interest, while with the reflection geometry the path length changes sharply for small angles.

(*c*) Since the attenuation is an exponential function of path length, there is a tendency for the diffuse background to rise rapidly at small scattering angles in the reflection geometry; therefore, the total signal is reduced by absorption but the diffuse background from the air/membrane and the membrane/solution interface is hardly attenuated, resulting in a low signal-to-background ratio at small angles. In the transmission geometry, the diffuse background is attenuated as much as the signal, and the signal-to-background ratio remains constant over the full angular range of interest. The low signal-to-background ratio is particularly important for in-plane scattering studies at glancing angles when the expected signal is very weak.

One apparent advantage of the reflection geometry is that there is no severe restriction on the size of the sample surface, while in the transmission case, the width of the cell must be optimized to the order of the absorption length of x rays, which is a few mm for readily available x-ray energies. However, sample size is hardly an advantage because of the small spot size of a focused synchrotron beam. Another apparent advantage of the reflection geometry is accessibility to high scattering angles. However, for most practical applications, the full hemispherical accessibility is really not needed because of the symmetry characteristics of the crystalline materials under investigation. For example, for a material with a three-fold symmetry axis, only a 120° section of the hemisphere (versus the full 360°) is required to obtain all the necessary information about the surface structure. Furthermore, transmission-geometry cell design with wide-angle accessibility is available.[245]

2. Problems with Solution Impurities at Beamlines

Charge transfer experiments can be very seriously affected by impurities in the test solution, often at levels as low as parts per billion. Keeping clean conditions is especially difficult under synchrotron-x-ray experiments. A main source of impurities is the x-ray radiation itself as it interacts with the test solution and structural parts of the cell. Some experiments indicated that bulk solution radiolysis can occur at high x-ray fluxes.[68,247] The degree of radiolysis and the seriousness of its consequences on the phenomena investigated strongly depend on the experimental conditions. This effect can be safely ignored for most measurements carried out at the *second-generation* bending-magnet beamlines. However, as higher-flux beamlines are developed at the *third-generation* synchrotrons, the x-ray interference with the system under investigations will have to be considered for every experiment. Radiolytic products of water (H_2 and H_2O_2, and a variety of intermediate radical species) will always be present, together with possible radiolytic products from the solutes. What effects these may have on the system investigated will have to be decided on a case-by-case basis. Just as in every case of impurity effects on electrochemical systems, there are no general rules. Some reactions on some electrodes will be sensitive to certain impurities but not to others, while other reactions or electrodes will have a different sensitivity to impurities. One either must investigate the individual effects, or assume the worst case and purify as much as possible. Admittedly, the problem is not easy to solve, but that is a poor reason for ignoring it. The degree of radiolysis will depend on the volume of solution irradiated; therefore, different cell designs will pose different problems. The reported experiments were carried out with transmission-type cells only, in which case the x rays traverse a large volume of solution. Thin-layer cells (reflection-geometry cells) contain much smaller irradiated volume; on the other hand, all radiolytic products are trapped close to the working-electrode interface. These systems will have to be investigated further. The only way to minimize the radiolytic effect is the continuous, high-speed flushing of the test solution to carry away most of the radiolytic products before they can diffuse to the interface (transmission-geometry cells are particularly suited for such experiments).

It was found in one set of expriments[247] that a rather simple, qualitative indicator of radiolysis was the open circuit potential shift upon irradiation of a platinum electrode immersed in perchloric acid solution. This was about 10 to 15 mV at a *bending-magnet* beamline (that, considering the much wider potential range used in the experiments, is easily negligible), while it increased to over 100 mV at a *wiggler* beamline. Figure 48 shows the potential shift at open-circuit potential and Fig. 49 shows the current measured when the electrode was potentiostatted at the open-circuit potential before application of the x-ray beam. There is a surprising scarcity of data about x-ray-induced radiolysis in the literature, however the general understanding of radiolysis of water was sufficient to permit at least an order-of-magnitude, or better, prediction of the interaction of the measuring x-ray probe with the system under investigation.[247]

Other workers[68] have reported that they encountered one major problem during x-ray measurements of copper bulk deposition: the strong interaction between the x-ray beam and the electrolyte seemed to create oxidizing free radicals that strongly interacted with the deposit and produced its dissolution. To check this x-ray effect, the beam was stopped for one minute during the measurements and this clearly demonstrated that the film did not dissolve

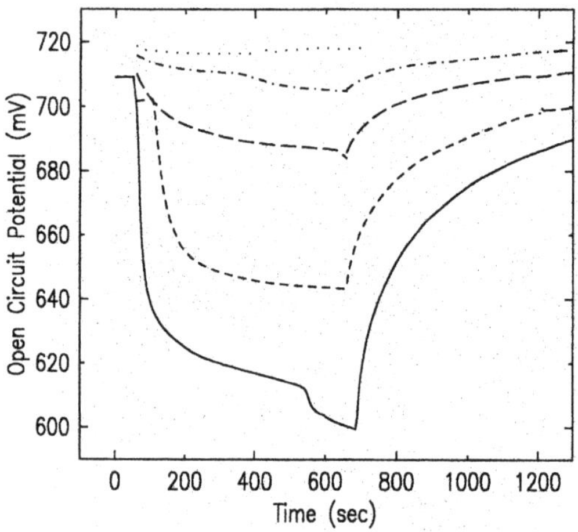

Figure 48. Shift of open-circuit potential (SCE) and its relaxation in 0.1 M $HClO_4$ solution. Beam directed at the center of the working electrode at an incident angle of 12° with an x-ray flux of 2×10^{11} (solid), 4.6×10^{10} (short dashed), 1.1×10^{10} (long dashed), 2.5×10^{9} (dot-dashed), and 5.7×10^{8} (dotted) photons per second.[247]

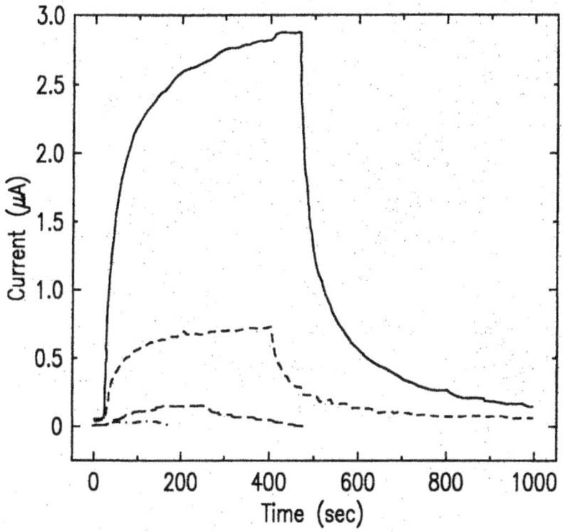

Figure 49. Current at open-circuit potential and its relaxation. Conditions as in Fig. 48.[247]

during the period when the x-ray beam was off (see Fig. 40). In another report, the radiolytic production of chloride ions from perchloric acid were assumed to occur.[43]

X rays can also damage structural parts of the cell, consequently, introduce decomposition products into the solution. The most noticeable damage is on the thin-film membranes that can easily be punctured by x rays. It was found that the most usable membrane is a sandwich of Kapton between two layers of Teflon.[212] This construction permitted the cleaning of the cell with strong acids because of the chemical inertness of the Teflon (Kapton would be strongly attacked under these conditions). Then, during experiments small holes were punctured through the Teflon layers by the x-ray beam, but this exposed only small areas of Kapton to the usually not-very-corrosive test solution. Unfortunately, this still introduced some of the decomposition products of the Teflon into the solution. The only way to minimize the effect was the flushing of the test solution.

A further difficulty is caused by the requirement of flexible connections between any solution reservoir or gas source and the cell due to the rather wide-angle movements of the x-ray goniometer. Similar connections in a laboratory are typically made of solid glass tubing, but at a beamline plastics are used. Teflon is desirable because of its chemical resistance and cleanability, but it is unfortunately very highly permeable to oxygen. Continuous or frequent flushing of the solution was again the only way to minimize the problem, unless other materials can be found that provide both chemical stability and oxygen impermeability (maybe Teflon tubing with a thin metal, or other oxygen impermeable, coating on the outside).

ACKNOWLEDGEMENT

This work was supported by the Office of Basic Energy Sciences, U.S. Department of Energy under contract no. W-31-109-ENG-38.

REFERENCES

[1]J. Robinson, in *Spectroscopic and Diffraction Techniques in Interfacial Electrochemistry,* C. Gutierrez and C. Melendres, eds. Kluwer, London, (1990) p 313.
[2]B. M. Ocko, in *Spectroscopic and Diffraction Techniques in Interfacial Electrochemistry,* C. Gutierrez and C. Melendres, eds. Kluwer, London, (1990) p 343.
[3]M. F. Toney and O. R. Melroy, in *Electrochemical Interfaces: Modern Techniques for In-Situ Interface Characterization,* H. D. Abruna, ed. Chapter 2, VCH, Weinheim, Germany (1991) p 55.
[4]H. You and Z. Nagy, in *Current Topics in Electrochemistry* **2** Research Trends, Council of Scientific Research Information, Trivandrum, India (1993) p 21.
[5]M. F. Toney and B. M. Ocko, *Synchrotron Radiation News* **6**(5) (1993) 28.
[6]C. Thompson, in *Synchrotron Techniques in Interfacial Electrochemistry,* C. A. Melendres and A. Tadjeddine, eds. Kluwer, London (1994) p 97.
[7]B. M. Ocko and J. Wang, in *Synchrotron Techniques in Interfacial Electrochemistry,* C. A. Melendres and A. Tadjeddine, eds. Kluwer, London (1994) p 127.
[8]J. X. Wang, R. R. Adzic, and B. M. Ocko, in *Interfacial Electrochemistry: Theory, Experiment, and Applications*, A. Wieckowski, ed. Marcel Dekker, NY (1999) p 175.
[9]N. M. Markovic and P. N. Ross, *Surf. Sci. Rep.* **45** (2002) 117.
[10]Z. Nagy and H. You, *Electrochim. Acta* **47** (2002) 3037.
[11]O. M. Magnussen, *Chem. Rev.* **102** (2002) 679.
[12]R. R. Adzic, J. X. Wang, B. M. Ocko, and J. McBreen, in *Handbook of Fuel Cells: Fundamentals, Technology and Applications Vol. 2 Electrocatalysis*, W. Vielstich, A. Lamm, and H. A. Gasteiger, eds. Wiley, Chichester, England (2003) p 279.

[13]C. A. Lucas and N. M. Markovic, in *Interfacial Kinetics and Mass Transport, Encyclopedia of Electrochemistry Vol. 2.* A. J. Bard, M. Stratmann, and E. J. Calvo, eds. Wiley-VCH, Weinheim, Germany (2003) p 295.

[14]P. Fenter and N. C. Sturchio, *Progr. Surf. Sci.* **77** (2004) 171.

[15]K.-C. Chang, A. Menzel, V. Komanicky, H. You, J. Inukai, A. Wieckowski, E. V. Timofeeva, and Y.V. Tolmachev, in *In-Situ Spectroscopic Studies of Adsorption at the Electrode and Electrocatalysis,* S.-G. Sun, P. A. Christensen, and A. Wieckowski, eds. Elsevier, to be published.

[16]G. G. Long and J. Kruger, in *Techniques for Characterization of Electrodes and Electrochemical Processes,* R. Varma and J. R. Selman, eds. Chapter 4, Wiley, New York, NY (1991) p167.

[17]H. D. Abruna, in *Electrochemical Interfaces: Modern Techniques for In-Situ Interface Characterization,* H. D. Abruna, ed. Chapter 1, VCH, Weinheim, Germany (1991) p 1.

[18]J. H. White, in *Electrochemical Interfaces: Modern Techniques for In-Situ Interface Characterization,* H. D. Abruna, ed. Chapter 3, VCH, Weinheim, Germany (1991) p 131.

[19]A. E. Russell and W. O'Grady, in *Synchrotron Techniques in Interfacial Electrochemistry,* C. A. Melendres and A. Tadjeddine, eds. Kluwer, London (1994) p421.

[20]B. W. Batterman and H. Cole, *Rev. Mod. Phys.* **36** (1964) 681.

[21]B. E. Warren, *X-Ray Diffraction* Addison-Wesley, Reading, Massachusetts, 1969.

[22]J. D. Jackson, *Classical Electrodynamics* Wiley, New York, 1974.

[23]H. You, C. A. Melendres, Z. Nagy, V. A. Maroni, W. Yun, and R. M. Yonco, *Phys. Rev.* **B45** (1992) 11288.

[24]A. Menzel, K.-C. Chang, V. Komanicky, H. You, Y. S. Chu, Y. V. Tolmachev, and J. J. Rehr, *Rad. Phys. Chem.* in press.

[25]Y. S. Chu, H. You, J. A. Tanzer, T. E. Lister, and Z. Nagy, *Phys. Rev. Lett.* **83** (1999) 552.

[26]A. Menzel, K.-C. Chang, V. Komanicky, Y. S. Chu, Y. V. Tolmachev, and J. J. Rehr, and H. You, *Europhys. Lett.* **74** (2006) 1032.

[27]G. L. Borges, M. G. Samant, and K. Ashley, *J. Electrochem. Soc.* **139** (1992) 1565.

[28]K. M. Robinson, W. E. O'Grady, and I. K. Robinson, in *X-Ray Methods in Corrosion and Interfacial Electrochemistry*, A. Davenport and J. G. Gordon, eds. Electrochemical Society Proceedings **92-1,** Pennington, NJ (1992) p 25.

[29]K. M. Robinson, I. K. Robinson, and W. E. O'Grady, *Surface Sci.* **262** (1992) 387.

[30]K. M. Robinson, W. E. O'Grady, and I. K. Robinson, in *Surface X-Ray and Neutron Scattering (Springer Proc. Phys. 61)* H. Zabel and I. K. Robinson, eds. Springer, Berlin (1992) p 175.

[31]K. M. Robinson, I. K. Robinson, and W. E. O'Grady, *Electrochim. Acta* **37** (1992) 2169.

[32]K. M.. Robinson and W. E. O'Grady, *J. Electroanal. Chem.* **384** (1995) 139.

[33]N. M. Markovic, B. N. Grgur, C. A. Lucas, and P. N. Ross, *Surf. Sci.* **384** (1997) L805.

[34]V. Komanicky, A. Menzel, K.-C. Chang, and H. You, *J. Phys. Chem.* **B109** (2005) 23543.

[35]V. Komanicky, A. Menzel, and H. You, in *Electrocatalysis*, R. Adzic, V. Birss, G. Brisard, and A. Wieckowski, eds. Electrochemical Society Proceedings **2005-11**, Pennington, NJ (2006) p 231.

[36]M. F. Toney, in *Synchrotron Techniques in Interfacial Electrochemistry,* C. A. Melendres and A. Tadjeddine, eds. Kluwer, London (1994) p109.
[37]M. F. Toney, J. N. Howard, J. Richer, G. L. Borges, J. G. Gordon, O. R. Melroy, D. G. Wiesler, D. Yee, and L. B. Sorensen, *Nature* **368** (1994) 444.
[38]M. F Toney, J. N. Howard, J. Richer, G. L. Borges, J. G. Gordon, O. R. Melroy, D. G. Wiesler, D. Yee, and L. B. Sorensen, *Surf. Sci.* **335** (1995) 326.
[39]J. .G. Gordon, O. R. Melroy, and M. F. Toney, *Electrochim. Acta* **40** (1995) 3.
[40]M. L. Foresti, A. Pozzi, M. Innocenti, G. Pezzatini, F. Loglio, E. Salvietti, A. Giusti, F. D'Anca, R. Felici, and F. Borgatti, *Electrochim. Acta* **51** (2006) 5532.
[41]B. M. Ocko, J. Wang, A. Davenport, and H. Isaacs, *Phys. Rev. Lett.* **65** (1990) 1466.
[42]B. M. Ocko and J. Wang, in *Structural Effects in Electrocatalysis and Oxygen Electrochemistry,* D. Scherson, D. Tryk, M. Daroux, and X. Xing, eds. Electrochemical Society Proceedings **92-11,** Pennington, NJ (1992) p 147.
[43]I. M Tidswell, N. M. Markovic, C. A. Lucas, and P. N. Ross, *Phys. Rev.* **B47** (1993) 16542.
[44]K. M. Robinson and W. E. O'Grady, *Rev. Sci. Instrum.* **64** (1993) 1061.
[45]K. M. Robinson and W. E. O'Grady, *Faraday Discuss.* **95** (1993) 55.
[46]K. M. Robinson and W. E. O'Grady, in *Synchrotron Techniques in Interfacial Electrochemistry,* C. A. Melendres and A. Tadjeddine, eds. Kluwer, London (1994) p 157.
[47]N. M. Markovic, I. M. Tidswell, and P. N. Ross, *Langmuir* **10** (1994) 1.
[48]I. M Tidswell, N. M. Markovic, and P. N. Ross, *Surf. Sci.* **317** (1994) 241.
[49]R. R. Adzic, J. X. Wang, O. M. Magnussen, and B. M. Ocko, *Langmuir* **12** (1996) 513.
[50]T. Wandlowski, J. X. Wang, O. M. Magnussen, and B. M. Ocko, *J. Phys. Chem.* **100** (1996) 10277.
[51]C. A. Lucas, *Electrochim. Acta* **47** (2002) 3065.
[52]B. B. Blizanac, C. A. Lucas, M. E. Gallagher, M. Arenz, P. N. Ross, and N. M. Markovic, *J. Phys. Chem.* **B108** (2004) 625.
[53]B. B. Blizanac, C. A. Lucas, M. E. Gallagher, P. N. Ross, and N. M. Markovic, *J. Phys. Chem.* **B108** (2004) 5304.
[54]M. E. Gallagher, B. B. Blizanac, C. A. Lucas, P. N. Ross, and N. M. Markovic, *Surf. Sci.* **582** (2005) 215.
[55]B. M. Ocko, G. Helgesen, B. Schardt, J. Wang, and A. Hamelin, *Phys. Rev. Lett.* **69** (1992) 3350.
[56]J. Wang, A. J. Davenport, H. S. Isaacs, and B. M. Ocko, *Science* **255** (1992) 1416.
[57]J. Wang, B. M. Ocko, A. J. Davenport, and H. S. Isaacs, in *X-Ray Methods in Corrosion and Interfacial Electrochemistry*, A. Davenport and J. G. Gordon, eds. Electrochemical Society Proceedings **92-1,** Pennington, NJ (1992) p 34.
[58]J. Wang, B. M. Ocko, A. J. Davenport, and H. S. Isaacs, *Phys. Rev.* **B46** (1992) 10321.
[59]B. M. Ocko, A. Gibaud, and J. Wang, *J. Vac. Sci. Technol.* **A10** (1992) 3019.
[60]J. Wang, G. M. Watson, and B. M. Ocko, *Physica* **A200** (1993) 679.
[61]B. M. Ocko, O. M. Magnussen, R. R. Adzic, J. X. Wang, Z. Shi, and J. Lipkowski, *J. Electroanal. Chem.* **376** (1994) 35.
[62]B. M. Ocko, O. M. Magnussen, J. X. Wang ,and R. R. Adzic, in *Nanoscale Probes of the Solid/Liquid Interface,* A. A. Gewirth and H. Siegenthaler, eds. Kluwer, London (1995) p 103.

[63]T. Wandlowski, B. M. Ocko, O. M. Magnussen, S. Wu, and J. Lipkowski, *J. Electroanal. Chem.* **409** (1996) 155.
[64]F. Brossard, V. H. Etgens, and A. Tadjeddine, *Nucl. Inst. Methods* **B129** (1997) 419.
[65]S. Wu, and J. Lipkowski, O. M. Magnussen, B. M. Ocko, and T. Wandlowski, *J. Electroanal. Chem.* **446** (1998) 67.
[66]J. X. Wang and R. R. Adzic, *J. Electroanal. Chem.* **448** (1998) 205.
[67]C. A. Lucas, *J. Phys.* **D32** (1999) A198.
[68]V. H. Etgens, M. C. M. Alves, and A. Tadjeddine, *Electrochim. Acta* **45** (1999) 591.
[69]M. Takahasi, Y. Hayashi, J. Mizuki, K. Tamura, T. Kondo, H. Naohara, and K. Uosaki, *Surf. Sci.* **461** (2000) 213.
[70]R. J. Nichols, T. Nouar, C. A. Lucas, W. Haiss, and W. A. Hofer, *Surf. Sci.* **513** (2002) 263.
[71]A. H. Ayyad, J. Stettner, and O. M. Magnussen, *Phys. Rev. Lett.* **94** (2005) 066106.
[72]M. J. Ball, C. A. Lucas, N. M. Markovic, V. Stamenkovic, and P. N. Ross, *Surf. Sci.* **540** (2003) 295.
[73]N. M. Markovic, C. A. Lucas, V. Climent, V. Stamenkovic, and P. N. Ross, *Surf. Sci.* **465** (2000) 103.
[74]C. A. Lucas, N. M. Markovic, M. Ball, V. Stamenkovic, V. Climent, and P. N. Ross, *Surf. Sci.* **479** (2001) 241.
[75]M. J. Ball, C. A. Lucas, N. M. Markovic, V. Stamenkovic, and P. N. Ross, *Surf. Sci.* **518** (2002) 201.
[76]I. M Tidswell, N. M. Markovic, and P. N. Ross, *Phys. Rev. Lett.* **71** (1993) 1601.
[77]N. M. Markovic, C. A. Lucas, B. N. Grgur, and P. N. Ross, *J. Phys. Chem.* **B103** (1999) 9616.
[78]N. M. Markovic and P. N. Ross, in *Interfacial Electrochemistry: Theory, Experiment, and Applications*, A. Wieckowski, ed. Marcel Dekker, NY (1999) p 815.
[79]C. A. Lucas, N. M. Markovic, and P. N. Ross, *Phys. Rev. Lett.* **77** (1996) 4922.
[80]B. N. Grgur, N. M. Markovic, C. A. Lucas, and P. N. Ross, *J. Serb. Chem. Soc.* **66** (2001) 785.
[81]I. M Tidswell, N. M. Markovic, and P. N. Ross, *J. Electroanal. Chem.* **376** (1994) 119.
[82]C. A. Lucas, N. M. Markovic, and P. N. Ross, *Phys. Rev.* **B55** (1997) 7964.
[83]C. A. Lucas, N. M. Markovic, and P. N. Ross, *Surf. Sci.* **425** (1999) L381.
[84]J. X. Wang, I. K. Robinson, B. M. Ocko, and R. R. Adzic, *J. Phys. Chem.* **B109** (2005) 24.
[85]V. R. Stamenkovic, M. Arenz, C. A. Lucas, M. E. Gallagher, P. N. Ross, and N. M. Markovic, *J. Am. Chem. Soc.* **125** (2003) 2736.
[86]M. E. Gallagher, C. A. Lucas, V. Stamenkovic, N. M. Markovic, and P. N. Ross, *Surf. Sci.* **544** (2003) L729.
[87]T. E. Lister, Y. Chu, W. Cullen, H. You, R. M. Yonco, J. F. Mitchell, and Z. Nagy, *J. Electroanal. Chem.* ***524-525*** (2002) 201.
[88]Y. S. Chu, T. E. Lister, W. G. Cullen, H. You, and Z. Nagy, *Phys. Rev. Lett.* **86** (2001) 3364.
[89]Z. Nagy, H. You, R. M. Yonco, C. A. Melendres, W. Yun, and V. A. Maroni, *Electrochim. Acta* **36** (1991) 209.
[90]M. F. Toney, J. G. Gordon, M. G. Samant, G. L. Borges, O. R. Melroy, L.-S. Kau, D. G. Wiesler, D. Yee, and L. B. Sorensen, *Phys. Rev.* **B42** (1990) 5594.

[91]J. Zegenhagen, F. U. Renner, A. Reitzle, T. L. Lee, S. Warren, A. Stierle, H. Dosch, G. Scherb, B. O. Fimland, and D. M. Kolb, *Surf. Sci.* **573** (2004) 67.
[92]J. Zegenhagen, A. Kazimirov, G. Scherb, D. M. Kolb, D. M. Smilgies, and R. Feidenhans'l, *Surf. Sci.* **352** (1996) 346.
[93]D. G. Wiesler, M. F. Toney, C. S. McMillan, and W. H. Smyrl, in *The Application of Surface Analysis Methods to Environmental/Material Interactions*, D. R. Bear, C. R. Clayton, and G. D. Davis, eds. Electrochemical Society Proceedings **91-7**, Pennington, NJ (1991) p 440.
[94]G. M. Bommarito, D. Acevedo, and H. D. Abruna, *J. Phys. Chem.* **96** (1992) 3416.
[95]H. You, D. J. Zurawski, Z. Nagy, and R. M. Yonco, *J. Chem. Phys.* **100** (1994) 4699.
[96]H. You and Z. Nagy, *Physica* **B198** (1994) 187.
[97]H. You, D. J. Zurawski, R. P. Chiarello, and Z. Nagy, unpublished results.
[98]G. Luo, S. Malkova, S. V. Pingali, D. G. Schultz, B. Lin, M. Meron, T. J. Graber, J. Gebhardt, P. Vanysek, and M. L. Schlossman, *Faraday Discuss.* **129** (2005) 23.
[99]G. Luo, S. Malkova, J. Yoon, D. G. Schultz, B. Lin, M. Meron, I. Benjamin, P. Vanysek, and M. L. Schlossman, *J. Electroanal. Chem.* **593** (2006) 142.
[100]G. Luo, S. Malkova, S. V. Pingali, D. G. Schultz, B. Lin, M. Meron, I. Benjamin, P. Vanysek, and M. L. Schlossman, *J. Phys. Chem.* **B110** (2006) 4527.
[101]G. Luo, S. Malkova, J. Yoon, D. G. Schultz, B. Lin, M. Meron, I. Benjamin, P. Vanysek, and M. L. Schlossman, *Science* **311** (2006) 216.
[102]J. Li, E. Herrero, and H. D. Abruna, *Colloid Surf.* **A134** (1998) 113.
[103]B. M. Ocko, J. X. Wang, and T. Wandlowski, *Phys. Rev. Lett.* **79** (1997) 1511.
[104]J. X. Wang, T. Wandlowski, and B. M. Ocko, in *The Electrochemical Double Layer,* C. Korzeniewski and B. E. Conway, eds. Electrochemical Society Proceedings **97-17**, Pennington, NJ (1997) p 293.
[105]R. R. Adzic, and J. X. Wang, *J. Phys. Chem.* **B104** (2000) 869.
[106]R.R. Adzic and J. X. Wang, *Electrochim. Acta* **45** (2000) 4203.
[107]T. Wandlowski, J. X. Wang, and B. M. Ocko, *J. Electroanal. Chem.* **500** (2001) 418.
[108]J. E. DeVilbiss, J. X. Wang, B. M. Ocko, K. Tamura, R. R. Adzic, I. A. Vartanyants, and I. K. Robinson, *Electrochim. Acta* **47** (2002) 3057.
[109]R. R. Adzic and J. X. Wang, *Solid State Ionics* **150** (2002) 105.
[110]B. M. Ocko, O. M. Magnussen, J. X. Wang, R. R. Adzic, and T. Wandlowski, *Physica* **B221** (1996) 238.
[111]B. M. Ocko, O. M. Magnussen, J. X. Wang, and T. Wandlowski, *Phys. Rev.* **B53** (1996) R7654.
[112]J. X. Wang, N. S. Marinkovic, and R. R. Adzic, *Colloid Surf.* **A134** (1998) 165.
[113]O. M. Magnussen, B. M. Ocko, R. R. Adzic, and J. X. Wang, *Phys. Rev.* **B51** (1995) 5510.
[114]O. M. Magnussen, B. M. Ocko, J. X. Wang, and R. R. Adzic, *J. Phys. Chem.* **100** (1996) 5500.
[115]E. Herrero, S. Glazier, and H. D. Abruna, *J. Phys. Chem.* **B102** (1998) 9825.
[116]E. Herrero, S. Glazier, L. J. Buller, and H. D. Abruna, *J. Electroanal. Chem.* **461** (1999) 121.
[117]N. M. Markovic, C. A. Lucas, H. A. Gasteiger, and P. N. Ross, *Surf. Sci.* **365** (1996) 229.
[118]C. A. Lucas, N. M. Markovic, and P. N. Ross, *Langmuir* **13** (1997) 5517.

[119]N. M. Markovic, B. N. Grgur, C. A. Lucas, and P. N. Ross, *J. Chem. Soc. Faraday Trans.* **94** (1998) 3373.
[120]I. M Tidswell, C. A. Lucas, N. M. Markovic, and P. N. Ross, *Phys. Rev.* **B51** (1995) 10205.
[121]C. A. Lucas, N. M. Markovic, and P. N. Ross, *Surf. Sci.* **340** (1995) L949.
[122]C. A. Lucas, N. M. Markovic, I. M. Tidswell, and P. N. Ross, *Physica* **B221** (1996) 245.
[132]R. R. Adzic and J. X. Wang, *J. Serb. Chem. Soc.* **62** (1997) 873.
[124]N. M. Markovic, C. A. Lucas, H. A. Gasteiger, and P. N. Ross, *Surf. Sci.* **372** (1997) 239.
[125]N. M. Markovic, C. A. Lucas, H. A. Gasteiger, and P. N. Ross, in *Solid-Liquid Electrochemical Interfaces,* G. Jerkiewicz, M. P. Soriaga, K. Uosaki, and A. Wieckowski, eds. ACS Symposium Series **656,** Washington, DC (1997) p 87.
[126]N. M. Markovic, B. N. Grgur, C. A. Lucas, and P. N. Ross, *J. Electroanal. Chem.* **448** (1998) 183.
[127]N. M. Markovic, C. A. Lucas, A. Rodes, V. Stamenkovic, and P. N. Ross, *Surf. Sci.* **499** (2002) L149.
[128]C. A. Lucas, N. M. Markovic, and P. N. Ross, *Phys. Rev.* **B57** (1998) 13184.
[129]N. M. Markovic, B. N. Grgur, C. A. Lucas, and P. N. Ross, *Electrochim. Acta* **44** (1998) 1009.
[130]C. A. Lucas, N. M. Markovic, and P. N. Ross, *Surf. Rev. Lett.* **6** (1999) 917.
[131]C. A. Lucas, N. M. Markovic, B. N. Grgur, and P. N. Ross, *Surf. Sci.* **448** (2000) 65.
[132]C. A. Lucas, N. M. Markovic, and P. N. Ross, *Surf. Sci.* **448** (2000) 77.
[133]N. M. Markovic, H. A. Gasteiger, C. A. Lucas, I. M. Tidswell, and P. N. Ross, *Surf. Sci.* **335** (1995) 91.
[134]N. M. Markovic, B. N. Grgur, C. A. Lucas, and P. N. Ross, *J. Phys. Chem.* **B103** (1999) 487.
[135]Y. V. Tolmachev, A. Menzel, A. V. Tkachuk, Y. S. Chu, and H. You, *Electrochem. Solid State Lett.* **7** (2004) E23.
[136]A. Menzel, K.-C. Chang, V. Komanicky, and H. You, *Phys. Rev. Lett.* submitted.
[137]A. Menzel, K.-C. Chang, V. Komanicky, Y. V. Tolmachev, A. Tkachuk, Y. S. Chu, and H. You, *Phys. Rev. B* submitted.
[138]M Ball, C. A. Lucas, N. M. Markovic, B. M. Murphy, P. Steadman, T. J. Schmidt, V. Stamenkovic, and P. N. Ross, *Langmuir* **17** (2001) 5943.
[139]J. X. Wang, G. M. Watson, and B. M. Ocko, *J. Phys. Chem.* **100** (1996) 6672.
[140]M. Cappadonia, K. M. Robinson, J. Schmidberger, and U. Stimming, *J. Electroanal. Chem.* **436** (1997) 73.
[141]A. C. Finnefrock, L. J. Buller, K. L. Ringland, J. D. Brock, and H. D. Abruna, *J. Am. Chem. Soc.* **119** (1997) 11703.
[142]H. D. Abruna, J. M. Feliu, J. D. Brock, L. J. Buller, E. Herrero, J. Li, R. Gomez, and A. Finnefrock, *Electrochim. Acta* **43** (1998) 2899.
[143]A. C. Finnefrock, K. L. Ringland, J. D. Brock, L. J. Buller, and H. D. Abruna, *Phys. Rev. Lett.* **81** (1998) 3459.
[144]E. Herrero, L. J. Buller, J. Li, A. C. Finnefrock, A. B. Salomon, C. Alonso, J. D. Brock, and H. D. Abruna, *Electrochim. Acta* **44** (1998) 983.
[145]X. Torrelles, C. Vericat, M. E. Vela, M. H. Fonticelli, M. A. D. Millone, R. Felici, T.-L. Lee, J. Zegenhagen, G. Munoz, J. A. Martın-Gago, and R. C. Salvarezza, *J. Phys. Chem.* **B110** (2006) 5586.
[146]B. M. Ocko, G. M. Watson, and J. Wang, *J. Phys. Chem.* **98** (1994) 897.

[147]E. Casero, C. Alonso, J. A. Martin-Gago, F. Borgatti, R. Felici, F. Renner, T.-L. Lee, and J. Zegenhagen, *Surf. Sci.* **507-510** (2002) 688.
[148]Y. S. Chu, Ph.D. Thesis, University of Illinois (1997).
[149]J. X. Wang, R. R. Adzic, and B. M. Ocko, *J. Phys. Chem.* **98** (1994) 7182.
[150]C. Vericat, M. E. Vela, G. A. Andreasen, R. C. Salvarezza, F. Borgatti, [RFelici], T.-L. Lee, F. Renner, J. Zegenhagen, and J. A. Martin-Gago, *Phys. Rev. Lett.* **90** (2003) 075506.
[151]C. Vericat, M. E. Vela, J. Gago, and R. C. Salvarezza, *Electrichim. Acta* **49** (2004) 3643.
[152]L Lee, J. X. Wang, R. R. Adzic, I. K. Robinson, and A. A. Gewirth, *J. Am. Chem. Soc.* **123** (2001) 8838.
[153]J. Li and H. D. Abruna, *J. Phys. Chem.* **B101** (1997) 244.
[154]E. Herrero, J. Li, and H. D. Abruna, *Electrochim. Acta* 1999 **44** (1999) 2385.
[155]C. A. Lucas, N. M. Markovic, and P. N. Ross, *Phys. Rev.* **B56** (1997) 3651.
[156]R. R. Adzic, J. X. Wang, O. M. Magnussen, and B. M. Ocko, *J. Phys. Chem.* **100** (1996) 14721.
[157]J. X. Wang, N. S. Marinkovic, H. Zajonz, B. M. Ocko, and R. R. Adzic, *J. Phys. Chem.* **B105** (2001) 2809.
[158]S. R. Brankovic, J. X. Wang, Y. Zhu, R. Sabatini, J. McBreen, and R. R. Adzic, *J. Electroanal. Chem.* **524** (2002) 231.
[159]N. M. Jisrawi, H. Wiesmann, M. W. Ruckman, T. R. Thurston, G. Reisfeld, B. M. Ocko, and M. Strongin, *J. Mater. Res.* **12** (1997) 2091.
[160]N. M. Jisrawi, H. Wiesmann, M. W. Ruckman, T. R. Thurston, G. Reisfeld, B. M. Ocko, and M. Strongin, *J. Mater. Res.* **13** (1998) 518.
[161]H. You and Z. Nagy, *Cornell University CHESS Newsletter* (1995) p 17.
[162]H. You, Z. Nagy, D. J. Zurawski, and R. P. Chiarello, in *6th International Symposium on Electrode Processes,* A. Wieckowski and K. Itaya, eds. Electrochemical Society Proceedings **96-8,** Pennington, NJ (1996) p 136.
[163]H. You and Z. Nagy, *MRS Bulletin* **24**(1) (1999) 36.
[164]K Tamura, J. X. Wang, R. R. Adzic, and B. M. Ocko, *J. Phys. Chem.* **B108** (2004) 1992.
[165]T. J. Schmidt, V. R. Stamenkovic, C. A. Lucas, N. M. Markovic, and P. N. Ross, *Phys. Chem. Chem. Phys.* **3** (2001) 3879.
[166]T. E. Lister, Y. V. Tolmachev, Y. Chu, W. G. Cullen, H. You, R. Yonco, and Z. Nagy, *J. Electroanal. Chem.* **554-555** (2003) 71.
[167]R. R. Adzic, J. Wang, and B. M. Ocko, *Electrochim. Acta* **40** (1995) 83.
[168]K. Tamura, B. M. Ocko, J. X. Wang, and R. R. Adzic, *J. Phys. Chem.* **B106** (2002) 3896.
[169]R. R. Adzic and J. X. Wang, in *Oxygen Electrochemistry*, R. R. Adzic, F. C. Anson, and K. Kinoshita, eds. Electrochemical Society Proceedings **95-26**, Pennington, NJ (1995) p 61.
[170]N. S. Marinkovic, J. X. Wang, and R.R. Adzic, in *The Electrochemical Double Layer,* C. Korzeniewski and B. E. Conway, eds. Electrochemical Society Proceedings **97-17**, Pennington, NJ (1997) p 251.
[171]R. R. Adzic and J. X. Wang, *J. Phys. Chem.* **B102** (1998) 8988.
[172]T. Kondo, J. Morita, M. Okamura, T. Saito, and K. Uosaki, *J. Electroanal. Chem.* **532** (2002) 201.
[173]J. X. Wang, B. M. Ocko, and R. R. Adzic, *Surf. Sci.* **540** (2003) 230.
[174]J. X. Wang, N. S. Marinkovic, R. R. Adzic, and B. M. Ocko, *Surf. Sci.* **398** (1998) L291.

[175]N. S. Marinkovic, J. X. Wang, J. S. Marinkovic, and R. R. Adzic, *J. Phys. Chem.* **B103** (1999) 139.
[176]M. F. Toney, J. G. Gordon, M. G. Samant, G. L. Borges, D. G. Wiesler, D. Yee, and L. B. Sorensen, *Langmuir* **7** (1991) 796.
[177]M. F. Toney, in *Structural Effects in Electrocatalysis and Oxygen Electrochemistry,* D. Scherson, D. Tryk, M. Daroux, and X. Xing, eds. Electrochemical Society Proceedings **92-11,** Pennington, NJ (1992) p 121.
[178]C.-h. Chien, K. D. Kepler, A. .A Gewirth, B. M. Ocko, and J. Wang, *J. Phys. Chem.* **97** (1993) 7290.
[179]H. Kawamura, M. Takahasi, and J. Mizuki, *J. Electrochem. Soc.* **149** (2002) C586.
[180]M. Cappadonia, K. M. Robinson, J. Schmidberger, and U. Stimming, *Surf. Rev. Lett.* **4** (1997) 1173.
[181]R. J. Randler, D. M. Kolb, B. M. Ocko, and I. K. Robinson, *Surf. Sci* **447** (2000) 187.
[182]M. F Toney, J. N. Howard, J. Richer, G. L. Borges, J. G. Gordon, O. R. Melroy, D. Yee, and L. B. Sorensen, *Phys. Rev. Lett.* **75** (1995) 4472.
[183]J. Li and H. D. Abruna, *J. Phys. Chem.* **B101** (1997) 2907.
[184]O. R. Melroy, M. F. Toney, G. L. Borges, M. G. Samant, J. B. Kortright, P. N. Ross, and L. Blum, *Phys. Rev.* **B38** (1988) 10962.
[185]M. G. Samant, M. F. Toney, G. L. Borges, L. Blum, and O. R. Melroy, *Surf. Sci.* **193** (1988) L29.
[186]M. G. Samant, M. F. Toney, G. L. Borges, L. Blum, and O. R. Melroy, *J. Phys. Chem.* **92** (1988) 220.
[187]M. F. Toney and O. R. Melroy, in *Synchrotron Radiation in Materials Research,* R. Clarke, J. Gland, and J. H. Weaver, eds. Materials Research Society Proceedings **143**, Pittsburgh, PA (1989) p 37.
[188]J. B. Kortright, P. N. Ross, O. R. Melroy, M. F. Toney, G. L. Borges, and M. G. Samant, *J. Phys. Colloq. No. C7* **50** Suppl. 10 (1989) p C7-153.
[189]O. R. Melroy, M. F. Toney, G. L. Borges, M. G. Samant, J. B. Kortright, P. N. Ross, and L. Blum, *J. Electroanal. Chem.* **258** (1989) 403.
[190]M. F. Toney, J. G. Gordon, M. G. Samant, G. L. Borges, O. R. Melroy, D. Yee, and L. B. Sorensen, *J. Phys. Chem.* **99** (1995) 4733.
[191]Y. S. Chu, I. K. Robinson, and A. A. Gewirth, *Phys. Rev.* **B55** (1997) 7945.
[192]R. R. Adzic, J. Wang, C. M. Vitus, and B. M. Ocko, *Surf. Sci. Lett.* **293** (1993) L876.
[193]J. X. Wang, I. K. Robinson, J. E. DeVilbiss, and R. R. Adzic, *J. Phys. Chem.* **B104** (2000) 7951.
[194]T. Kondo, K. Tamura, M. Takahasi, J.-i. Mizuki, and K. Uosaki, *Electrochim. Acta* **47** (2002) 3075.
[195]J. X. Wang, R. R. Adzic, O. M. Magnussen, and B.M. Ocko, *Surf. Sci.* **344** (1995) 111.
[196]M. F. Toney, in *The Application of Surface Analysis Methods to Environmental/Material Interactions*, D. R. Bear, C. R. Clayton, and G. D. Davis, eds. Electrochemical Society Proceedings **91-7**, Pennington, NJ (1991) p 200.
[197]M. F. Toney, J. G. Gordon, M. G. Samant, G. L. Borges, O. R. Melroy, D. Yee, and L. B. Sorensen, *Phys. Rev.* **B45** (1992) 9362.
[198]M. F. Toney, in *X-Ray Methods in Corrosion and Interfacial Electrochemistry*, A. Davenport and J. G. Gordon, eds. Electrochemical Society Proceedings **92-1,** Pennington, NJ (1992) p 1.

[199]J. X. Wang, R. R. Adzic, O. M. Magnussen, and B.M. Ocko, *Surf. Sci.* **335** (1995) 120.
[200]R. R. Adzic and J. X. Wang, *J. Phys. Chem.* **B102** (1998) 6305.
[201]J. X. Wang, I. K. Robinson, and R. R. Adzic, *Surf. Sci.* **412/413** (1998) 374.
[202]K. Krug, J. Stettner, and O. M. Magnussen, *Phys. Rev. Lett.* **96** (2006) 246101.
[203]J. C. Ziegler, A. Reitzle, O. Bunk, J. Zegenhagen, and D. M. Kolb, *Electrochim. Acta* **45** (2000) 4599.
[204]T. Koop, W. Schindler, A. Kazimirov, G. Scherb, J. Zegenhagen, T. Schultz, R. Feidenhans'l, and J. Kirschner, *Rev. Sci. Instr.* **69** (1998) 1840.
[205]W. Schindler, T. Koop, A. Kazimirov, G. Scherb, J. Zegenhagen, T. Schultz, R. Feidenhans'l, and J. Kirschner, *Surf. Sci.* **465** (2000) L783.
[206]B. M. Ocko, I. K. Robinson, M. Weinert, R. J. Randler, and D. M. Kolb, *Phys. Rev. Lett.* **83** (1999) 780.
[207]D.-M Smilgies, R. Feidenhans'l, G. Scherb, D. M. Kolb, A. Kazimirov, and J Zegenhagen, *Surf. Sci.* **367** (1996) 40.
[208]M. J. Armstrong, G. M. Whitney, and . M. F. Toney, in *X-Ray Methods in Corrosion and Interfacial Electrochemistry*, A. Davenport and J. G. Gordon, eds. Electrochemical Society Proceedings **92-1,** Pennington, NJ (1992) p 62.
[209]C. K. Rhee, M. Wakisaka, Y. V. Tolmachev, C. M. Johnston, R. Haasch, K. Attenkofer, G. Q. Lu, H. You, and A. Wieckowski, *J. Electroanal. Chem.* **554-555** (2003) 367.
[210]J. C. Ziegler, G. Scherb, O. Bunk, A. Kazimirov, L. X. Cao, D. M Kolb, R. L. Johnson, and J. Zegenhagen, *Surf. Sci.* **452** (2000) 150.
[211]H. You, Y. S. Chu, T. E. Lister, and Z. Nagy, *Jpn. J. Appl. Phys.* **38**(Suppl. 38-1) (1999) 239.
[212]H. You, Z. Nagy, C. A. Melendres, D. J. Zurawski, R. P. Chiarello, R. M. Yonco, H. K. Kim, and V. A. Maroni, in *X-Ray Methods in Corrosion and Interfacial Electrochemistry*, A. Davenport and J. G. Gordon, eds. Electrochemical Society Proceedings **92-1,** Pennington, NJ (1992) p 73.
[213]C. A. Melendres, H. You, V. A. Maroni, Z. Nagy, and W. Yun, *J. Electroanal. Chem.* **297** (1991) 549.
[214]H. You, Y. S. Chu, Z. Nagy, and V. Parkhutik, in *Proceedings of The Eighth International Symposium on the Passivity of Metals and Semiconductors,* M. B. Ives, J. L. Lou, and J. R. Rodda, eds. Electrochemical Society Proceedings **99-42,** Pennington, NJ (2001) p 357.
[215]Y. S. Chu, I. K. Robinson, and A. A. Gewirth, *J. Chem. Phys.* **110** (1999) 5952.
[216]D. H. Kim, S. S. Kim, H. H. Lee, H. W. Jang, J. W. Kim, M. Tang, K. S. Liang, S. K. Sinha, and D. Y. Noh, *J. Phys. Chem.* **B108** (2004) 20213.
[217]M. F. Toney, A. J. Davenport, L. J. Oblonsky, M. P. Ryan, C. M. Vitus, *Phys. Rev. Lett.* **79** (1997) 4282.
[218]A. J. Davenport, L. J. Oblonsky, M. P. Ryan, and M. F. Toney, *J. Electrochem. Soc.* **147** (2000) 2162.
[219]O. M. Magnussen, J. Scherer, B. M. Ocko, and R. J. Behm, *J. Phys. Chem.* **B104** (2000) 1222.
[220]J. Scherer, B. M. Ocko, and O. M. Magnussen, *Electrochim. Acta* **48** (2003) 1169.
[221]S. L. Medway, C. A. Lucas, A. Kowal, R. J. Nichols, and D. Johnson, *J. Electroanal. Chem.* **587** (2006) 172.
[222]H. You, Y. S. Chu, T. E. Lister, Z. Nagy, A. L. Ankudiniv, and J. J. Rehr, *Physica* **B283** (2000) 212.

[223]V. Parkhutik, Y. Chu, H. You, Z. Nagy, and P. A. Montano, *J. Porous Mater.* **7** (2000) 27.
[224]D. H. Kim, H. H. Lee, S. S. Kim, H. C. Kang, D. Y. Noh, H. Kim, and S. K. Sinha, *Appl. Phys. Lett.* **85** (2004) 6427.
[225]K. Uosaki, M. Koinuma, T. Kondo, S. Ye, I. Yagi, H. Noguchi, K. Tamura, K. Takeshita, and T. Matsushita, *J. Electroanal. Chem.* **429** (1997) 13.
[226]G. Scherb, A. Kazimirov, J. Zegenhagen, T. Schultz, R. Feidenhans'l, and B. O. Fimland, *Appl. Phys. Lett.* **71** (1997) 2990.
[227]G. Sherb, A. Kazimirov, and J. Zegenhagen, *Rev. Sci. Instr.* **69** (1998) 512.
[228]F. U. Renner, A. Stierle, H. Dosch, D. M. Kolb, T.-L. Lee, and J. Zegenhagen, *Nature* **439** (2006) 707.
[229]C. A. Melendres, Y. P. Feng, D. D. Lee, and S. K. Sinha, *J. Electrochem. Soc.* **142** (1995) L19.
[230]Y. P. Feng, S. K. Sinha, C. A. Melendres, and D. D. Lee, *Physica* **B221** (1996) 251.
[231]R. De Marco, Z.-T. Jiang, B. Pejcic, and E. Poinen, *J. Electrochem. Soc.* **152** (2005) B389.
[232]H. You, K. Huang, S. S. Yoo, and Z. Nagy, in *International Symposium on Advanced Luminescent Materials,* D. J. Lockwood, P. M. Fauchet, N. Koshida, and S. R. J. Brueck, eds. Electrochemical Society Proceedings **95-25,** Pennington, NJ (1995) p 230.
[233]H. You, Z. Nagy, and K. Huang, *Phys. Rev. Lett.* **78** (1997) 1367.
[234]H. You, J. A. Tanzer, Z. Nagy, Z. Gaburro, and D. Babic, in *Pits and Pores: Formation, Properties, and Significance for Advanced Luminescent Materials,* P. Schmuki, D. J. Lockwood, H. Isaacs, and A. Bsiesy, eds. Electrochemical Society Proceedings **97-7,** Pennington, NJ (1997) p 215.
[235]L. Bosio, R. Cortes, M. Denoziere, and G. Folcher, *J. Phys. Colloq. No. C7* **50** Suppl. 10 (1989) p C7-23.
[236]L. Bosio, R. Cortes, G. Folcher, and M. Froment, in *X-Ray Methods in Corrosion and Interfacial Electrochemistry,* A. Davenport and J. G. Gordon, eds. Electrochemical Society Proceedings **92-1,** Pennington, NJ (1992) p 49.
[237]L. Bosio, R. Cortes, G. Folcher, and M. Froment, *J. Electrochem. Soc.* **139** (1992) 2110.
[238]Z. Nagy, in *Historical Perspectives on the Evolution of Electrochemical Tools,* J. Leddy, V. Birss, and P. Vanysek, eds. Electrochemical Society Special Volume **SV2002-29,** Pennington, NJ (2004) p 235.
[239]G. G. Long, J. Kruger, and M. Kuriyama, in *Passivity of Metals and Semiconductors,* M. Froment, ed. Elsevier, Amsterdam (1983) p 139.
[240]L. Bosio, R. Cortes, A. Defrain, M. Froment, and A. M. Lebrun, in *Passivity of Metals and Semiconductors,* M. Froment, ed. Elsevier, Amsterdam (1983) p 131.
[241]L. Bosio, R. Cortes, and M. Froment, in *EXAFS Near Edge Struct. 3 Springer Proc. Phys. 2),* K. Hodgson, B. Hedman, and J. E. Penner-Hahn, eds. Springer, Berlin (1984) p 484.
[242]M. Fleishmann, A. Oliver, and J. Robinson, *Electrochim. Acta* **31** (1986) 899.
[243]M. G. Dowsett and A. Adriaens, *Anal. Chem.* **78** (2006) 3360.
[244]Z. Nagy, R. M. Yonco, H. You, and C. A. Melendres, U. S. Patent 5,141,617 (1992).
[245]Z. Nagy, H. You, and R. M. Yonco, *Rev. Sci. Instrum.* **65** (1994) 2199.
[246]Z. Nagy, H. You, Y. V. Tolmachev, and R. M. Yonco, unpublished results.
[247]Z. Nagy and H. You, *J. Electroanal. Chem.* **381** (1995) 275.

Index

Cathode reduction(s) of nitrate, 1
Chronoamperometry(ic), 107, 108, 110
Corrosion, 308
  pitting, 309
  metal dissolution, 309
Crystal surface, 259
  double layer, 270
  ionic disturbance, 276
  metal surface preparation, 259
  oxidation, 299, 302
  reconstruction of metal surfaces, 260
    platinum, 262
    gold, 263
  relaxation of metal surfaces, 260
    platinum, 260
    platinum oxidation, reduction, 264
  restructuring, 259
  roughening, 259
  copper passivation, 268
  roughness, 308
  surface roughness measurements, 264
Current density, 1, 4, 10, 12-14, 16, 17, 20, 21, 24- 27, 30-33, 37-39, 43, 44, 46, 47, 50, 53, 55
  detection, 35
  electrolyte, 235
  gas diffusion, 37, 41
  horizontal, 37
  membrane assemblies (MEA), 48
  metal, 9
  microelectrode, 36, 38, 45
  parallel plate, 48
  platinum, 24
  potential, 1, 13
  single crystal, 273
  solid, 34
  surface(s), 9, 15, 21, 26
  titanium, 12
Deposition, 288
  copper, 298
  gold, 188, 298
  monolayer, 288
  multilayer, 288
  palladium, 290
  physical vapor, 185
  platinum, 288, 290
  silver, 288, 293
  submonolayer, 288
  thallium, 293
Electrocatalysis, 283
Electrochemical/chemical synthesis, 195
Electrochemical interphase, 247, 249
  active surface sites, 248
  atomic level, 248
  buried interface, 248
  molecular level, 248

R.E. White (ed.), *Modern Aspects of Electrochemistry, No. 45*, Modern Aspects of Electrochemistry 45, DOI 10.1007/978-1-4419-0655-7_BM2,

other than metal, electrolyte, 313
reproducible surface preparation techniques, 248
ruthenium dioxide surface, 273
silver, 272
surface effects, 248
water, 272, 273
Electrochemical step edge decoration, 175, 177
advantages, 202
canonical, 179, 188
limitations, 202
weaknesses, 200
Electrodeposition of metals, 177, 178, 182,
cyclic/stripping, 190, 192
Electrode(s), 1, 14, 28
anode, 223
bipolar, 47
cathode, 234
copper, 10, 30, 46
corrosion, 43
counter, 91, 133
Electrode surfaces, 277
absorption, 277, 282
adsorption, 277, 280
Electrolysis, 207
ammonia, 209, 218, 219, 221, 222
ammonia as hydrogen carrier, 209
cracking, 219, 221, 222
electrocatalysts, 227
efficiency, 23 0, 243
Faradaic efficiency, 237
oxidation, 223, 226
production, 209
separator membrane, 238
source of ammonia, 217
substrates, 231, 232
Electrolyzer, 239
ammonia, 239
Electron transfer steps, 1
Electrooxidation, 200
Fuel cells, 208
ammonia (hydrogen carrier), 209
source (hydrogen), 208
transportation and storage, 208
Glass frit, 91, 92
Growth law for nanowires, 180
Highly oriented pyrolytic graphite (HOPG), 177, 178, 188
Hydrogen production, 211
ammonia, 217
coal gasification, 212, 213
efficiency of methods, 215, 216
partial oxidation reforming, 212
steam reforming (internal reforming), 211, 212
water electrolysis, 213, 214, 215
Ionic liquid(s), 63
cations and anions, 64
general properties, 64
molecular solvent, 64
non-haloaluminate, 66, 75, 121
Kinetic, 85, 108, 123, 283

current, 85, 106
data, 85
oxygen reduction, 283, 286
Molten or fused salts, 63
Nanowire(s), 175, 177, 178
base metal, 182
copper, 182
fabricating, 175
gold, 188
manganese, 185
molybdenum, 184
platinum, 182
palladium, 182
semiconducting, 189
Oxides, electrodepositing nanowires, 197
Reduction of nitrate (cathodic), 8, 83
direct, 8, 17, 19-23, 25, 29, 34
indirect, 8, 17, 19-21, 34
parallel mechanisms, 1
Reference electrode, 92, 93, 99
Renewable fuels/sources, 207
ammonia, 209
fossil, 208
hydroelectric, 207
nuclear, 207
solar, 208
thermal, 207
wind, 208
Room-temperature ionic liquids and melts (RTIL(s)) 63, 64
dialkylimidazolium chlorides, 75
dialkylimidazolium salts with fluorohydrogenate anions, 76
dialkylimidazolium salts with fluorocomplex anions, 77
non-chloroaluminate, 121, 132, 134, 142, 153
tetraalkylammonium salts with bis(trifluoromethly)-sulfonylimide anions, 78
Rotating disk electrode (RDE), 103, 104, 105, 113
Rotating ring-disk electrode (RRDE), 103, 104, 107
Room-temperature melts, 64
RTIL properties,
chloride ion contamination, 84
conductance measurements at high frequencies, 88
other considerations, 89
oxygen contamination, 85
thermal stability, 79, 82
water contamination, 82
Silver, 182
template synthesis, 176
Solvents, 66, 75, 76, 79, 85, 89, 93, 98, 107, 113, 133, 134, 136, 137, 142, 147, 149, 153
Synchrotron x-ray scattering, 247
Voltammetric methods, 83
Voltammetry
cyclic, 90, 99, 188
hydrodynamic, 103, 104

Voltammograms, 91, 105, 113, 182
X-ray scattering, 250
  crystal truncation rods, 250, 257
  general description, 250
  glancing angle in-plane x-ray diffraction, 257
  resonance anomalous, 259
  resonance surface scattering, 307
  reflectivity, 250, 257
  structure factor, 250
  usefulness of techniques, 315
X-ray electrochemical cell designs, 316
  comparison of cell designs, 321
  reflection geometry, 317
  solution impurities, 322
  transmission geometry, 318
  hanging drop transmission geometry, 319

MIX
Papier aus verantwortungsvollen Quellen
Paper from responsible sources
FSC® C105338

If you have any concerns about our products,
you can contact us on
ProductSafety@springernature.com

In case Publisher is established outside the EU,
the EU authorized representative is:
Springer Nature Customer Service Center GmbH
Europaplatz 3, 69115 Heidelberg, Germany

Printed by Libri Plureos GmbH
in Hamburg, Germany